Lecture Notes in Computer Science

Commenced Publication in 1973
Founding and Former Series Editors:
Gerhard Goos, Juris Hartmanis, and Jan van Leeuwen

Cristian S. Calude   Jarkko Kari
Ion Petre   Grzegorz Rozenberg (Eds.)

Unconventional Computation

10th International Conference, UC 2011
Turku, Finland, June 6-10, 2011
Proceedings

 Springer

Volume Editors

Cristian S. Calude
The University of Auckland, Dept. of Computer Science, Science Centre
38 Princes Street, Auckland 1142, New Zealand
E-mail: cristian@cs.auckland.ac.nz

Jarkko Kari
University of Turku, Department of Mathematics
20014 Turku, Finland
E-mail: jkari@utu.fi

Ion Petre
Åbo Akademi University, Dept. of Information Technologies, ICT building
Joukahaisenkatu 3-5 A, 20520 Turku, Finland
E-mail: ipetre@abo.fi

Grzegorz Rozenberg
Leiden University, Leiden Institute of Advanced Computer Science (LIACS)
Niels Bohrweg 1, 2333 CA Leiden, The Netherlands
E-mail: rozenber@liacs.nl

ISSN 0302-9743   e-ISSN 1611-3349
ISBN 978-3-642-21340-3   e-ISBN 978-3-642-21341-0
DOI 10.1007/978-3-642-21341-0
Springer Heidelberg Dordrecht London New York

Library of Congress Control Number: Applied for

CR Subject Classification (1998): F.1, F.2, I.1, C.1.3, C.1, J.2

LNCS Sublibrary: SL 1 – Theoretical Computer Science and General Issues

Typesetting: Camera-ready by author, data conversion by Scientific Publishing Services, Chennai, India

Printed on acid-free paper

Springer is part of Springer Science+Business Media (www.springer.com)

Preface

The 10th International Conference on Unconventional Computation, UC 2011, was organized under the auspices of EATCS and Academia Europaea, by the Department of Mathematics of the University of Turku (Turku, Finland), and the Center for Discrete Mathematics and Theoretical Computer Science (Auckland, New Zealand). The event was held in Turku, Finland, during June 6–10, 2011. The conference venues were the Calonia and Arcanum buildings of the university.

The city of Turku was founded in the thirteenth century, which makes it the oldest town in Finland. For centuries, it remained the capital of Finland, until 1812 when the capital was moved to Helsinki. Turku is situated by the Baltic Sea and surrounded by one of the largest and most beautiful archipelagoes of the world. The archipelago consists of thousands of small islands and provides a unique and spectacular natural environment for travelers to enjoy. Turku was European Capital of Culture in 2011, and many cultural events were organized in the city throughout the year, also during the Unconventional Computation conference.

The International Conference on Unconventional Computation (UC) series is devoted to all aspects of unconventional computation theory as well as experiments and applications. Typical, but not exclusive, topics are: natural computing including quantum, cellular, molecular, membrane, neural, and evolutionary computing, as well as chaos and dynamical system-based computing, and various proposals for computational mechanisms that go beyond the Turing model. The first venue of the Unconventional Computation Conference (formerly called Unconventional Models of Computation) was Auckland, New Zealand, in 1998. Subsequent sites of the conference were Brussels, Belgium, in 2000, Kobe, Japan, in 2002, Seville, Spain, in 2005, York, UK, in 2006, Kingston, Canada, in 2007, Vienna, Austria, in 2008, Ponta Delgada, Portugal, in 2009, and Tokyo, Japan, in 2010.

The six keynote speakers of the 2011 conference were:

- Samson Abramsky (University of Oxford, UK): "The Logic and Topology of Non-locality and Contextuality"
- Bastien Chopard (University of Geneva, Switzerland): "A Framework for Multiscale and Multiscience Modeling Based on the Cellular Automata and Lattice Boltzmann Approaches"
- David Corne (Heriot-Watt University, UK): "Unconventional Optimizer Development"
- Juhani Karhumäki (University of Turku, Finland): "Weighted Finite Automata: Computing with Different Topologies"
- Gheorghe Păun (Institute of Mathematics of the Romanian Academy, Romania): "Membrane Computing at Twelve Years (Back to Turku)"
- Grzegorz Rozenberg (Leiden University, The Netherlands): "A Formal Framework for Bioprocesses in Living Cells"

The conference also included three tutorials:

- Mika Hirvensalo (University of Turku, Finland): "Quantum Information"
- Nicolas Ollinger (Aix-Marseille University, France): "Cellular Automata"
- Ignacio Pérez-Hurtado, Mario J. Pérez-Jiménez, Agustín Riscos-Núñez, and Francisco J. Romero-Campero (University of Seville, Spain): "Membrane Computing"

In addition to the main UC2010 conference, four workshops were also hosted: "Physics and Computation" and "Hypercomputation" were organized by Mike Stannett (University of Sheffield, UK), "Language Theory in Biocomputing" was organized by Tero Harju (University of Turku, Finland), and "Summer Solstice Conference on Discrete Models of Complex Systems" was organized by Danuta Makowiec (University of Gdansk, Poland), Anna Lawniczak (University of Guelph, Canada), and Bruno Di Stefano (Nuptek Systems Ltd., Canada).

The Program Committee selected 17 papers (out of 33 submissions) to be presented at the conference. This volume includes 6 (extended) abstracts of invited talks, 3 (extended) abstracts of tutorials, and 17 regular papers. The papers presented at the workshops appeared in a separate proceedings volume, published by the Turku Centre for Computer Science (TUCS). The conference also hosted a poster session.

The editors are grateful to all the contributors to the scientific content of the conference. We thank especially the invited speakers, tutorial speakers, all authors of contributed papers, and the organizers of the satellite workshops. We are indebted to the Program Committee and the additional reviewers for their help in selecting the papers. We extend our thanks to the members of the local Organizing Committee. We are also grateful for the support by the Federation of Finnish Learned Societies, Finnish Academy of Science and Letters, Turku Centre for Computer Science, the University of Turku and the City of Turku. Finally, we acknowledge the excellent cooperation from the *Lecture Notes in Computer Science* team of Springer for their help in producing this volume in time for the conference.

March 2011

Cristian S. Calude
Jarkko Kari
Ion Petre
Grzegorz Rozenberg

Organization

Program Committee

Selim Akl	Queen's University, Canada
Olivier Bournez	Ecole Polytechnique, France
Thomas Bäck	Leiden University, The Netherlands
Adan Cabello	University of Seville, Spain
Barry Cooper	University of Leeds, UK
José Félix Costa	IST – Technical University of Lisbon, Portugal
Nachum Dershowitz	Tel Aviv University, Israel
Eric Goles	Universidad Adolfo Ibáñez, Chile
Shan He	University of Birmingham, UK
Mika Hirvensalo	University of Turku, Finland
Natasha Jonoska	University of South Florida, USA
Jarkko Kari	University of Turku, Finland (Co-chair)
Giancarlo Mauri	University of Milano-Bicocca, Italy
Kenichi Morita	Hiroshima University, Japan
Ion Petre	Åbo Akademi University, Finland (Co-chair)
Kai Salomaa	Queen's University, Canada
Hava Siegelmann	University of Massachusetts Amherst, USA
Susan Stepney	University of York, UK
Fumiaki Tanaka	University of Tokyo, Japan
Jon Timmis	University of York, UK

Organizing Committee

Pierre Guillon	Juhani Karhumäki	Petri Salmela
Tero Harju	Jarkko Kari (Chair)	Charalampos Zinoviadis
Mika Hirvensalo	Arto Lepistö	
Timo Jolivet	Ion Petre	

External Referees

Scott Aaronson	Claudio Ferretti	Gheorghe Păun
Alastair Abbott	Yoichi Hirai	Vladimir Rogojin
Alhazov Artiom	Katsunobu Imai	Ville Salo
Sepinoud Azimi	Timo Jolivet	Siamak Taati
Robert Brijder	Ibuki Kawamata	Tarmo Uustalu
Cris Calude	Alberto Leporati	Akihiro Yamashita
Salimur Choudhury	Yusuke Matsumura	Charalampos Zinoviadis
Erzsébet Csuhaj-Varjú	Victor Mitrana	
Eugen Czeizler	Simon O'Keefe	

Table of Contents

Invited Lectures and Tutorials

Regular Contributions

The Logic and Topology of Non-locality and Contextuality

Samson Abramsky

Oxford University Computing Laboratory
Wolfson Building, Parks Road Oxford OX1 3QD, U.K.

Bell's theorem famously shows that no local theory can account for the predictions of quantum mechanics; while the Kochen-Specker theorem shows the same for non-contextual theories. Non-locality, and increasingly also contextuality, play an important role as computational resources in current work on quantum information. Much has been written on these matters, but there is surprisingly little unanimity even on basic definitions or the inter-relationships among the various concepts and results. We use the mathematical language of sheaves and monads to give a very general and mathematically robust description of the behaviour of systems in which one or more measurements can be selected, and one or more outcomes observed. In particular, we give a unified account of contextuality and non-locality in this setting.

- A central result is that an empirical model can be extended to all sets of measurements if and only if it can be realized by a factorizable hidden-variable model, where factorizability subsumes both non-contextuality and Bell locality. Thus the existence of incompatible measurements is the essential ingredient in non-local and contextual behavior in quantum mechanics.
- We give hidden-variable-free proofs of Bell style theorems.
- We identify a notion of strong contextuality, with surprising separations between non-local models: Hardy is not strongly contextual, GHZ is.
- We interpret Kochen-Specker as a generic (model-independent) strong contextuality result.
- We give general combinatorial and graph-theoretic conditions, independent of Hilbert space, for such results.

C.S. Calude et al. (Eds.): UC 2011, LNCS 6714, p. 1, 2011.

A Framework for Multiscale and Multiscience Modeling and Numerical Simulations

Bastien Chopard[1], Jean-Luc Falcone[1], Alfons G. Hoekstra[2],
and Joris Borgdorff[2]

[1] University of Geneva, Switzerland
[2] University of Amsterdam, The Netherlands

Abstract. The Complex Automata (CxA) methodology offers a new framework to develop multiscale and multiscience numerical simulations. The CxA approach assumes that a multiscale model can be formulated in terms of several coupled single-scale submodels. With concepts such as the scale separation map, the generic submodel execution loop and the coupling templates, one can define a multiscale modeling language which is a bridge between the application design and the computer implementation.

1 Introduction

Multiscale applications are an important challenge for computational science and many domains of research. Many phenomena involve many spatial or temporal scales, and the interaction of various physical processes. Biomedical applications are an illustration of a multiscale, multiphysics problem, where biology is coupled to fluid mechanics. For instance, in the problem of restenosis [7], blood flow is coupled to the growth of smooth muscle cells. Haemodynamics is a fast varying process, acting over spatial scales ranging from micrometers to centimeters. On the other hand, smooth muscle cells evolve at a much slower time scale.

Although the term "multiscale modeling" is commonly used in many research fields, there is only a few methodological papers [6,13,5] offering a conceptual framework, or a general theoretical approach. As a result, in most of the mutliscale applications found in the literature, methodology is entangled with the specificity of the problem and researchers keep reinventing similar strategies, but with different names. It is yet clear that a good methodology is quite important when developing an interdisciplinary application within a group of people with different scientific backgrounds and different geographical locations. A multiscale modeling framework and the corresponding modeling language is an important step in this direction. It allows one to clearly describe a multiscale, multiphysics phenomena, separating the problem specific components from the strategy used to deal with a large range of scales. It gives a blueprint of a complex application, offering a way to co-develop a global numerical solution within a large team. A good matching between the application design and its implementation on a computer is central for an incremental development and its long term sustainability.

C.S. Calude et al. (Eds.): UC 2011, LNCS 6714, pp. 2–8, 2011.

2   The Complex Automata Approach

Cellular automata (CA) and Lattice Boltzmann (LB) models (see for instance [4]) are powerful tools to model various complex systems, in particular biological processes and fluid flows. In terms of scales, each CA or LB model is characterized by its lattice spacing Δx, its lattice size L, its time step Δt and its simulation time T. These quantities specifies the range of spatial and temporal scales that can be resolved through a numerical simulation.

Recently, we have developed a theoretical framework for multiscale modeling, based on the concept of Complex Automata (CxA) [11,10,12]. A CxA is a set of "single-scale" coupled LB and/or CA systems, each representing a different physical process . Here the term "single-scale" means that each CA or LB submodel only resolves a limited range of spatial and temporal scales.

This framework gave rise to MUSCLE, a general coupling software [9,1] and a Multiscale Modeling Language (MML) [8]. They are now further developed within the European project MAPPER [2], with the goal of reformulating various existing multiscale applications in a formalism that would allow their execution of a distributed HPC platform.

Below, we briefly describe the CxA methodology for multiscale numerical simulations. We refer the reader to [12] for more detail and to [7,3] for the application of this approach to a biomedical problem.

2.1   The Scale Separation Map

A multiscale systems can be represented on a scale separation map (SSM), i.e. a map that describes the range of spatial and temporal scales that need to be resolved in order to solve the problem at hand. This range of scales is shown as a rectangle, as indicated in Fig. 1 left. In order to simulate such a process on a computer, one has to find a strategy to "split" these scales and keep only the relevant ones. Otherwise, the CPU time will become prohibitive. Concretely, this process amounts to the identification of submodels acting on a reduced range of scales, and their mutual couplings, as illustrated in Fig. 1 right. Such a splitting is however likely to affect the quality of the result and the art of multiscale modeling is to propose a good compromise between CPU performance and accuracy.

The arrows shown in Fig. 1 represent the coupling between the submodel that arise due to the splitting of the scales. They correspond to an exchange of data, often supplemented by a transformation to match the difference of scales at both extremities. In the proposed methodology, these couplings are implemented as software components, coined *smart conduits* that completely take care of the mapping of the data from one submodel to the other. Therefore, in the CxA approach, submodels are autonomous solvers that are not aware of the scales of the other submodels and that can be substituted anytime by a better version of the code or the algorithm.

The relation between two submodels can be described through their respective position in the SSM. The scales of the two submodels either overlap or can be separated. Since scale overlap or scale separation concerns space or time, there

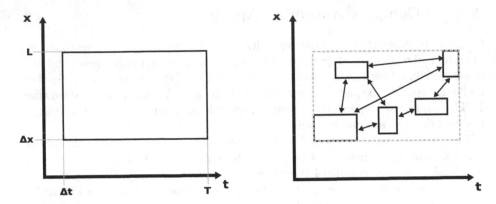

Fig. 1. Illustration of the process of "scale splitting": a multiscale models is decomposed into several "single-scale" coupled submodels

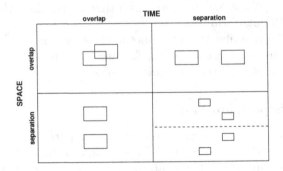

Fig. 2. The possible relations between two submodels in the scale separation map

are in total five possible relations, as illustrated in Fig. 2. In addition to their respective position in the SSM, two interacting submodels are characterized by the relation between their computational domains. Both submodels can share the same domain (situation termed *single-domain*). Otherwise, the submodels have different or overlapping computational domains. This case is termed *multi-domain*.

The above features (respective position in the SSM and domain relation) offer a way to classified the interactions between two coupled submodels.

Another benefit of the SSM is to give a quick estimate of the CPU gained by the scale splitting process. The CPU time of a submodel goes as $(L/\Delta x)^d (T/\Delta t)$ where d is the spatial dimension of the model. Therefore, the computational time of the system on the left panel of Fig. 1 is likely to be much larger than the one of the right panel.

It also turns out that the SSM is a powerful way to describe a multiscale, multiscience problem. The coupling can be annotated with the quantity that is exchanged between each pairs of submodels. Fig. 3 illustrates this point in

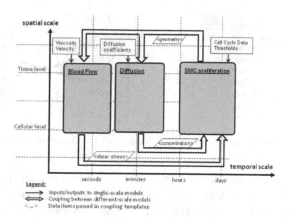

Fig. 3. Scale separation map (SSM) of the in-stent restenosis application described in [7,3]

the case of the in-stent restenosis simulation [7,3]. After stenting of a coronary artery, the smooth muscle cells (SMC) are likely to proliferate into the lumen, causing again a stenosis. Drug eluting stents can be used to slow down the SMC growth. From this picture we see that, contrary to many situations reported in the literature, multiscale modeling is more than the coupling of just two submodels, one at a microscopic scales and the other at a macroscopic scale.

2.2   Submodel Execution Loop and Coupling Templates

A second ingredient of the CxA methodology is to express CA and LB models with a generic, abstract execution temporal loop. We term this algorithmic structure the *submodel execution loop* (SEL). It reflects the fact that, during the time iterations of the CA of LB model, a reduced set of generic operations have to be performed over and over. The most important one is the S operation, where S stands for *Solver*. For a CA or LB model, S is the composition of the so-called *collision* and *propagation* phases [4]. A second important step of the SEL is the B operation, where B means *Boundary*. The execution of B amounts to specifying the boundary conditions for the computation. The repetition of S and B is the core of the submodel time loop. In addition, in order to initialize the process, an other operation has to be specified. In our approach we term it f_{init} to reflect that the variables of the model (often denoted $f_i(r,t)$ in a LB model) need to be given an initial value. This initialization phase also specifies the computational domain and possibly some termination condition for the time loop.

Finally, we define two observation operators, O_i and O_f, which compute some desired quantities from the model variables. The subscript i and f are for intermediate and final observations, respectively. Note that the algorithmic structure of the SEL also applies to other numerical schemes than CA and LB models. It is actually quite general and is now used for other types of solvers in the MAPPER project [2].

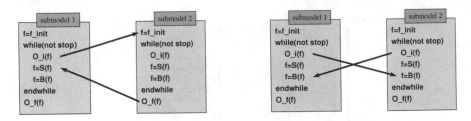

Fig. 4. Generic submodel execution loop and two examples of coupling

Imposing the above generic structure to the evolution loop limits the ways to couples two submodels. A coupling amounts to an exchange of data between a pair of operators belonging to the SEL of the two submodels. According to our definitions, the sender of information is either O_i or O_f. And the receiving operators can only be S, B or f_{init}. This is illustrated in Fig. 4. We call *coupling templates* the different possible pairs of input/output operators in a coupling. Therefore the coupling templates are $O_i \rightarrow S$, $O_i \rightarrow B$, $O_i \rightarrow f_{init}$, and $O_f \rightarrow S$, $O_f \rightarrow B$, $O_f \rightarrow f_{init}$. Names can be given to each of these coupling templates: $O_i \rightarrow S/B$ is called *interact*, $O_i \rightarrow f_{init}$ is termed *call*, $O_f \rightarrow S/B$ is termed *release* and $O_f \rightarrow f_{init}$ is called *dispatch*.

It is quite interesting to notice that these coupling templates reflects very closely the relative position of the two submodels in the SSM and the relation between their computational domains. From analyzing several multiscale systems and the way their submodel are mutually coupled, we reach the conclusion that the relations shown in table 1 hold, between any two coupled submodels, X and Y, with a single domain relation. In case where X and Y have a multi-domain relation, the same table holds but the operator S is replaced by B. To illustrate the meaning of table 1 we can consider a system of particles transported in a fluid flow. One submodel X is the fluid solver and the second submodel Y is an advection-diffusion solver. This is a typical single-domain situation with overlapping temporal scales. The observation O_i^X of the flow velocity is needed to compute the advection process. So $O_i^X \rightarrow S^Y$. In return, $O_i^Y \rightarrow S^X$ because the density of transported particles may affect the viscosity of the fluid.

In the example of the growth of biological cells subject to the blood flow shear stress, there is a clear time scale separation between the two processes (see Fig 3). Therefore, the converged flow field is first sent to the biological model to define

Table 1. Relation between coupling templates and position in the scale separation map for two submodels X and Y sharing the same computational domain

name	coupling	temporal scale relation
interact	$O_i^X \rightarrow S^Y$	overlap
call	$O_i^X \rightarrow f_{init}^Y$	X larger than Y
release	$O_f^Y \rightarrow S^X$	Y smaller than X
dispatch	$O_f^X \rightarrow f_{init}^Y$	any

the cell growth rate $(O_f^X \to S^Y)$. Then, the new geometry of the cells induces a new boundary condition for the flow, which must be recomputed $(O_i^X \to f_{init}^Y)$.

2.3 Conclusion

We have presented a methodological framework to develop and design a multiscale, multiscience application. This framework has been called Complex Automata (CxA) because it was initially meant to describe a system of coupled Cellular Automata and Lattice Boltzmann models with different scales [12]. It turns out that other types of solver can also be included in the same formalism. The proof of concept of this approach has been demonstrated for a challenging biomedical application, namely the simulation of restenosis in stented arteries [7,3]

Through the concepts of the scale separation map, submodel execution loops and coupling templates, the proposed approach offers a new programming model and a new description language [8] coined Multiscale Modeling Language (MML). From this representation, important submodel scheduling information can be inferred, leading to the possibility of automatic code generation.

Within the European project MAPPER [2], we are developing xMML, a XML-like version of MML, bridging the application design to its implementation and execution on distributed HPC platforms. The concept is applied to a wide range of multiscale applications.

References

1. http://developer.berlios.de/projects/muscle/
2. http://www.mapper-project.eu
3. Caiazzo, A., Evans, D., Falcone, J.-L., Hegewald, J., Lorenz, E., Stahl, B., Wang, D., Bernsdorf, J., Chopard, B., Gunn, J., Hose, R., Krafczyk, M., Lawford, P., Smallwood, R., Walker, D., Hoekstra, A.: A complex automata approach for in-stent restenosis: two-dimensional multiscale modeling and simulations. J. of. Computational Sciences (2010), doi:10.1016/j.jocs.2010.09.002
4. Chopard, B., Droz, M.: Cellular Automata Modeling of Physical Systems. Cambridge University Press, Cambridge (1998)
5. Dada, J.O., Mendes, P.: Multi-scale modelling and simulation in systems biology. Integrative Biology (2011)
6. Weinan, E., Li, X., Ren, W., Vanden-Eijnden, E.: Heterogeneous multiscale methods: A review. Commun. Comput. Phys. 2, 367–450 (2007)
7. Evans, D., Lawford, P.-V., Gunn, J., Walker, D., Hose, D.-R., Smallwood, R.H., Chopard, B., Krafczyk, M., Bernsdorf, J., Hoekstra, A.: The application of multiscale modelling to the process of development and prevention of stenosis in a stented coronary artery. Phil. Trans. Roy. Soc. (2008) (in press)
8. Falcone, J.-L., Chopard, B., Hoekstra, A.: Mml: towards a multiscale modeling language. Procedia Computer Science 1(11), 819–826 (2010)
9. Hegewald, J., Krafczyk, M., Tölke, J., Hoekstra, A., Chopard, B.: An agent-based coupling platform for complex automata. In: Bubak, M., van Albada, G.D., Dongarra, J., Sloot, P.M.A. (eds.) ICCS 2008, Part II. LNCS, vol. 5102, pp. 227–233. Springer, Heidelberg (2008)

10. Hoekstra, A.G., Falcone, J.-L., Caiazzo, A., Chopard, B.: Multi-scale modeling with cellular automata: The complex automata approach. In: Umeo, H., Morishita, S., Nishinari, K., Komatsuzaki, T., Bandini, S. (eds.) ACRI 2008. LNCS, vol. 5191, pp. 192–199. Springer, Heidelberg (2008)
11. Hoekstra, A., Lorenz, E., Falcone, J.-L., Chopard, B.: Towards a complex automata formalism for multuscale modeling. Int. J. Multiscale Computational Engineering 5(6), 491–502 (2008)
12. Hoekstra, A.G., Caiazzo, A., Lorenz, E., Falcone, J.-L., Chopard, B.: Complex automata: multi-scale modeling with coupled cellular automata. In: Hoekstra, A., Kroc, J., Sloot, P. (eds.) Modeling Complex Systems with Cellular Automata, ch. 3, Springer, Heidelberg (2010)
13. Ingram, G.D., Cameron, I.T., Hangos, K.M.: Classification and analysis of integrating frameworks in multiscale modelling. Chemical Engineering Science 59, 2171–2187 (2004)

Unconventional Optimizer Development

David Corne

School of MACS, Heriot-Watt University
Edinburgh EH14 4AS, UK

The fruits of bio-inspired approaches to optimisation include several techniques that are now commonly used in practice to address real-world problems. A common situation is as follows: an organisation has a regularly occurring problem to solve (typically a logistics problem), and they engage a research group or a consultancy to deliver an optimizer that can then be used as they regularly solve instances of that problem. The research group will then spend perhaps several months developing the optimizer, and this will almost always involve:

(i) deciding to use a specific algorithm framework (e.g. tabu search or evolutionary search);
(ii) tuning an algorithm over many problem instances in the space of interest, towards getting the best results achievable in a given time (perhaps minutes).

I argue that this typical approach should, in many, arguably most cases, be changed completely. First, the client does not need a slow algorithm that delivers great solutions - they need a very fast algorithm that delivers acceptable solutions. Second, there are many drawbacks and uncertainties in the enterprise of algorithm tuning; it would be good to mitigate these uncertainties via a different approach. Third, to spend several months designing and tuning an algorithm that solves instances seems like a great waste of time when, in several cases, it may be possible to simply use this time to solve all of the instances the company is likely to face! In this talk I therefore discuss the ingredients of the unconventional approach.

C.S. Calude et al. (Eds.): UC 2011, LNCS 6714, p. 9, 2011.

A Formal Framework for Bioprocesses in Living Cells

Andrzej Ehrenfeucht[1] and Grzegorz Rozenberg[1,2]

[1] Department of Computer Science, University of Colorado at Boulder
430 UCB, Boulder, CO 80309, USA
[2] Leiden Institute of Advanced Computer Science, Leiden University
Niels Bohrweg 1, 2333 CA Leiden, The Netherlands

Natural Computing is an interdisciplinary field of research that investigates human-designed computing inspired by nature as well as computation taking place in nature. In other words, Natural Computing investigates models, computational techniques, and computational technologies inspired by nature as well as it investigates, in terms of information processing, phenomena/processes taking place in nature.

One of the fascinating research areas of Natural Computing is the computational nature of biochemical reactions taking place in living cells. It is hoped that this line of research may contribute to a computational understanding of the functioning of a living cell. An important step towards this goal is understanding interactions between biochemical reactions. These reactions and their interactions are regulated, and the main regulation mechanisms are facilitation/acceleration and inhibition/retardation. The interactions between individual reactions take place through their influence on each other, and this influence happens through the two mechanisms.

In our lecture we present a formal framework for the investigation of processes carried by biochemical reactions in living cells. We motivate this framework by explicitly stating a number of assumptions that hold for a great number of biochemical reactions, and we point out that these assumptions are very different from the ones underlying traditional models of computation. We discuss some basic properties of processes carried out by biochemical reactions, and demonstrate how to capture and analyze, in our formal framework, some biochemistry-related notions.

Besides providing a formal framework for reasoning about processes instigated by biochemical reactions, the models discussed in the lecture are novel and attractive from the computer science point of view.

The lecture is of a tutorial style and self-contained. In particular no knowledge of biochemistry is required.

C.S. Calude et al. (Eds.): UC 2011, LNCS 6714, p. 10, 2011.

Quantum Information – A Tutorial

Mika Hirvensalo

Department of Mathematics, University of Turku, FI-20014 Turku, Finland
TUCS – Turku Centre for Computer Science
mikhirve@utu.fi

1 Introduction

Information processing, information transmission, and information security are everyday notions of modern society. But what exactly is information? This seems to be quite a hard question. Analogous complication arises in physical sciences when asking what exactly energy is. A naive approach to define information is to define information as a message contained in a string of symbols, but naturally enough, a similar question about the meaning of "message" arises. In this presentation all potential societal and qualitative connotations of information are stripped away and we will restrict only to the quantitative mathematical aspects of information.

20th century witnessed the birth of quantum mechanics, a novel theory establishing a united way to treat two apparently distinct aspects of microsystems: undulatory and corpuscular. Quantum mechanics did not only bring unification, but also severe philosophical problems on the nature of reality.

2 Bell Inequalities

Even without a definition of information it is possible to give a relative description of quantum information as information represented in quantum systems; physical systems so small that they governed by quantum mechanics rather than classical mechanics. This description is already sufficient for learning what might become different when entering the realm of quantum information: A brief but a very famous observation [9] published in 1982 shows that in general, quantum information cannot be cloned – a task supposed evident for classical information.

Much stranger behaviour was discovered by Albert Einstein & al. when trying to establish the incompleteness of quantum mechanics as a physical theory [4]. So-called EPR-paradox raised a long-lasting debate between Niels Bohr and Albert Einstein, but the paradox was resolved in satisfactory manner only decades later [1]. Quantum systems can be shown to violate Bell inequalities, which demonstrates that the quantum systems carrying information are different from classical ones in a very fundamental sense.

C.S. Calude et al. (Eds.): UC 2011, LNCS 6714, pp. 11–13, 2011.

3   Von Neumann and Shannon Entropy

The problem of defining information is equivalent to defining the nature of *en-tropy*. Intuitively, entropy is some sort of measure of disorder: Entropy is small if the system is in order (in some sense), and large if the system under investigation is in disorder. A notable early approach to quantify entropy was taken by Ludwig Boltzmann [3] in 1872.

John von Neumann was an ingenious mathematician whose contribution to the mathematical structures of quantum mechanics is beyond comparison. He performed a Gedanken Experiment to find a formula

$$S(\rho) = - \operatorname{Tr} \rho \log \rho, \tag{1}$$

for the entropy of a quantum state described by a density matrix ρ. Von Neumann entropy formula is perfectly consistent with the notion presented later by Claude Shannon [8]: Von Neumann entropy of a mixed quantum system coincides with the Shannon entropy

$$H(X) = -(p_1 \log p_1 + \ldots + p_n \log p_n) \tag{2}$$

of a random variable X corresponding to the quantum state in a very natural way. Equation (2) is worth reconsidering, as it clearly measures the uncertainty of the random variable X: Minimum entropy 0 is obtained with simple distribution $p_i = 1$ for some i, and maximum entropy with the uniform distribution. The unit of the entropy depends on the choice of logarithm base: Choosing base 2 corresponds to measuring the entropy in bits.

The conditional entropy of X, when the value of another random variable Y is known to be y_j is defined via conditional probabilities:

$$H(X \mid y_j) = -(p(x_1 \mid y_j) \log p(x_1 \mid y_j) + \ldots + p(x_n \mid y_j) \log p(x_n \mid y_j)),$$

and the conditional entropy of X when Y is known is defined as the expected value

$$H(X \mid Y) = p(y_1)H(X \mid y_1) + \ldots + p(y_m)H(X \mid y_m).$$

Conditional entropy thus measures the uncertainty of X, when another random variable Y is known. Now, the information of X when Y is known is defined as

$$I(X \mid Y) = H(X) - H(X \mid Y),$$

which makes perfectly sense: The information that Y gives is shown as the difference in the uncertainties.

An important part of quantum information theory concentrates on cases where X and Y are obtained from a quantum system via measurement, and a famous classical result governing the gained information, so-called Holevo Bound [5] must be mentioned:

$$I(X \mid Y) \leq S(p_1\rho_1 + \ldots + p_n\rho_n) - (p_1 S(\rho_1) + \ldots + p_n S(\rho_n)),$$

where X is random variable with potential values $\{1, \ldots, n\}$, and Y corresponds to an observable.

4   Quantum Cryptography

In 1984, Charles Bennett and Gilles Brassard presented a quantum information -based scheme BB84 for secure communication [2]. The protocol BB84 is actually a key exchange protocol, and the secure communication can thereafter be established by one-time pad encryption. However, it turned out to be a quite difficult task to establish the absolute security of BB84 protocol, and Mayers [6] was apparently the first one to provide the unconditional security proof.

References

1. Bell, J.S.: On the Einstein-Podolsky-Rosen paradox. Physics 1, 195–200 (1964)
2. Bennett, C.H., Brassard, G.: Quantum cryptography: public key distribution and coin tossing. In: Proceedings of IEEE Conference on Computers, Systems, and Signal Processing, Bangalore, India, pp. 175–179 (1984)
3. Boltzmann, L.: Weitere Studien über das Wärmegleichgewicht unter Gasmolekülen. Wiener Berichte 66, 275–370 (1872)
4. Einstein, A., Podolsky, B., Rosen, N.: Can quantum-mechanical description of physical reality be considered complete? Physical Review 47, 777–780 (1935)
5. Holevo, A.S.: Statistical Problems in Quantum Physics. In: Murayama, G., Prokhorov, J.V. (eds.) Proceedings of the Second Japan-USSR Symposium on Probability Theory, pp. 104–109. Springer, Heidelberg (1973)
6. Mayers, D.: Unconditional security in quantum cryptography. Journal of the ACM 48(3), 351–406 (2001)
7. von Neumann, J.: Thermodynamik quantummechanischer Gesamheiten. Nachrichten von der Gesellschaft der Wissenschaften zu Göttingen 1, 273–291 (1927)
8. Shannon, C.E.: A Mathematical Theory of Communication. Bell System Technical Journal 27, 379–423, 623–656 (1948)
9. Wootters, W.K., Zurek, W.H.: A single quantum cannot be cloned. Nature 299, 802–803 (1982)

Weighted Finite Automata: Computing with Different Topologies*

Juhani Karhumäki and Turo Sallinen

Department of Mathematics and
Turku Centre for Computer Science (TUCS)
University of Turku
FIN-20014 Turku, Finland
{karhumak,thtsal}@utu.fi

Abstract. We use a very conventional model of computation to define unconventional computational processes. This leads to an easily computable class of real functions, however, this class is very different to those of nicely behaving real functions in a classical sense. All this is based on the fact that the topology of the unit interval is very different to that of infinite words representing numbers in that interval. In addition, the very inherent recursive structure of finite automata is central here.

1 Introduction

Finite automata are among the simplest models of conventional computing. They operate on words over a finite alphabet. If we associate to automata also outputs, that is consider finite transducers, we obtain a simple classical model of computing *word functions*: $A^* \to B^*$.

However, we can proceed also in a different way. Let us take an ordinary finite nondeterministic automaton $\mathcal{A}$. We associate to each (accepted) word the number how many times it is accepted. As an extension of this we can use instead of ordinary nondeterministic automata those with weights, that is each transition is labelled not only by a symbol but also its weight. This yields to functions $f_{\mathcal{A}} : A^* \to \mathbb{R}_+$ first introduced by Schützenberger [9] and later referred to as *rational functions*, e.g., in [5]. Their behavior is broadly studied, e.g., in [5], [1] and [8].

Now, let us fix $A = \{0, 1\}$, that is a binary alphabet. Then there exists the well-known bijection between the unit interval $[0, 1)$ and the set of infinite binary words in $A^\omega \setminus A^*1^\omega$, that is between infinite binary words containing infinitely many zeros. We make use of this bijection to associate to each nondeterministic (weighted) automaton $\mathcal{A}$ two functions:

$$f_{\mathcal{A}} : A^\omega \to \mathbb{R}_+ \tag{1}$$

---

* Work supported by the Academy of Finland project 121419.

C.S. Calude et al. (Eds.): UC 2011, LNCS 6714, pp. 14–33, 2011.

and

$$\hat{f}_{\mathcal{A}} : [0, 1) \to \mathbb{R}_+, \tag{2}$$

where the values of the functions on $w \in A^\omega$ and $\hat{w} \in [0, 1)$ represented by w, respectively, are the multiplicities (or weights) given by $\mathcal{A}$ on the input w. The values of these functions are obtained as infinite products of matrices, for rational functions the products being only finite.

The above requires some considerations. In general, these lead to very challenging and only poorly understood questions. However, we avoid these problems by using only very special classes of automata, so called *level automata*, which are acyclic automata except that cycles from a state to itself are allowed, but even then their weights are either strictly smaller than 1 or equal to 1 depending on whether the state is non-final or final, respectively.

This is the way we use finite automata to compute functions. We obtain functions $f_{\mathcal{A}}$ of (1) and $\hat{f}_{\mathcal{A}}$ of (2). An important observation here is that these functions behave topologically very differently. Namely, for our model of level automata, the functions $f_{\mathcal{A}}$ are always continuous, in fact even uniformly continuous (see Theorem 1), while those of $\hat{f}_{\mathcal{A}}$ are very rarely so. We shall characterize in Theorem 2 when this happens.

The explanation to above comes from the fact that the topologies of domain spaces of $f_{\mathcal{A}}$ and $\hat{f}_{\mathcal{A}}$ are very different. Indeed, in A^ω the points

$$w10^\omega \text{ and } w01^\omega \quad \text{for } w \in A^+,$$

are not close to each other, while their numerical interpretations coincide.

The above explains our title, as well as justifies our presentation in this meeting: we use a conventional model of computing in an unconventional way. Moreover, we are trying to convince the reader that this approach reveals interesting phenomena. More concretely, it allows to compute some in a classical sense very complicated functions in a very simple way, as well as to compute only very few nicely behaving classical functions. The former statement is supported by our construction of a continuous nowhere differentiable function – and still its computation is not more demanding than that of a cubic polynomial! The latter observation is based on our analysis of automata computing the parabola $f(x) = x^2$.

In our presentation, we restrict our considerations to functions $[0, 1] \to \mathbb{R}_+$. Extensions to higher dimension, that is to functions $[0, 1]^2 \to \mathbb{R}_+$, in other words to real images, might be more interesting from the point of view of applications. However, the theoretical phenomena we want to stress are clearer in one dimension, we believe.

More details of this topic can be found in [6] and references given therein.

2   Preliminaries

In this section we define the basic notions and fix the terminology of our presentation. This requires, e.g., to interrelate quite different areas of mathematical research. On the one hand we are dealing with very concrete objects like natural (and artificial) images, or their one-dimensional counterparts, functions, and how they can be represented in terms of words. On more theoretical side, we are interrelating automata theory to that of classical analysis, that is the theory of functions – a connection which is rather rarely analyzed. Crucial notions here are matrices and their products, allowing the use of powerful theory of linear algebra.

We denote by A a finite alphabet. In our considerations it is typically binary, say $A_2 = \{0,1\}$, or quartic, that is $A_4 = A_2 \times A_2$, which we rename as: $(0,0) \leftrightarrow 0$, $(0,1) \leftrightarrow 1$, $(1,0) \leftrightarrow 2$ and $(1,1) \leftrightarrow 3$. An *image* F is a formal power series

$$F : A^n \to S\,,$$

where S is a semiring. Semirings considered here are those of number semirings, such as $\mathbb{R}$, $\mathbb{R}_+$, $\mathbb{Q}$ or $\mathbb{Q}_+$. The power series F, for $A = A_4$, can be interpreted as an n-resolution picture, and hence referred to as an n-*image*, as described below.

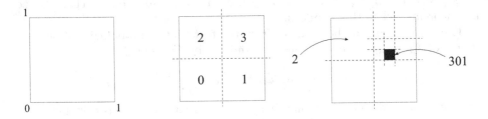

Fig. 1. The unit square, its division into four subsquares and the addresses of the pixels

Consider the unit square on Euclidean space, and its division into four equal subsquares shown in Fig. 1. The subsquares are addressed by elements of A_4. We repeat the process n times. Then any element of A^n can be viewed as an *address* of a pixel in this division as shown in Fig. 1. As a conclusion the formal power series F is a multiresolution representation of a picture, where for each $w \in A_4^n$, $F(w)$ tells the value of the pixel addressed by w in the $2^n \times 2^n$ -resolution image.

The above representation is very flexible. There is no reason to consider only the (unit) square as the shape of the picture. This can equally well be a rectangular or, for example, a triangle or a line as illustrated in Fig. 2.

The division pattern, and hence also the size of the address alphabet seems to be irrelevant (up to technicalities) for any presentation. The choice of the values of F allows to consider in the same framework

- *black and white* pictures, when S is the Boolean set $\{0,1\}$,
- *greytone* pictures, when S is $\mathbb{R}$ or its subset,
- *colour* pictures, when S is $\mathbb{R}^3$ or its subset.

In the last case the three components of the image tell, in a natural way, the *intensity* of the three main colour components.

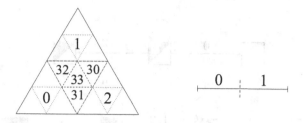

Fig. 2. Alternative division patterns

Fig. 3. Sierpinski's triangle

We continue with an example. Consider the picture of a (modified) Sierpinski's triangle depicted in Fig. 3. The above division argument yields subpictures

and the first automaton depicted in Fig. 4 telling how these subpictures are related to the addresses. After the second step of divisions we obtain the second automaton depicted in Fig. 4. From now on the process gets stabilized, no new states are introduced in divisions. The automaton becomes completed when the original picture is chosen to initial state of the automaton, as marked by ↑. Now, we obtain an approximation, an $2^n \times 2^n$ -resolution image, of Sierpinski's triangle by associating to each address $w \in A_4^n$ the subimage given by that state of the automaton where w leads from the initial state. Interestingly, in this example all resolutions are computed by the same automaton – a property which, of course, is not true in general.

In the above example the value of a pixel was determined by the state reached by the address of that pixel in an automaton. An alternative way of defining this is to associate outputs to each transition of the automaton. When doing this we have to calculate which "amount of darkness" goes with each transition. Easy calculations show that this leads from the second automaton of Fig. 4 to that of third automaton of Fig. 4, where the numbers after semicolons tell these portions.

In this automaton the darkness of a pixel is obtained as the product of weights of its address. Relative darknesses of pixels of any level are in correct portion. However, the image is getting paler and paler when the resolution is increased – indeed, the weight of any infinite computation tends to zero. In order to avoid

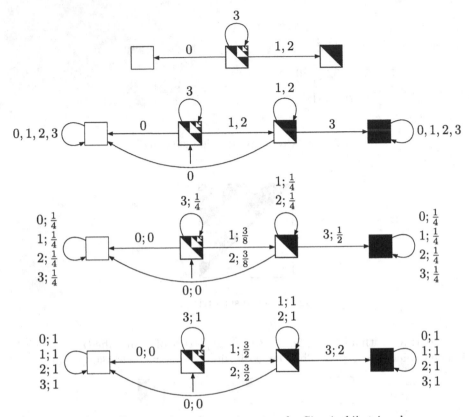

Fig. 4. Construction of an automaton for Sierpinski's triangle

this we *scale* the automaton as follows. Since in each division step we introduce four times more pixels, we multiply all the weights by four. This yields out final (weighted) automaton approximating the Sierpinski's triangle.

Actually the second and fourth automata of Figure 4 compute the same $2^n \times 2^n$ -resolution approximations of our image. The difference being that in the second automaton the value of a pixel is determined by the end state, while in the fourth it is obtained as the product of all weights of the computation. A few first approximations are shown in Fig. 5.

Before going into more formal definitions we want to make a few remarks. In above we were very detailed not only to make it clear what we mean by automata-theoretic representation of images, but to emphasize the diversified nature of this approach. Indeed from above it should be clear that the dimension of the picture, that it whether it is one, two or three, makes only notational differences and difficulties.

We continue by giving our formal definitions. A *weighted finite automaton* (WFA for short) is a 5-tuple $\mathcal{A} = (Q, A, W, I, T)$, where

- Q is a finite set of states;
- A is a finite input alphabet;

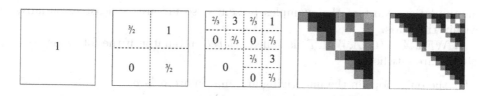

Fig. 5. Approximation of Fig. 3 by automata of Fig. 4

- $W : Q \times A \times Q \to \mathbb{R}$ is a weight function;
- $I : Q \to \mathbb{R}$ is the initial distribution and;
- $T : Q \to \mathbb{R}$ is the final distribution.

The weight function W decomposes in a natural way to functions $W_a : Q \times Q \to \mathbb{R}$ for $a \in A$, and hence can be viewed as a set of matrices over $\mathbb{R}$ indexed by elements of A. A WFA is a natural extension of the notion of *nondeterministic* finite automaton (NFA for short). Indeed, by setting

$$E = \{(p, a, q) \mid W_a(p, q) \neq 0\},$$
$$i = \{q \in Q \mid I(q) \neq 0\}, \text{ and}$$
$$t = \{q \in Q \mid T(q) \neq 0\}$$

we obtain from $\mathcal{A}$ an NFA (Q, A, E, i, t), so-called *underlying automaton* of WFA $\mathcal{A}$. It is worth mentioning already here that it does not make any difference whether W is total or partial, as long as the zero is allowed as a weight.

The theory of weighted finite automata (or $\mathbb{R} - A$ -automata in terms of Eilenberg) is a well studied research area, see, e.g., [5,8,1], usually referred to as the theory of *rational formal power series*. The two aspects which make this chapter justified are the following two features. We use the theory to a particular application, namely to deal with images and real functions, and we consider these automata on infinite inputs.

Let $\mathcal{A} = (Q, A, W, I, T)$ be a weighted finite automaton. It defines the functions

$$F_{\mathcal{A}} : A^* \to \mathbb{R}, \quad F_{\mathcal{A}}(w) = I \cdot W_w \cdot T$$

and

$$f_{\mathcal{A}} : A^\omega \to \mathbb{R}, \quad f_{\mathcal{A}}(w) = \lim_{n \to \infty} I \cdot W_{\mathrm{pref}_n(w)} \cdot T,$$

where pref_n denotes the prefix of length n and the matrix W_w for $w = a_1, \ldots a_t$ is defined by $W_w = W_{a_1} \cdots W_{a_t}$. As we noted (due to zero weights) $F_{\mathcal{A}}$ can be assumed to be always defined while $f_{\mathcal{A}}$ is only a partial function – if the limit on the right does not exist then the value of $f_{\mathcal{A}}$ is undefined. We call $F_{\mathcal{A}}$ and $f_{\mathcal{A}}$ as *word and ω-word functions* defined by $\mathcal{A}$, respectively.

The question whether or not $f_{\mathcal{A}}$ is a total function seems to be highly non-trivial, in general. This is one reason why we want to introduce a special class of WFA such that this problem does not occur. The other reason is that this special class is large enough to illustrate the power of automata theory in image manipulation. Most of the examples presented so far are, at least essentially, within this class.

A WFA $\mathcal{A} = (Q, A, W, I, T)$ is called a *level* automaton if the following conditions are satisfied:

  (i) the only loops in the underlying automaton are $p \xrightarrow{a} p$;

  (ii) $W(p, a, q) \geq 0$ for all $a \in A$ and $p, q \in Q$;

 (iii) for every $p \in Q$:
- $W(p, a, p) < 1$ for all $a \in A$, if there exists $q \neq p$ and $b \in A$ such that $W(p, b, q) \neq 0$;
- $W(p, a, p) = 1$ for all $a \in A$, otherwise;

 (iv) $I \in \mathbb{R}_+^t$ and $T \in \mathbb{R}_+^t$;

  (v) the underlying automaton of $\mathcal{A}$ is reduced.

In terms of classical theory, word functions computed by level automata are $\mathbb{R}_+$-rational. In addition to that the condition (iii) is crucial. It guarantees, as we shall soon see, that an ω-word function is always defined. It also allows to define the *degree* of a state of a WFA as follows: The states with loops of weight one are of degree 0, and a state q is of degree i if there exists a transition from that state to a state of degree $i - 1$ with a nonzero weight, and moreover all transitions from it with nonzero weights go to states of degree at most $i - 1$. The degree of the level automaton is defined to be the maximum of the degrees of its states.

Part of the motivation of introducing the above restricted class of WFA is demonstrated in the following lemma and its consequences. A further motivation comes from the fact, as we shall see, that already this class allows a number of nontrivial and illustrative results and examples.

Let $\mathcal{A} = (Q, A, W, I, T)$ be a level automaton. For any $q \in Q$, let T_q denote the final distribution where state q is assigned value 1 and all other states value 0. We define an ω-word function $f_q : A^\omega \to \mathbb{R}_+$ by

$$f_q(w) = \lim_{n \to \infty} I W_{\mathrm{pref}_n(w)} T_q$$

and prove

Lemma 1. *The functions f_q are always defined. Moreover, if q is of positive degree then f_q is the zero function.*

Proof. A simple argumentation from analysis, see e.g. [6].　　　　　□

As an immediate consequence of Lemma 1 we formulate

Corollary 1. *The ω-word function of a level automaton is always defined.*

Actually, we can state even a stronger statement. We recall that the set of all infinite words A^ω becomes a metric space when we define distance function $d : A^\omega \to \mathbb{R}_+$ as follows:

$$d(u, v) = 2^{-|u \wedge v|},$$

where the function $\wedge$ defines the maximal common prefix of infinite words. By convention $d(u, u) = 0$. Now we can formulate

Theorem 1. *For each level automaton $\mathcal{A}$ the ω-word function $f_{\mathcal{A}}$ is continuous, in fact, even uniformly continuous.*

Proof. Follows from considerations of Lemma 1 and the triangular inequality:

$$\left| f_{\mathcal{A}}(u) - f_{\mathcal{A}}(v) \right| \leq \left| f_{\mathcal{A}}(u) - F_{\mathcal{A}}(u \wedge v) \right| + \left| f_{\mathcal{A}}(v) - F_{\mathcal{A}}(u \wedge v) \right| . \qquad \square$$

The above shows that WFA, or at least their restriction to level automata, behave quite smoothly as generators of ω-word functions. The picture changes drastically when they are viewed as tools to define images, or to compute real functions. This is an essential point of this chapter.

As shown earlier we can transform a four letter alphabet into coordinates of a plane, or accordingly a binary alphabet into digits of numbers. We are mainly dealing with the binary case and hence analyzing how automata can be used to compute real functions.

Let $A = \{0, 1\}$ be a binary alphabet and $\mathcal{A}$ a WFA with A as the alphabet. An infinite word $\omega = a_1 a_2 \ldots$ over A is viewed as a binary representation

$$0.a_1 a_2 \ldots$$

of a real number. This representation is not unique since for finite word $u \in A^*$ the words

$$u10^{\omega} \text{ and } u01^{\omega}$$

represent the same number. The uniqueness is achieved by restricting to infinite words in the set

$$X = A^{\omega} \setminus A^* 1^{\omega} = \{ w \in A^{\omega} \mid w \text{ contains infinitely many 0's} \} .$$

Now, every real $x \in [0, 1)$ possesses the unique binary representation $\text{bin}(x) \in X$. In other words $\text{bin}(x) = a_1 a_2 \ldots$ is the unique word in X satisfying

$$x = \sum_{i=1}^{\infty} a_i 2^{-i} .$$

For a binary infinite word ω in X we denote by $\hat{\omega}$ the unique binary number it represents, that is $\omega = \text{bin}(\hat{\omega})$. A class of real numbers, namely those which possess a finite binary representation, are of special interest in our considerations. They are all rational numbers referred to as *dyadic rationals*. Clearly, each dyadic rational $r \in [0, 1)$ is of the form $r = \frac{m}{2^{\ell}}$ for some nonnegative integers m and ℓ with $m < 2^{\ell}$.

The real function $\hat{f}_{\mathcal{A}} : [0, 1) \to \mathbb{R}$ computed by WFA $\mathcal{A}$ is defined by

$$\hat{f}_{\mathcal{A}}(x) = f_{\mathcal{A}}\big(\text{bin}(x)\big) .$$

Of course, the above allows $\hat{f}_{\mathcal{A}}$ to be only partial, although for level automaton it is always total by Corollary 1. Central questions of our considerations will be those where $\hat{f}_{\mathcal{A}}$ is either *continuous* or even *smooth*. We recall that a real function is smooth if and only if it has all derivatives on the considered interval.

Theorem 1 has the following interesting corollary stating that functions $\hat{f}_{\mathcal{A}}$ are continuous everywhere except possibly in the denumerable set of dyadic points.

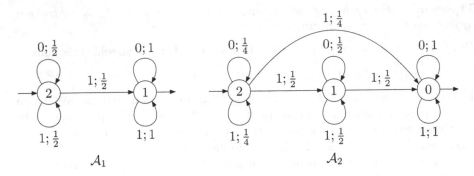

Fig. 6. Two level automata

Corollary 2. *For any level automaton $\mathcal{A}$, the real function $\hat{f}_\mathcal{A}$ is continuous at any non-dyadic $x \in [0,1)$, and right continuous at every dyadic point $x \in [0,1)$.*

Proof. Let $w = \text{bin}(x)$. If x is non-dyadic, w does not end in 0^ω or 1^ω. Let d be the metric of A^ω defined earlier, and d_E the normal Euclidean metric in the interval $[0,1)$. It is easy to verify (relying on the fact that w does not end in 0^ω or 1^ω) that for any $\varepsilon > 0$ there exists $\delta > 0$ such that

$$d_E(x,y) < \delta \Longrightarrow d\big(\text{bin}(x), \text{bin}(y)\big) < \varepsilon. \qquad (3)$$

The result then immediately follows from the continuity of $f_\mathcal{A}$ (Theorem 1). The proof of the right continuity at dyadic rationals $x \in [0,1)$ is similar, except that the implication (3) is only guaranteed to hold for $y \geq x$. $\qquad\square$

We call the subinterval

$$\left[\frac{m}{2^\ell}, \frac{m+1}{2^\ell}\right) \subseteq [0,1)$$

a *dyadic interval*, and consistently to our earlier explanations define its *address* to be the prefix of $\text{bin}\left(\frac{m}{2^\ell}\right)$ of length ℓ. In other words the address of the above interval is the length ℓ binary expansion of the integer $m < 2^\ell$.

Example 1. Consider the WFA depicted in Figure 6. In these illustrations we use incoming arrows to represent initial distributions, in this example $(1,0)$ and $(1,0,0)$, respectively, and outgoing arrows to represent final distributions, here $(0,1)$ and $(0,0,1)$, respectively.

Let us analyze first the automaton $\mathcal{A}_1$. Clearly, the weight of $w = 0110^\omega$ is

$$\frac{1}{4} + \frac{1}{8} = \frac{3}{8}.$$

In other words $\hat{f}_{\mathcal{A}_1}\left(\frac{3}{8}\right) = \frac{3}{8}$. Interestingly, we also have

$$f_{\mathcal{A}_1}(0101^\omega) = \frac{1}{4} + \frac{1}{16}\sum_{i=0}^{\infty}\frac{1}{2^i} = \frac{1}{4} + \frac{1}{8} = \frac{3}{8}.$$

Actually, simple general computations show that $\hat{f}_{A_1}(x) = x$ for all $x \in [0,1)$. Similarly, we can conclude that $\hat{f}_{A_2}(x) = x^2$ for all $x \in [0,1)$. By construction of A_2,

$$f_{A_2}(0w) = \frac{1}{4}f_{A_2}(w)$$

and

$$f_{A_2}(1w) = \frac{1}{4}f_{A_2}(w) + \frac{1}{2}f_{A_1}(w) + \frac{1}{4}.$$

Since $\widehat{0w} = \frac{1}{2}x$ and $\widehat{1w} = \frac{1}{2}x + \frac{1}{2}$, we have

$$\hat{f}_{A_2}\left(\frac{1}{2}x\right) = \frac{1}{4}f_{A_2}(x) \tag{4}$$

and

$$\hat{f}_{A_2}\left(\frac{1}{2}x + \frac{1}{2}\right) = \frac{1}{4}f_{A_2}(x) + \frac{1}{2}f_{A_1}(x) + \frac{1}{4}. \tag{5}$$

Moreover

$$\hat{f}_{A_2}(1) = 1. \tag{6}$$

Formulas (4) and (5) hold true for all x in $[0,1]$. So together with (6) they imply that

$$\hat{f}_{A_2}(z) = z^2$$

for all z having a finite binary representation.

In the above examples the functions computed by WFA were not only continuous but also smooth. As we shall see in the Corollary 5, and will illustrate in the next example, this happens very rarely.

As we hinted in the above example, a necessary condition for the continuity of a function $\hat{f}_A$ at point $x = \frac{1}{2}$ is the equality

$$f_A(10^\omega) = f_A(01^\omega).$$

The next theorem formulates the continuity of $\hat{f}_A$ at dyadic rationals in terms of simple conditions on the ω-word function f_A. Later we see that these conditions can be effectively checked.

Theorem 2. *For any level automaton A the following conditions are equivalent:*

(i) $\hat{f}_A$ *is continuous on the interval $[0,1)$;*
(ii) $f_A(u01^\omega) = f_A(u10^\omega)$ *for all words $u \in A^*$.*

Proof. If $\hat{f}_A$ is continuous then for every $u \in A^*$

$$f_A(u10^\omega) = \hat{f}_A(\widehat{u10^\omega})$$
$$= \lim_{n \to \infty} \hat{f}_A(\widehat{u01^n0^\omega})$$
$$= \lim_{n \to \infty} f_A(u01^n0^\omega)$$
$$= f_A(u01^\omega).$$

Conversely, assume (ii). Continuity of $\hat{f}_A$ at non-dyadic rationals and 0 was extablished in Corollary 2. Consider then an arbitrary positive dyadic rational $x = \widehat{u10^\omega}$. As in the proof of Corollary 2, for any $\varepsilon > 0$ there exists $\delta > 0$ such that

$$d_E(x, y) < \delta \implies d(\mathrm{bin}(y), w) < \varepsilon \text{ for } w = u01^\omega \text{ or } w = u10^\omega.$$

The result now follows from the continuity of f_A and the assumption that $f_A(u01^\omega) = f_A(u10^\omega)$. □

We call a WFA *strongly* continuous, if it computes a continuous function for all initial distributions. For a WFA $\mathcal{A}$, we denote by $\mathcal{A}_I$ the automaton obtained by changing the initial distribution into I. The following corollary is obtained from Theorem 2, by considering all initial distributions.

Corollary 3. *For any level automaton $\mathcal{A}$ the following conditions are equivalent:*

(i) *$\mathcal{A}$ is strongly continuous on the interval $[0, 1)$;*
(ii) *$f_{\mathcal{A}_I}(01^\omega) = f_{\mathcal{A}_I}(10^\omega)$ for all initial distributions I.*
(iii) *$f_{\mathcal{A}_I}(01^\omega) = f_{\mathcal{A}_I}(10^\omega)$ for all unit coordinate vectors $I = (\ldots 0, 1, 0, \ldots)$.*

Proof. Implication (i) $\implies$ (ii) follows from Theorem 2, and (ii) $\implies$ (iii) is trivial. Let us prove that (iii) implies (i). Let I be an arbitrary initial distribution. By Theorem 2, to prove (i) it is enough to show that $f_{\mathcal{A}_I}(u01^\omega) = f_{\mathcal{A}_I}(u10^\omega)$ for all $u \in A^*$. Denote $IW_u = (a_1, a_2, \ldots a_n)$, and let I_i be the i-th unit coordinate vector. Then, by (iii),

$$f_{\mathcal{A}_I}(u01^\omega) = \sum_{i=1}^{n} a_i f_{\mathcal{A}_{I_i}}(01^\omega) = \sum_{i=1}^{n} a_i f_{\mathcal{A}_{I_i}}(10^\omega) = f_{\mathcal{A}_I}(u10^\omega). \qquad □$$

After the above results, it is illustrative to consider the following general example.

Example 2. Consider the level automaton shown in Figure 7. Here, we have assumed, consistently to our definition of level automata, that $\alpha, \beta \in [0, 1)$, $\gamma, \delta \in \mathbb{R}_+$, the weights of the loops on state q_1 are equal to 1, and that q_0 and q_1 correspond to initial and final states, respectively.

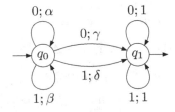

Fig. 7. A simple two state level automaton

In this special case the continuity and the strong continuity clearly coinside. By (iii) of Corollary 3 we see that $\hat{f}_A$ is continuous if and only if

$$f_A(01^\omega) = f_A(10^\omega),$$

which can be rewritten as

$$\gamma + \alpha\delta \sum_{i=0}^{\infty} \beta^i = \delta + \beta\gamma \sum_{i=0}^{\infty} \alpha^i,$$

or equivalently, as

$$(\alpha + \beta - 1)\big(\delta(1 - \alpha) - \gamma(1 - \beta)\big) = 0.$$

As a conclusion we have shown that the automaton $\mathcal{A}$ of Figure 7 computes a continuous function if and only if one of the following conditions is satisfied:

(i) $\alpha + \beta = 1$;

or

(ii) $\delta(1 - \alpha) = \gamma(1 - \beta)$.

At this point a few comments are in order. First, for a random choice of parameters α, β, γ and δ the probability that $\mathcal{A}$ computes a continuous function is zero! This justifies our view that WFA compute very rarely nicely behaving functions of analysis. Secondly, one can continue the above analysis, for example, by requiring that $\hat{f}_A$ would have a derivative at point $x = \frac{1}{2}$. A necessary condition would be $\alpha = \beta = \frac{1}{2}$, see [2]. An example of such an automaton is that shown in Figure 6, where $\alpha = \beta = \frac{1}{2}$, $\gamma = 0$ and $\delta = \frac{1}{2}$. That computed the identity function. In Figure 8 two other functions computed by automaton $\mathcal{A}$ of Figure 7 are shown. The first one, with $\alpha = \frac{3}{4}$, $\beta = \frac{1}{4}$, $\gamma = 0$ and $\delta = 1$ computes a continuous function while the second one, with $\alpha = \frac{1}{2}$, $\beta = \frac{1}{4}$, $\gamma = 1$ and $\delta = 0$, a noncontinuous one. Further, as analyzed in [2], the latter function has no derivative in nondenumerably many points.

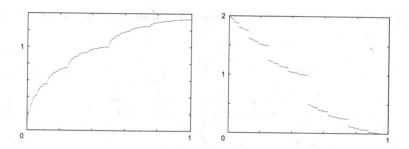

Fig. 8. Two examples of functions computed by the automaton of Figure 7

3   Computing the Parabola

In this section we focus on computing the parabola.

Example 3. Consider the three state automaton $\mathcal{A}$ illustrated in Figure 9. Now, since $\gamma + (1 - \gamma) = 1$, the automaton $\mathcal{A}_1$, that is $\mathcal{A}$ with only state 1 as initial one, computes a continuous function. Further the inequality $0 = f_{\mathcal{A}_1}(0^\omega) \neq f_{\mathcal{A}_1}(1^\omega)$ guarantees that there exists the unique hyperplane in $\mathbb{R}^2$, that is a line going through the origin, such that $\mathcal{A}$ defines a continuous function if and only if the pair (x, y) is on this line. In other words, $f_{\mathcal{A}}$ is continuous if and only if $\frac{x}{y}$ is fixed.

If we choose $\alpha = \beta = \frac{1}{4}$ and $\gamma = \frac{1}{2}$, the above ratio yields the value $\frac{1}{2}$. If we further fix $y = \frac{1}{2}$, then necessarily $x = \frac{1}{4}$ and we are in the automaton $\mathcal{A}_2$ of Figure 6. As we saw, this computes the parabola $f(x) = x^2$. Now, after fixing α, β and γ as above, the requirement that $\mathcal{A}$ computes a continuous function leads to the unique, up to the ratio of x and y, automaton $\mathcal{A}_2$. This indeed is the unique minimal level automaton for the function $f(x) = x^2$. This is justified as follows. Any such automaton has three states, as a consequence of Corollary 5 below. Moreover, by considerations in Section 4, necessarily $\alpha = \beta = \frac{1}{4}$ and $W(1,0) = W(0,0) = \frac{1}{2}$.

A striking phenomenon occurs when we try to decompose the above (essentially) unique level automaton computing the parabola $f(x) = x^2$. The only, at least automata theoretic, way of doing this is to compute $\hat{f}_{\mathcal{A}_2}$ as the sum of two functions

$$\hat{f}_{\mathcal{A}_2} = \hat{f}_{\mathcal{A}_2'} + \hat{f}_{\mathcal{A}_1'},$$

where $\mathcal{A}_2'$ is obtained from $\mathcal{A}_2$ by deleting the edge $2 \xrightarrow{1;\frac{1}{4}} 0$ and $\mathcal{A}_1'$ is obtained from $\mathcal{A}_2$ by deleting the state 1. It follows that while $\hat{f}_{\mathcal{A}_2}$ is very nicely behaving, both $f_{\mathcal{A}_2'}$ and $f_{\mathcal{A}_1}$ are complicated noncontinuous fractal type functions, as illustrated in Figure 10.

We continue by outlining the justification to the conclusions of Example 3. We associate to any WFA $\mathcal{A} = (Q, A, W, I, T)$ the linear subspaces

$$\mathcal{L} = \langle IW_u \mid u \in A^* \rangle, \text{ and}$$
$$\mathcal{R} = \langle W_u T \mid u \in A^* \rangle$$

of $\mathbb{R}^{1 \times n}$ and $\mathbb{R}^{n \times 1}$, respectively, called the *left* and the *right* generated subspace.

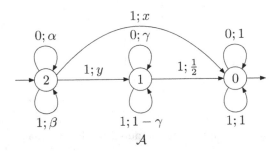

Fig. 9. A level automaton of degree two

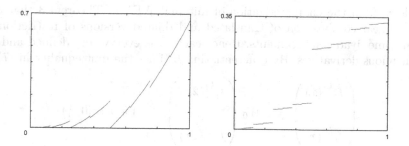

Fig. 10. Functions computed by WFA's $\mathcal{A}'_2$ and $\mathcal{A}'_1$

The following application of the generated subspaces $\mathcal{L}$ and $\mathcal{R}$ allows us to minimize a given WFA. Here the minimality is with respect to the number of states, as usual. We call a WFA $\mathcal{A}$ *minimal* if there exist no *equivalent* WFA with a fewer number of states. The equivalence is interpreted here as the equivalence of the word function $F_{\mathcal{A}}$ computed by the automata. The WFA considered are not necessarily level automata.

Theorem 3. *WFA $\mathcal{A}$ is minimal if and only if the dimensions of $\mathcal{L}$ and $\mathcal{R}$ are equal to n, the number of states. One can effectively test minimality and find an equivalent minimal WFA for any given WFA.*

The proof of Theorem 3 is an application of linear algebra, see, e.g., [6]. We have analyzed above when WFA, and in particular the level variants, define continuous functions. In the next results we require not only the continuity, but even more regularity. Namely, we characterize when a WFA defines a *smooth* function. We start with the following result from [4].

Theorem 4. *Let $\mathcal{A} = (Q, A, W, I, T)$ be a minimal (not necessarily level) WFA whose ω-word function $f_{\mathcal{A}}$ is everywhere defined and possesses continuous k first derivatives. If the k'th derivative is not zero, then W_0 has $k + 1$ distinct eigenvalues $\frac{1}{2^i}$ for $i = 0, 1, \ldots k$.*

Proof. Let $Q = \{q_1, q_2, \ldots, q_n\}$. For any $q_i \in Q$ we denote by $\hat{f}_i$ the real function computed by the WFA obtained from $\mathcal{A}$ by making q_i the only initial state. The weight matrices W_0 and W_1 state a mutually recursive linear decomposition of functions $\hat{f}_i$:

$$
\begin{pmatrix} \hat{f}_1(x) \\ \vdots \\ \hat{f}_n(x) \end{pmatrix} = \begin{cases} W_0 \begin{pmatrix} \hat{f}_1(2x) \\ \vdots \\ \hat{f}_n(2x) \end{pmatrix} & \text{for } x \in [0, \tfrac{1}{2}) \ , \\[4ex] W_1 \begin{pmatrix} \hat{f}_1(2x - 1) \\ \vdots \\ \hat{f}_n(2x - 1) \end{pmatrix} & \text{for } x \in [\tfrac{1}{2}, 1) \ . \end{cases} \tag{7}
$$

It follows from the characterization of minimal WFA in Theorem 3 that each $\hat{f}_i$ is a linear composition of translated and dilated versions of restrictions of $\hat{f}_A$ on some dyadic subsegments. Hence each $\hat{f}_i$ is everywhere defined and has k continuous derivatives. By differentiating k times the first equality in (7) we obtain

$$\begin{pmatrix} \hat{f}_1^{(k)}(x) \\ \vdots \\ \hat{f}_n^{(k)}(x) \end{pmatrix} = 2^k W_0 \begin{pmatrix} \hat{f}_1^{(k)}(2x) \\ \vdots \\ \hat{f}_n^{(k)}(2x) \end{pmatrix} \qquad \text{for } x \in [0, \tfrac{1}{2}).$$

Substituting $x = 0$ gives

$$Z = 2^k W_0 Z, \tag{8}$$

where

$$Z = \begin{pmatrix} \hat{f}_1^{(k)}(0) \\ \vdots \\ \hat{f}_n^{(k)}(0) \end{pmatrix}.$$

If Z is the zero vector then $\hat{f}_i^{(k)}(0) = 0$ for $i = 1, 2, \ldots, n$. This implies that $\hat{f}_A^{(k)}(x) = 0$ for every dyadic rational x. Dyadic rationals are dense in $[0, 1)$, so the k'th derivative of $\hat{f}_A$ is zero, contradicting the assumption in the theorem statement.

We conclude that Z is an eigenvector of W_0 corresponding to eigenvalue $\frac{1}{2^k}$. This reasoning can be applied with any $i \leq k$ in place of k.   □

Since an $n \times n$ matrix can have at most n eigenvalues we have

Corollary 4. *Each WFA computing a real function with non-vanishing k'th derivative contains at least $k + 1$ states.*

If an n-state WFA computes a smooth function then, by the corollary, the n'th derivative of the function must be zero, so the function must be a polynomial of degree at most $n - 1$.

Corollary 5. *Any smooth function computed by a WFA is a polynomial. A WFA that computes a polynomial of degree k must have at least $k + 1$ states.*

4 A Monster Function

The goal of this section is to further emphasize how simple WFA can define complicated functions – at least in the spirit of classical analysis. More precisely we introduce a four state level automaton which defines a continuous but nowhere derivable real function. The material of this section comes from [3] and [7].

We consider the automaton $\mathcal{A}(t)$ of Figure 11. We first note that $\mathcal{A}(t)$ is strongly continuous. Indeed subautomaton constituting of states 0 and 1 (or 0 and 1') defines a continuous function by Example 2. And the whole automaton does the same by symmetry. Our second observation is that the subautomaton,

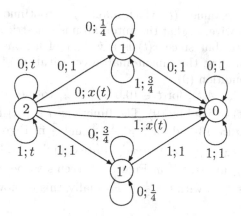

Fig. 11. A four state level automaton $\mathcal{A}(t)$

say $\mathcal{A}_1$, constituting of states 0, 1 and 2 is strongly continuous, and hence by Corollary 3, the continuity of $\hat{f}_{\mathcal{A}_1}$ is equivalent to the condition

$$f_{\mathcal{A}_1}(10^\omega) = f_{\mathcal{A}_1}(01^\omega).$$

Simple calculations show that this is equivalent to

$$t\left(\frac{1}{1-t}x(t) + \frac{1}{1-t}\cdot 1 \cdot \frac{4}{3}\right) = x(t),$$

and further to

$$x(t) = \frac{4t}{2t-1}, \quad \text{provided } t \neq \frac{1}{2}. \tag{9}$$

The behavior of the function $x(t)$ is depicted in Figure 12. As a conclusion, for $t = \frac{1}{2}$, $\mathcal{A}_1$ never defines a continuous function but for other values of $t \in (0,1)$,

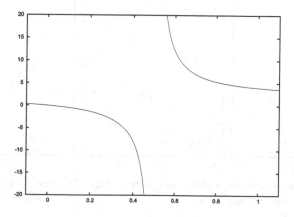

Fig. 12. A graph of the function $x(t)$

there exists the unique value $x(t)$ which makes $\hat{f}_{A_1}$ continuous. Of course, this value can be also negative, so that the automaton is not strictly within the class of our level automata, but since $x(t)$ is a weight of a noncycle this is just a technicality. By symmetry, the subautomaton constituting of states 0, 1′ and 2 leads to the same condition (9).

From now on we fix $\mathcal{A}(t)$, for $t \in (0,1)$, $t \neq \frac{1}{2}$, to mean the automaton of Figure 11, where $x(t)$ is fixed by (9). The functions $\hat{f}_{A(t)}(x)$ for $t = \frac{1}{5}$, $\frac{1}{4}$ and $\frac{1}{3}$ are illustrated in Figure 13 and for $t = \frac{11}{15}$, $\frac{3}{4}$ and $\frac{7}{9}$ in Figure 15. While functions $\hat{f}_{A(1/4)}(x)$ and $\hat{f}_{A(3/4)}(x)$ are already quite complicated looking fractaltype functions, $\hat{f}_{A(2/3)}(x)$, illustrated in Figure 14, seems to be even more chaotic, likewise all functions $\hat{f}_{A(t)}$ with $t \neq \frac{1}{4}, \frac{3}{4}$. Formally, this is shown in the following results.

Theorem 5. *Functions $\hat{f}_{A(1/4)}$ and $\hat{f}_{A(3/4)}$ are continuous, but they possess a derivative equal to zero at dyadic rational points.*

Theorem 6. *The function $\hat{f}_{A(t)}$, for $t \in (0,1)$, $t \neq \frac{1}{4}, \frac{3}{4}$, is continuous, but does not have a derivative at any point.*

Proof. We outline the proof of Theorem 6 – that of Theorem 5 is implicitly in these considerations.

Let $t \neq \frac{1}{4}, \frac{3}{4}$ and $x(t)$ be fixed. We choose an arbitrary $w \in \{0,1\}^\omega \setminus \{0,1\}^*1^\omega$ and show that, for some constant c, there exist infinitely many words $w_i \in \{0,1\}^\omega \setminus \{0,1\}^*1^\omega$ such that

$$|f_{A(t)}(w) - f_{A(t)}(w_i)| \geq c|\hat{w} - \hat{w}_i|. \qquad (10)$$

This indeed proves the theorem.

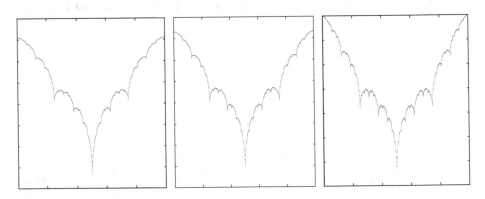

Fig. 13. Functions computed by the automaton $\mathcal{A}(t)$ of Figure 11 for values $t = \frac{1}{5}$, $t = \frac{1}{4}$ and $t = \frac{1}{3}$, respectively

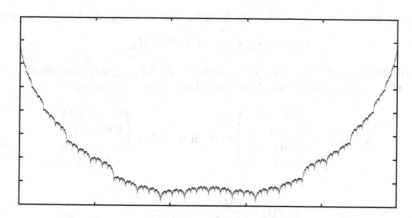

Fig. 14. An example of a continuous nowhere differentiable function computed by the automaton $\mathcal{A}(t)$ of Figure 11 for value $t = \frac{2}{3}$

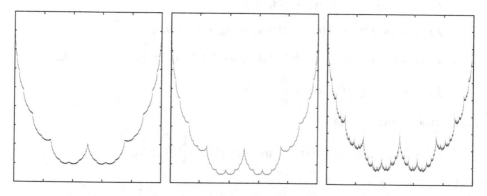

Fig. 15. Functions computed by the automaton $\mathcal{A}(t)$ of Figure 11 for values $t = \frac{11}{15}$, $t = \frac{3}{4}$ and $t = \frac{7}{9}$, respectively

Let $w_n = \mathrm{pref}_n(w)$ and define the following four sequences of words:

$$w_0(n) = w_n 0^\omega, \qquad w_1(n) = w_n 10^\omega,$$
$$w_2(n) = w_n 110^\omega, \qquad w_3(n) = w_n 1^\omega.$$

Actually, the word $w_3(n)$ is illegal but we can use it since the function is continuous. It follows that

$$|\hat{w} - \widehat{w_i(n)}| \leq \frac{1}{2^n} \quad \text{for } i = 0, 1, 2, 3.$$

The weight matrices associated to $\mathcal{A}(t)$ are

$$M_0 = \begin{pmatrix} t & 1 & 0 & x(t) \\ 0 & \frac{1}{4} & 0 & 1 \\ 0 & 0 & \frac{3}{4} & 1 \\ 0 & 0 & 0 & 1 \end{pmatrix} \text{ and } M_1 = \begin{pmatrix} t & 0 & 1 & x(t) \\ 0 & 0 & \frac{1}{4} & 1 \\ 0 & \frac{3}{4} & 0 & 1 \\ 0 & 0 & 0 & 1 \end{pmatrix},$$

and let us denote $(\alpha_n, \beta_n, \gamma_n, \delta_n)$ the distribution given by the prefix w_n, that is

$$(\alpha_n, \beta_n, \gamma_n, \delta_n) = (1, 0, 0, 0) M_{w_n} .$$

In particular, $\alpha_n = t^n$ for all n. We need the limits $\lim_{n \to \infty} M_0^n$ and $\lim_{n \to \infty} M_1^n$. Straightforward computations show that these exist, and moreover

$$\lim_{n \to \infty} M_0^n = \begin{pmatrix} 0 & 0 & 0 & r(t) \\ 0 & 0 & 0 & \frac{4}{3} \\ 0 & 0 & 0 & 0 \\ 0 & 0 & 0 & 1 \end{pmatrix} \quad \text{and} \quad \lim_{n \to \infty} M_1^n = \begin{pmatrix} 0 & 0 & 0 & r(t) \\ 0 & 0 & 0 & 0 \\ 0 & 0 & 0 & \frac{4}{3} \\ 0 & 0 & 0 & 1 \end{pmatrix} ,$$

where

$$r(t) = \frac{x(t)}{1 - t} - \frac{4}{3(t - 1)} .$$

With the help of above we can compute

$$f_{\mathcal{A}(t)}(w_0(n)) = r(t) \cdot \alpha_n + \frac{4}{3} \beta_n + \delta_n ,$$

$$f_{\mathcal{A}(t)}(w_1(n)) = (t \cdot r(t) + x(t)) \alpha_n + \beta_n + \gamma_n + \delta_n ,$$

$$f_{\mathcal{A}(t)}(w_2(n)) = (t^2 \cdot r(t) + t \cdot x(t) + x(t) + 1) \alpha_n + \frac{3}{4} \beta_n + \frac{5}{4} \gamma_n + \delta_n ,$$

$$f_{\mathcal{A}(t)}(w_3(n)) = r(t) \cdot \alpha_n + \frac{4}{3} \gamma_n + \delta_n ,$$

and further conclude that

$$f_{\mathcal{A}(t)}(w_0(n)) - f_{\mathcal{A}(t)}(w_3(n)) = \frac{4}{3}(\beta_n - \gamma_n) ,$$

and

$$f_{\mathcal{A}(t)}(w_1(n)) - f_{\mathcal{A}(t)}(w_2(n)) = \left(\frac{4}{3} t - 1 \right) \alpha_n + \frac{1}{4}(\beta_n - \gamma_n) .$$

Consequently,

$$4\Big(f_{\mathcal{A}(t)}(w_1(n)) - f_{\mathcal{A}(t)}(w_2(n)) \Big)$$
$$= \frac{3}{4}\Big(f_{\mathcal{A}(t)}(w_0(n)) - f_{\mathcal{A}(t)}(w_3(n)) \Big) + 4 \left(\frac{4}{3} t - 1 \right) t^n ,$$

implying that

$$\max\left\{ \left| f_{\mathcal{A}(t)}(w_1(n)) - f_{\mathcal{A}(t)}(w_2(n)) \right|, \left| f_{\mathcal{A}(t)}(w_0(n)) - f_{\mathcal{A}(t)}(w_3(n)) \right| \right\}$$
$$\geq \frac{1}{2} \left| \frac{4}{3} t - 1 \right| t^n .$$

This, in turn, means that there exists an infinite subset $I_0 \subseteq \mathbb{N}$ such that

$$\frac{\left| f_{\mathcal{A}(t)}(w) - f_{\mathcal{A}(t)}(w_i(n)) \right|}{|\hat{w} - \widehat{w_i(n)}|} \geq \left| \frac{1}{3} t - \frac{1}{4} \right| (2t)^n \quad \text{for all } i \in I_0 .$$

This confirms (10) when $t > \frac{1}{2}$ and $t \neq \frac{3}{4}$. The case $t < \frac{1}{2}$, with $t \neq \frac{1}{4}$, is similar, in fact, symmetric. The above reasoning also explains why the theorem does not hold for $t = \frac{3}{4}$, and even with some extra efforts why Theorem 5 holds true. □

As the final remark we emphasize the following. The functions $\hat{f}_{\mathcal{A}(t)}$ we defined are certainly quite complicated in terms of classical analysis. However, their automata-theoretic representation is very simple, and also provides a fast method to compute their values at any point. Since $\mathcal{A}(t)$ contains only four states, which is the minimal number of states needed to compute a cubic polynomial, as we have seen, computationally $\hat{f}_{\mathcal{A}(t)}$ is not harder than any cubic polynomial in our formalism.

References

1. Berstel, J., Reutenauer, C.: Noncommutative rational series with applications. Encyclopedia of Mathematics and its Applications, (137). Cambridge University Press, Cambridge (2010)
2. Culik II, K., Karhumäki, J.: Finite automata computing real functions. SIAM J. Comput. 23, 789–814 (1994)
3. Derencourt, D., Karhumäki, J., Latteux, M., Terlutte, A.: On continuous functions computed by finite automata. RAIRO-Theor. Inf. Appl. 29, 387–403 (1994)
4. Droste, M., Kari, J., Steinby, P.: Observations on the smoothness properties of real functions computed by weighted finite automata. Fund. Inform. 73(1,2), 99–106 (2006)
5. Eilenberg, S.: Automata, languages and machines. Academic Press, London (1974)
6. Karhumäki, J., Kari, J.: Finite automata, image manipulation and automatic real functions. In: Handbook of Automata. European Mathematical Society (to appear)
7. Sallinen, T.: Reaalifunktioiden laskennasta automaateilla. Master's thesis, University of Turku (2009)
8. Salomaa, A., Soittola, M.: Automata-theoretic aspects of formal power series. Springer, Heidelberg (1978)
9. Schützenberger, M.P.: On the definition of a family of automata. Information and Control 4, 245–270 (1961)

Tutorial on Cellular Automata

Nicolas Ollinger

Laboratoire d'informatique fondamentale de Marseille (LIF),
Aix-Marseille Université, CNRS,
39 rue Joliot-Curie, 13 013 Marseille, France
Nicolas.Ollinger@lif.univ-mrs.fr

This tutorial surveys computational aspects of cellular automata, a discrete dynamical model introduced by S. Ulam and J. von Neumann in the late 40s: a regular grid of finite state cells evolving synchronously according to a common local rule described by a finite automaton.

Formally, a *cellular automaton* is a tuple (d, S, N, f) where $d \in \mathbb{N}$ is the dimension of the cellular space, S is the finite set of states, $N \subseteq_{\text{finite}} \mathbb{Z}^d$ is the finite neighborhood and $f : S^N \to S$ is the local rule of the cellular automaton. A *configuration* $c \in S^{\mathbb{Z}^d}$ is a coloring of the cellular space by states.

The *global transition function* $G : S^{\mathbb{Z}^d} \to S^{\mathbb{Z}^d}$ applies f uniformly according to N, *i.e.* for every configuration $c \in S^{\mathbb{Z}^d}$ and every position $z \in \mathbb{Z}^d$ it holds

$$G(c)(z) = f(c(z + v_1), \ldots, c(z + v_m)) \quad \text{where } N = \{v_1, \ldots, v_m\}.$$

A *space-time diagram* $\Delta \in S^{\mathbb{Z}^d \times \mathbb{N}}$ is obtained by piling successive configurations of an orbit, *i.e.* for every time step $t \in \mathbb{N}$ it holds $\Delta_{t+1} = G(\Delta_t)$.

Computing inside the cellular space. The first part of the tutorial considers cellular automata as a universal model of computation. Several notions of universality are discussed: boolean circuit simulation, Turing universality, intrinsic universality. Special abilities of cellular automata as a model of massive parallelism are then investigated.

Computing properties of cellular automata. The second part of the tutorial considers properties of cellular automata and their computation. De Bruijn diagrams and associated regular languages are introduced as tools to decide injectivity and surjectivity of the global transition function in the one-dimensional case. Both immediate and dynamical properties are introduced, in particular the notion of limit set.

Computation and reduction: undecidability results. The last part of the tutorial considers computing by reduction to establish undecidability results on some properties of cellular automata: injectivity and surjectivity of the global transition function in higher dimensions, nilpotency and intrinsic universality in every dimension, a Rice's theorem for limit sets.

C.S. Calude et al. (Eds.): UC 2011, LNCS 6714, pp. 34–35, 2011.

References

1. Delorme, M.: An introduction to cellular automata: some basic definitions and concepts. In: Delorme, M., Mazoyer, J. (eds.) Cellular Automata, Saissac, pp. 5–49. Kluwer Acad. Publ., Dordrecht (1999)
2. Kari, J.: Theory of cellular automata: A survey. Theoretical Computer Science 334, 3–33 (2005)
3. Rozenberg, G., Baeck, T., Kok, J. (eds.): Handbook of Natural Computing. Springer, Berlin (2011, to appear)

Membrane Computing at Twelve Years (Back to Turku)

Gheorghe Păun

Institute of Mathematics of the Romanian Academy
P.O. Box 1-764, 014700 Bucureşti, Romania, and
Department of Computer Science and Artificial Intelligence
University of Sevilla
Avda. Reina Mercedes s/n, 41012 Sevilla, Spain
gpaun@us.es

The talk is a quick introduction to membrane computing, by briefly presenting twelve basic ideas (in the development of which the author was involved – several other ideas deserve to be mentioned), with some emphasis on two recently investigated notions, the *spiking neural P systems* (SN P systems, for short), inspired from neural biology, and the *dP systems*, a distributed class of P systems (initially introduced for so-called symport-antiport P systems, but extended also to SN P systems, a case which is discussed here in some details).

Membrane computing is a branch of natural computing whose initial aim was to abstract computing ideas (data structures, operations about them and ways to control the operations, computing architectures, models of computation) from the organization and the functioning of living cells, considered alone or in higher order structures, such as tissues, populations of cells (bacteria), neural networks. In short, multisets of objects (symbols in an alphabet, strings of symbols or even more complex data) are processed in the compartments defined by a membrane structure, which can be hierarchically organized, as in a cell, or can have the membranes placed in the nodes of a graph (like, for instance, the cells in a tissue). Various types of object processing rules are considered, with biological inspiration or with mathematical/computer science inspiration. Several ways of using the rules (semantics) were considered: synchronized (maximally or minimally parallel, with a bounded parallelism, etc.) or asynchronous. In many cases, the computing power of the obtained devices equals the power of Turing machines; sets of numbers or languages can be generated or recognized (the computing devices can be used in the generative, the accepting, or the computing modes). If ways of producing an exponential working space in a polynomial time is provided (e.g., by means of the biological operation of membrane division), then polynomial (parallel time) solutions are obtained to computationally hard (e.g., **NP**-complete) problems. Many applications were carried out, in biology and biomedicine, ecosystems study, economics, as well as in approximate optimization and computer science in general. Various software packages were produced and used in applications (including a programming language, *P-lingua*).

A state-of-the-art of membrane computing can be found in the recent Handbook, published in 2010 by Oxford University Press, [6].

C.S. Calude et al. (Eds.): UC 2011, LNCS 6714, pp. 36–37, 2011.
© Springer-Verlag Berlin Heidelberg 2011

The twelve ideas which will be discussed (chosen to also remind the number of years – actually, slightly more than twelve – since membrane computing was initiated: this happened in the autumn of the year 1998, in Turku, see [4]) are: cell-like P systems, string objects, symport-antiport rules (computing by communication), active membranes (especially, membrane division and membrane creation), tissue-like P systems, using P systems in the accepting mode, trace languages, numerical P systems, P systems with objects on membranes (brane calculi inspired P systems), P colonies, spiking neural P systems, dP systems. Other topics will be touched, without however exhausting the "dictionary" of research directions in membrane computing: controls on rule application, minimal parallelism, asynchronous systems, array objects, trees as a result of computations, population P systems, conformon objects, probabilistic-stochastic P systems, MP systems, complexity approaches, and so on. Also, some open problems and research topics will be mentioned.

Some more details – definitions, examples, and results – are given for SN P systems and for dP systems, as well as for the bridge between these two types of devices, the SN dP systems.

Also, some hints will be given concerning applications, especially in modeling biological processes, but also in studying-simulating ecosystems; applications in economics, computer graphics, approximate optimization, etc., will be also mentioned.

Only very general bibliographical information will be given below; a comprehensive source of information is the website of membrane computing, from http://ppage.psystems.eu. Many technical details can be found in the proceedings of the two yearly meetings in the area, the Brainstorming Week on Membrane Computing (held in Sevilla, in the beginning of February) and the Conference on Membrane Computing (each year, in the second part of August; until 2010, the conference was called a workshop), in the PhD theses with membrane computing subject, and in the special issues of journals devoted to membrane computing; references about all these can be found in the above mentioned website.

References

1. Ciobanu, G., Păun, G., Pérez-Jiménez, M.J. (eds.): Applications of Membrane Computing. Springer, Berlin (2006)
2. Frisco, P.: Computing with Cells. Advances in Membrane Computing. Oxford Univ. Press, Oxford (2009)
3. Păun, A.: Computability of the DNA and Cells. Splicing and Membrane Computing. SBEB Publ., Choudrant (2008)
4. Păun, G.: Computing with membranes. Turku Center for Computer Science (TUCS) Report 208 (November 1998), http://www.tucs.fi; published in J. Computer and System Sci. 61, 108–143 (2000)
5. Păun, G.: Membrane Computing. An Introduction. Springer, Berlin (2002); Chinese translation, Wuhan Univ. (2011)
6. Păun, G., Rozenberg, G., Salomaa, A. (eds.): Handbook of Membrane Computing. Oxford University Press, Oxford (2010)

Membrane Computing (Tutorial)

Ignacio Pérez-Hurtado, Mario J. Pérez-Jiménez, Agustín Riscos-Núñez,
and Francisco J. Romero-Campero

Research Group on Natural Computing
Department of Computer Science and Artificial Intelligence
University of Sevilla
Avda. Reina Mercedes s/n, 41012, Sevilla, Spain
{perezh,marper,ariscosn,fran} @us.es

The aim of the tutorial is to give a general overview of the Membrane Computing paradigm [2,5]. Membrane Computing is a quite active research field, initiated by Gh. Păun in 1998 [3]. It is a theoretical machine-oriented model, where the computational devices (known as *P systems*) are in some sense an abstraction of a living cell. There exist a large number of different definitions of P systems, but most of them share some common features: a *membrane structure* (defining in a natural way a number of regions or compartments), and an alphabet of *objects* that are able to evolve and/or move within the membrane structure according to a *set of rules* (emulating the way substances undergo biochemical reactions in a cell).

Many of the first P systems specifications that were investigated proved to be *universal* or *computationally complete* (i.e. equivalent to Turing Machines). Besides, the quest for *efficiency* has been another research direction, yielding in many cases cellular solutions to **NP**-complete problems working in polynomial time (making a space-time trade-off and using the inherent massive parallelism of P systems). The first session of the tutorial will formalize these concepts, presenting a computational complexity theory for Membrane Computing.

Another completely different approach is to concentrate on the evolution of the P systems itself, instead of focusing on the output of the computation and the number of steps. This research direction is getting an increasing attention recently. For instance, P systems have been used to model biological phenomena within the framework of cellular systems and population biology presenting models of oscillatory systems, signal transduction, gene regulation control, quorum sensing, metapopulations, and real ecosystems.

The second session of this tutorial will introduce a P systems based general framework for modelling complex dynamical phenomena.

Phenomena under investigation are described by means of multienvironment P systems consisting of a finite number of environments, each of them having a specific P system with active membranes. Each rule has associated a computable function which depends on the left-hand side of the rule and the run time. The inherent randomness and uncertainty of biological processes is captured by using probabilistic or stochastic strategies.

C.S. Calude et al. (Eds.): UC 2011, LNCS 6714, pp. 38–39, 2011.

Applications of Membrane Systems in Computational Systems and Synthetic Biology

As mentioned above, P systems are being used as tools for modelling purposes, adapting their semantics in an appropriate way. Works on this direction rely on the development of associated simulation software, since P systems have not yet been implemented neither in hardware nor in biological means.

It is interesting to note that some biological parameters of the modeled processes can be obtained experimentally by biologists or ecologists (and therefore they can be taken into account when designing the model), while some other relevant constants may be unknown. Software tools are thus necessary to enable virtual experimentation, as well as for the process of model validation.

Specifically, membrane computing, offers a modelling framework for multi-compartmental stochastic and discrete systems that allows us to design and assess bacterial synthetic gene circuits operating at the multi-cellular level [4]. The third session of this tutorial will present a specification/modelling language based on P systems that allows researchers to design synthetic gene circuits in a parsimonious and incremental manner. In this respect, membrane computing assists in one of the goals of synthetic biology, the incorporation of engineering principles into the pipelines used in the design and in vivo implementation of gene circuits exhibiting desirable behaviour or phenotypes.

The last session of the tutorial will illustrate the general modelling framework for systems biology and population dynamics based on P systems, by displaying some real case studies, and their associated software tools [1].

Acknowledgments. The authors acknowledge the support of the projects TIN2008-04487-E and TIN2009–13192 of the *Ministerio de Ciencia e Innovación* of Spain, cofinanced by FEDER funds, and the support of the Project of Excellence with *Investigador de Reconocida Valía* of the *Junta de Andalucía*, grant P08-TIC-04200.

References

1. García-Quismondo, M., Gutiérrez-Escudero, R., Pérez-Hurtado, I., Pérez-Jiménez, M.J., Riscos-Núñez, A.: An overview of P-lingua 2.0. In: Păun, G., Pérez-Jiménez, M.J., Riscos-Núñez, A., Rozenberg, G., Salomaa, A. (eds.) WMC 2009. LNCS, vol. 5957, pp. 264–288. Springer, Heidelberg (2010)
2. Păun, G., Rozenberg, G., Salomaa, A. (eds.): The Oxford Handbook of Membrane Computing. Oxford University Press, Oxford (2010)
3. Păun, G.: Computing with membranes. Journal of Computer and System Sciences 61(1), 108–143 (2000); and Turku Center for Computer Science-TUCS Report No 208
4. Smaldon, J., Romero-Campero, F.J., Fernandez-Trillo, F., Gheorghe, M., Alexander, C., Krasnogor, N.: A computational study of liposome logic: towards cellular computing from the bottom up. Systems and Synthetic Biology 4(3), 157–179 (2010)
5. The P systems Web Page, http://ppage.psystems.eu

Von Neumann Normalisation and Symptoms of Randomness: An Application to Sequences of Quantum Random Bits

Alastair A. Abbott* and Cristian S. Calude**

Department of Computer Science, University of Auckland,
Private Bag 92019, Auckland, New Zealand
aabb009@aucklanduni.ac.nz, cristian@cs.auckland.ac.nz
www.cs.auckland.ac.nz

Abstract. Due to imperfections in measurement and hardware, the flow of bits generated by a quantum random number generator (QRNG) contains bias and correlation, two symptoms of non-randomness. There is no algorithmic method to eliminate correlation as this amounts to guaranteeing incomputability. However, bias can be mitigated: QRNGs use normalisation techniques such as von Neumann's method—the first and simplest technique for reducing bias—and other more efficient modifications.

In this paper we study von Neumann un-biasing normalisation for an ideal QRNG operating 'to infinity', i.e. producing an infinite bit-sequence. We show that, surprisingly, von Neumann un-biasing normalisation can both increase or *decrease* the (algorithmic) randomness of the generated sequences. The impact this has on the quality of incomputability of sequences of bits from QRNGs is discussed.

A successful application of von Neumann normalisation—in fact, any un-biasing transformation—does exactly what it promises, *un-biasing*, one (among infinitely many) symptoms of randomness; it will not produce 'true' randomness, a mathematically vacuous concept.

1 Introduction

The outcome of some individual quantum-mechanical events cannot in principle be predicted, so they are thought as ideal sources of random numbers. An incomplete list of quantum phenomena used for random number generation include nuclear decay radiation sources [24], the quantum mechanical noise in electronic circuits known as shot noise [25] or photons travelling through a semi-transparent mirror [15,19,23,26,27,29].

Due to imperfections in measurement and hardware, the flow of bits generated by a quantum random number generator (QRNG) contains bias and correlation, two symptoms of non-randomness [7]. In this paper we study the first and simplest technique for reducing bias: von Neumann normalisation [31]. We specifically investigate the effect this has on the quality of randomness of infinite

---

* AA was in part supported by the CDMTCS.
** CC was in part supported by the CDMTCS and UoA R&SL grant.

C.S. Calude et al. (Eds.): UC 2011, LNCS 6714, pp. 40–51, 2011.

sequences of quantum random bits. Although some of the mathematical results we present apply to any RNG, our approach is intimately motivated by the operation of photon-based QRNGs, more specifically, by their mechanisms [3] and the quality of randomness they produce [2].

Von Neumann's method considers pairs of bits, and takes one of three actions: a) pairs of equal bits are discarded; b) the pair 01 becomes 0; c) the pair 10 becomes 1. Contrary to wide spread claims, the technique works for some sources of bits, but not for all. The output produced by a source of independent and constantly biased bits is transformed into a flow of bits in which the frequency of 0's and 1's are equal: 50% for each.

Mathematically, the notion of 'true randomness' is vacuous, so we can investigate only symptoms of randomness. Mathematical arguments show that we have to study infinite sequences of bits: various forms of algorithmic randomness [10] are each defined by an infinity of conditions, some 'statistical' (like bias), some 'non-statistical' (like lack of computable correlations). The symptoms of 'randomness' often emphasised for the source of a QRNG are unpredictability and uniformity of distribution. In fact, here are three—out of an infinity of—symptoms of randomness which can be used to understand quantum randomness:

1) *Unpredictability*, which is a manifestation of the strong incomputability of the bits [9] (an infinite sequence is strongly incomputable if it is provable that no Turing machine or equivalent formalism can compute and certify more than finitely many scattered bits of the sequence). For quantum bits, no bit at all can be provably computed in advance [2]. This type of unpredictability is strong—it is not due to ignorance of the system but is a fundamental feature of the system.

2) *Uniform distribution* of the generated bits (not just of individual bits, but of all n-bit strings), a manifestation of the Borel normality of a sequence. Unlike strong incomputability, this is not known to be guaranteed for quantum bits, but as we shall see, under certain conditions the sequence is Borel normal with probability one[1].

3) *Lack of patterns*, a manifestation of algorithmic randomness (incompressibility) of a sequence of bits. As for normality, this is not known to be guaranteed for quantum bits, but again a measure-theoretical argument can show such sequences are algorithmically random with probability one if normalisation can be successfully conducted.

Mathematically, it is well known that 3) → 1) and 2), but the converse implications are false. Further, both implications 1) → 2) and 2) → 1) are false [7]. While computability implies predictability, unpredictability depends not just on incomputability, but the *strength* of incomputability. Further, no sequence is absent of all possible patterns so only computable patterns can be excluded. This is one of the reasons for the impossibility of 'true randomness'.

---

[1] It is important to note the subtle theoretical difference between a *probability-one* event and a *provably guaranteed* event in the probability space of infinite sequences: in contrast with a provably guaranteed event whose complement is empty, the complement of a probability-one event can be not only non-empty, but even infinite [11].

Up until now, QRNGs have been given largely the same mathematical treatment used for pseudo-random number generators, focusing on producing uniformly distributed bits. For real devices this primarily entails the use of randomness extractors—which von Neumann's procedure is one of—to make the source as close as possible to the uniform distribution [30]. But as we have seen, uniformity does not imply unpredictability. Indeed, a device outputting successive bits of a predetermined Borel normal sequence will appear uniformly distributed. However, the strength of QRNGs is in the incomputability of the bits, and this requires computability analysis rather than probabilistic treatment. Until now, the unpredictability has been assumed an intrinsic feature of quantum bits and has escaped rigorous treatment [26]. A step in the right direction was made in [23], where violation of Bell inequalities was used to try and verify unpredictability. However, Bell tests, like the probabilistic treatment of the distribution, are unable to say anything about the computability of the bits [3].

Since uniformity of distribution is a necessary requirement for a good QRNG in addition to incomputability, techniques such as von Neumann's will need to be used for real devices where bias and correlation is inevitable. In this paper we study the effect of von Neumann normalisation on infinite sequences with a particular focus on the effect this has on the symptoms of randomness within the sequences. We focus on von Neumann normalisation because it is simple, easy to implement, and (along with the more efficient iterated version due to Peres [22] for which the results will also apply) is widely used by current proposals for QRNGs [19,20,13,26].

The main results of this paper are the following. In the 'ideal case', the von Neumann normalised output of an independent constantly biased QRNG is the probability space of the uniform distribution (un-biasing). We treat only the case of an infinite sequence of bits, but the corresponding result for finite strings also holds [1]. It is important to note that independence in the mathematical sense of multiplicity of probabilities is a model intended to correspond to the physical notion of independence of outcomes [16]. In order to study the theoretical behaviour of QRNGs, which are based on the *assumption of physical independence of measurements*, we must translate this appropriately into our formal model. In [1] we carefully defined independence of QRNGs to achieve this aim.

We then examine the effect von Neumann normalisation has on various properties of infinite sequences. In particular, Borel normality and (algorithmic) randomness are invariant under normalisation, but suprisingly for ε-random sequences with $0 < \varepsilon < 1$, normalisation can both decrease or increase the randomness of the source. Full proofs of all results presented in this paper can be found in [1].

2   Notation

We present the main notation used throughout the paper. By 2^X we denote the power set of X. By $|X|$ we denote the cardinality of the set of X. Let $B = \{0, 1\}$ and denote by B^* the set of all bit-strings (λ is the empty string). If $x \in B^*$

and $i \in B$ then $|x|$ is the length of x and $\#_i(x)$ represents the number of i's in x. By B^n we denote the finite set $\{x \in B^* \mid n = |x|\}$. The concatenation product of two subsets X, Y of B^* is defined by $XY = \{xy \mid x \in X, y \in Y\}$. If $X = \{x\}$ then we write xY instead of $\{x\}Y$. By B^ω we denote the set of all infinite binary sequences. For $\mathbf{x} \in B^\omega$ and natural n we denote by $\mathbf{x}(n)$ the prefix of $\mathbf{x}$ of length n. We write $w \sqsubset v$ or $w \sqsubset \mathbf{x}$ in case w is a prefix of the string v or the sequence $\mathbf{x}$.

A prefix-free (Turing) machine is a Turing machine whose domain is a prefix-free set of strings [7]. The prefix complexity of a string, $H_W(\sigma)$, induced by a prefix-free machine W is $H_W(\sigma) = \min\{|p| : W(p) = \sigma\}$. Fix a computable ε with $0 < \varepsilon \leq 1$. An ε–universal prefix-free machine U is a machine such that for every machine W there is a constant c (depending on U and W) such that $\varepsilon \cdot H_U(\sigma) \leq H_W(\sigma) + c$, for all $\sigma \in B^*$. If $\varepsilon = 1$ then U is simply called a universal prefix-free machine. A sequence $\mathbf{x} \in B^\omega$ is called ε–random if there exists a constant c such that $H_U(\mathbf{x}(n)) \geq \varepsilon \cdot n - c$, for all $n \geq 1$. Sequences that are 1–random are simply called random.

A sequence $\mathbf{x}$ is called Borel m–normal ($m \geq 1$) if for every $1 \leq i \leq 2^m$ one has: $\lim_{n \to \infty} N_i^m(\mathbf{x}(n))/\lfloor \frac{n}{m} \rfloor = \lim_{n \to \infty} \mathcal{N}_i^m(\mathbf{x}(n))/n = 2^{-m}$; here $N_i^m(y)$ ($\mathcal{N}_i^m(y)$) counts the number of non-overlapping (respectively, overlapping [18]) occurrences of the ith (in lexicographical order) binary string of length m in the string y. The sequence $\mathbf{x}$ is called Borel normal if it is Borel m–normal, for every natural $m \geq 1$.

A probability space is a measure space such that the measure of the whole space is equal to one [5]. More precisely, a (Kolmogorov) probability space is a triple consisting of a sample space Ω, a σ–algebra $\mathcal{F}$ on Ω, and a probability measure P, i.e. a countably additive function defined on $\mathcal{F}$ with values in $[0, 1]$ such that $P(\Omega) = 1$.

3   Von Neumann Normalisation

We define the mapping $F : B^2 \to B \cup \{\lambda\}$ as

$$F(x_1 x_2) = \begin{cases} \lambda & \text{if } x_1 = x_2, \\ x_1 & \text{if } x_1 \neq x_2, \end{cases}$$

and $f : B \to B^2$ as $f(x) = x\bar{x}$, where $\bar{x} = 1 - x$. Note that for all $x \in B$ we have $F(f(x)) = x$ and, for all $x_1, x_2 \in B$ with $x_1 \neq x_2$, $f(F(x_1 x_2)) = x_1 x_2$.

For $m \leq \lfloor n/2 \rfloor$ we define the normalisation function $VN_{n,m} : B^n \to \left(\bigcup_{k \leq m} B^k \right) \cup \{\lambda\}$ as

$$VN_{n,m}(x_1 \ldots x_n) = F(x_1 x_2)F(x_3 x_4) \cdots F\left(x_{(2\lfloor \frac{m}{2} \rfloor - 1)} x_{2\lfloor \frac{m}{2} \rfloor}\right).$$

Consider a string of n independent bits produced by a constantly biased QRNG. Let p_0, p_1 be the probability that a bit is 0 or 1, respectively, with $p_0 + p_1 = 1$, $p_0, p_1 \leq 1$.

The probability space of bit-strings produced by the QRNG is $(B^n, 2^{B^n}, P_n)$ where $P_n : 2^{B^n} \to [0,1]$ is defined by

$$P_n(X) = \sum_{x \in X} p_0^{\#_0(x)} p_1^{\#_1(x)}, \tag{1}$$

for all $X \subseteq B^n$.

The space $(B^n, 2^{B^n}, P_n)$ is just the n-fold product of the single bit probability space $(B, 2^B, P_1)$, and for this reason it is often called an 'independent identically-distributed bit source'. The outcome of successive context preparations and measurements (which photon-based QRNGs consist of) are postulated to be independent of previous and future outcomes [14]. This means there must be no causal link between one measurement and the next within the system (preparation and measurement devices included) so that the system has no memory of previous or future events. It is this physical understanding which is behind the modelling of QRNGs as independent bit sources.

The above assumption needs to be made clear as in high bit-rate experimental configurations to generate QRNs with photons, its validity may not always be clear. If the wave-functions of successive photons 'overlap' the assumption no longer holds and (anti)bunching phenomena may play a role. This is an issue that needs to be more seriously considered in QRNG design and will only become more relevant as the bit-rate of QRNGs is pushed higher and higher [3].

The von Neumann normalisation function $VN_{n,m}$ transforms the source probability space $(B^n, 2^{B^n}, P_n)$ into the target probability space $(B^m, 2^{B^m}, P_{n \to m})$. The target space of normalised bit-strings of length $1 < m \le \lfloor n/2 \rfloor$ associated to the source probability space $(B^n, 2^{B^n}, P_n)$ is the space $(B^m, 2^{B^m}, P_{n \to m})$, where $P_{n \to m} : 2^{B^m} \to [0,1]$ is defined for all $Y \subseteq B^m$ by the formula:

$$P_{n \to m}(Y) = \frac{P_n\left(VN_{n,m}^{-1}(Y)\right)}{P_n\left(VN_{n,m}^{-1}(B^m)\right)}.$$

The von Neumann procedure transforms the source probability space with constant bias into the probability space with the uniform distribution over B^m, i.e. the target probability space $(B^m, 2^{B^m}, P_{n \to m})$ has $P_{n \to m} = U_m$, the uniform distribution. Simple examples show that independence and the constant bias of P_n are essential hypotheses in the following result.

Theorem 1 (von Neumann, [1,31]). *Assume that $1 < m \le \lfloor n/2 \rfloor$. In the target probability space $(B^m, 2^{B^m}, P_{n \to m})$ associated to the source probability space $(B^n, 2^{B^n}, P_n)$ we have $P_{n \to m}(Y) = U_m(Y) = |Y| \cdot 2^{-m}$, for every $Y \subseteq B^m$.*

4   Infinite Von Neumann Normalisation

We extend von Neumann normalisation to infinite sequences of bits to study those produced by QRNGs. First, we extend the definition of the normalisation function $VN_{n,m}$ to sequences. We define $VN : B^\omega \to B^\omega \cup B^*$ as

$$VN(\mathbf{x} = x_1 \ldots x_n \ldots) = F(x_1 x_2) F(x_3 x_4) \cdots F(x_{2\lfloor \frac{n}{2} \rfloor - 1} x_{2\lfloor \frac{n}{2} \rfloor}) \cdots .$$

For convenience we also define $VN_n : B^\omega \to \left(\bigcup_{k \leq n} B^k \right) \cup \{\lambda\}$ as

$$VN_n(\mathbf{x}) = F(x_1 x_2) F(x_3 x_4) \cdots F(x_{2\lfloor \frac{n}{2} \rfloor - 1} x_{2\lfloor \frac{n}{2} \rfloor}) = VN_{n,n}(x_1 \ldots x_n).$$

Secondly, we introduce the probability space of infinite sequences as in [7]. Let $A_Q = \{a_1, \ldots, a_Q\}$, $Q \geq 2$ be an alphabet with Q elements. We let $\mathcal{P} = \{xA_Q^\omega \mid x \in A_Q^*\} \cup \{\emptyset\}$ and $\mathcal{C}$ be the class of all finite mutually disjoint unions of sets in $\mathcal{P}$; the class $\mathcal{P}$ can be readily shown to generate a σ-algebra $\mathcal{M}$. Using Theorem 1.7 from [7], the probabilities on $\mathcal{M}$ are characterised by the functions $h : A_Q^* \to [0,1]$ satisfying the following two conditions: 1. $h(\lambda) = 1$, 2. $h(x) = h(x_{a_1}) + \cdots + h(x_{a_Q})$, for all $x \in A_Q^*$.

If $Q = 2$ so $A_2 = B$, and for $x \in B^n$ we take $h(x) = P_n(\{x\})$ with P_n as defined in (1), then the above conditions are satisfied. This induces our probability measure μ_P on $\mathcal{M}$, which satisfies $\mu_P(XB^\omega) = P_n(X)$ for $X \subseteq B^n$. Hence the suitable extension of the finite case probability space to infinite generated sequences is the space $(B^\omega, \mathcal{M}, \mu_P)$. In the special case when $p_0 = p_1$ we get the Lebesgue probability $\mu_{P_L}(XB^\omega) = \sum_{x \in X} 2^{-|x|}$.

In general, if $Q \geq 2$, $p_i \geq 0$ for $i = 1, \ldots, Q$ are reals in $[0,1]$ such that $\sum_{i=1}^{Q} p_i = 1$, we can take $h_Q(x) = p_1^{\#_{a_1}(x)} \ldots p_Q^{\#_{a_Q}(x)}$ to obtain the probability space $(A_Q^\omega, \mathcal{M}, \mu_{P_Q})$ in which $\mu_{P_Q}(xA_Q^\omega) = h_Q(x)$, for all $x \in A_Q^*$.

Theorem 2. *For every string* $y \in B^*$ *there exists an uncountable set* $R \subset B^\omega$ *of* μ_P *measure zero such that for all* $\mathbf{x} \in R$, $VN(\mathbf{x}) = y$.

Proof. Let $y = y_1 \ldots y_n \in B^*$ and $D = \{00, 11\}$, the two-bit blocks which are deleted by von Neumann normalisation and $y' = f(y_1) \ldots f(y_n)$. Then every sequence $\mathbf{x} \in y'D^\omega$ satisfies $VN(\mathbf{x}) = VN_{2n}(\mathbf{x})VN(x_{2n+1}x_{2n+2}\ldots) = y$ since $VN_{2n}(\mathbf{x}) = VN_{2n,2n}(y') = y$ and for all $\mathbf{z} \in D^\omega$ we have $VN(\mathbf{z}) = \lambda$. Obviously, the set $R = y'D^\omega$ is uncountable and has μ_P measure zero as the set of Borel normal sequences has measure one [7]. $\square$

Corollary 1. *The set* $Q = \{\mathbf{x} \in B^\omega \mid VN(x) \in B^*\}$ *has* μ_P *measure zero.*

It is interesting to note that the 'collapse' in the generated sequence produced by von Neumann normalisation in Theorem 2 is not due to computability properties of the sequence. In particular, there are random sequences that collapse to any string, so to strings which are not Borel normal (see [7] for the definition of normality for strings).

In the following we need a measure-theoretic characterisation of random sequences, so we present a few facts from constructive topology and probability.

Consider the compact topological space (A_Q^ω, τ) in which the basic open sets are the sets wA_Q^ω, with $w \in A_Q^*$. Accordingly, an open set $G \subset A_Q^\omega$ is of the form $G = VA_Q^\omega$, where $V \subset A_Q^*$.

From now on we assume that the reals $p_i, 1 \leq i \leq Q$ which define the probability μ_{P_Q} are all computable. A constructively open set $G \subset A_Q^\omega$ is an open set $G = VA_Q^\omega$ for which $V \subset A_Q^*$ is computably enumerable (c.e.). A constructive sequence of constructively open sets, c.s.c.o. sets for short, is a sequence

$(G_m)_{m\geq 1}$ of constructively open sets $G_m = V_m A_Q^\omega$ such that there exists a c.e. set $X \subset A_Q^* \times \mathbf{N}$ with $V_m = \{x \in A_Q^* \mid (x,m) \in X\}$, for all natural $m \geq 1$. A constructively null set $S \subset A_Q^\omega$ is a set for which there exists a c.s.c.o. sets $(G_m)_{m\geq 1}$ with $S \subset \bigcap_{m\geq 1} G_m$, $\mu_{P_Q}(G_m) \leq 2^{-m}$. A sequence $\mathbf{x} \in A_Q^\omega$ is random in the probability space $(A_Q^\omega, \mathcal{M}, \mu_{P_Q})$ if $\mathbf{x}$ is not contained in any constructively null set in $(A_Q^\omega, \mathcal{M}, \mu_{P_Q})$. For the case of the Lebesgue probability μ_{P_L} the measure-theoretic characterisation of random sequences holds true: $\mathbf{x}$ is random if and only if $\mathbf{x}$ is not contained in any constructively null set of $(A_Q^\omega, \mathcal{M}, \mu_{P_L})$ [7,21].

We continue with another instance in which von Neumann normalisation decreases randomness.

Proposition 1. *There exist (continuously many) infinite ε-random sequences $\mathbf{x} \in B^\omega$ such that $VN(\mathbf{x}) = 000\ldots 00\ldots$ for any computable $0 < \varepsilon < 1$.*

We follow this with instances for which the converse is true: von Neumann normalisation conservers or increases randomness.

Proposition 2. *There exist (continuously many) infinite ε-random sequences $\mathbf{x} \in B^\omega$ such that $VN(\mathbf{x})$ is random for any computable $0 < \varepsilon < 1$.*

Theorem 3. *Let $\mathbf{x} \in B^\omega$ be Borel normal in $(B^\omega, \mathcal{M}, \mu_{P_L})$. Then $VN(\mathbf{x})$ is also Borel normal in $(B^\omega, \mathcal{M}, \mu_{P_L})$.*

Proof. Note that $VN(\mathbf{x}) \in B^\omega$ because $\mathbf{x}$ contains infinitely many occurrences of 01 on even/odd positions. Let $D = \{00, 11\}$, $\mathbf{x}^*(n) = VN_{n,n}(\mathbf{x}(n))$, $n' = |\mathbf{x}^*(n)|$. We have

$$\lim_{n'\to\infty} \frac{N_i^m(\mathbf{x}^*(n))}{n'} = \lim_{n'\to\infty} \left(\frac{n}{n'}\right)\left(\frac{N_i^m(\mathbf{x}^*(n))}{n}\right),$$

but as $n \to \infty$, $n' \to \infty$. We thus have $\lim_{n'\to\infty} \frac{n'}{n} = 2^{-1}$, by the normality of $\mathbf{x}$. The number of occurrences of each $i = i_1 \ldots i_m \in B^m$ in $\mathbf{x}^*(n)$ is the number of occurrences of $i' = f(i_1)y_1 f(i_2) \ldots y_{m-1}f(i_m)$ in $\mathbf{x}(n)$, summed over all $y_1, \ldots, y_{m-1} \in D^*$. Viewing i' as a string over B^2 we have:

$$\lim_{n'\to\infty} \frac{N_i^m(\mathbf{x}^*(n))}{n} = \lim_{n\to\infty} \frac{\sum_{y_1,\ldots,y_{m-1}} N_{i'}^{|i'|}(\mathbf{x}(n))}{n} = 2^{-(m+1)}.$$

Hence, both limits exist and we have

$$\lim_{n'\to\infty} \frac{N_i^m(\mathbf{x}^*(n))}{n'} = \lim_{n'\to\infty} \left(\frac{n}{n'}\right)\left(\frac{N_i^m(\mathbf{x}^*(n))}{n}\right) = 2^{-m}.$$

Since this holds for all m, i, we have that $VN(\mathbf{x})$ is Borel normal. □

Let $A_Q = \{a_1, \ldots, a_Q\}$, $Q \geq 3$. Let $\sum_{i=1}^Q p_i = 1$ where $p_i \geq 0$ for $i = 1, \ldots, Q$ and $(A_Q^\omega, \mathcal{M}, \mu_{P_Q})$ be the probability space defined by the probabilities p_i. Let

$A_{Q-1} = \{a_1, \ldots, a_{Q-1}\}$ and $(A_{Q-1}^\omega, \mathcal{M}, \mu_{P_{Q-1}^T})$ be the probability space defined by the probabilities

$$p_i^T = p_i \left(1 + \frac{p_Q}{\sum_{j=1}^{Q-1} p_j}\right) = \frac{p_i}{1 - p_Q},$$

with $1 \le i \le Q - 1$. Let $T : A_Q^* \to A_{Q-1}^*$ be the monoid morphism defined by $T(a_i) = a_i$ for $1 \le i \le Q - 1$, $T(a_Q) = \lambda$; $T(x) = T(x_1)T(x_2) \cdots T(x_n)$ for $x \in A_Q^n$. As T is prefix-increasing we naturally extend T to sequences to obtain the function $T : A_Q^\omega \to A_{Q-1}^\omega$ given by $T(\mathbf{x}) = \lim_{n \to \infty} T(\mathbf{x}(n))$ for $\mathbf{x} \in A_Q^\omega$.

Lemma 1. *The transformation T is $(\mu_{P_Q}, \mu_{P_{Q-1}^T})$-preserving, i.e. for all $w \in A_{Q-1}^*$ we have $\mu_{P_Q}\left(T^{-1}(wA_{Q-1}^\omega)\right) = \mu_{P_{Q-1}^T}\left(wA_{Q-1}^\omega\right)$.*

Proof. Take $w = w_1 \ldots w_m \in A_{Q-1}^\omega$. We have:

$$\mu_{P_Q}\left(T^{-1}(wA_{Q-1}^\omega)\right) = \mu_{P_Q}\left(\{\mathbf{x} \in A_Q^\omega \mid w \sqsubset T(\mathbf{x})\}\right) = \mu_{P_{Q-1}^T}\left(wA_{Q-1}^\omega\right). \qquad \square$$

Proposition 3. *If $\mathbf{x} \in A_Q^\omega$ is random in $(A_Q^\omega, \mathcal{M}, \mu_{P_Q})$ and T is the transformation defined above, then $T(\mathbf{x})$ is random in $(A_{Q-1}^\omega, \mathcal{M}, \mu_{P_{Q-1}^T})$.*

Proof. We generalise a result in [8] stating that, for the Lebesgue probability, measure-preserving transformations preserve randomness. Assume that $\mathbf{x}$ is random in $(A_Q^\omega, \mathcal{M}, \mu_{P_Q})$ but $T(\mathbf{x})$ is not random in in $(A_{Q-1}^\omega, \mathcal{M}, \mu_{P_{Q-1}^T})$, i.e. there is a constructive null set $R = (G_m)_{m \ge 1}$ containing $T(\mathbf{x})$. Assume that $G_m = X_m A_{Q-1}^\omega$, where $X_m \subset A_{Q-1}^\omega$ is c.e. and has the measure $\mu_{P_{Q-1}^T}(X_m A_{Q-1}^\omega)$ smaller than 2^{-m}. Define $S_m = T^{-1}(X_m A_{Q-1}^\omega) \subset A_Q^\omega$ and note that S_m is open because it is equal to $\bigcup_{w \in X_m} V_w A_Q^\omega$ with $V_w = \{v \in A_Q^\omega \mid w \sqsubset T(v)\}$ and, using Lemma 1, has the measure smaller than 2^{-m}:

$$\mu_{P_Q}(S_m) = \mu_{P_Q}\left(\bigcup_{w \in X_m} V_w A_Q^\omega\right) \le 2^{-m}.$$

We have proved that $\mathbf{x}$ is not random in $(A_Q^\omega, \mathcal{M}, \mu_{P_Q})$, a contradiction. $\qquad \square$

Let us define $VN^{-1} : 2^{B^*} \to 2^{B^*}$ for $x = x_1 \ldots x_m \in B^m$ as

$$VN^{-1}(x) = \{y \mid y = u_1 f(x_1) u_2 \ldots u_m f(x_m) u_{m+1} v \text{ and}$$
$$u_i \in \{00, 11\}^* \text{ for } 1 \le i \le m, v \in B \cup \{\lambda\}\}$$
$$= \bigcup_{n=0}^{\infty} VN_{n+2m, m}^{-1}(x),$$

and for $X \subseteq B^*$ as

$$VN^{-1}(X) = \bigcup_{x \in X} VN^{-1}(x).$$

For all $x \in B^*$ and $\mathbf{y} \in VN^{-1}(x)B^\omega$ we then have $x \sqsubset VN(\mathbf{y})$.

For the cases that $VN(\mathbf{x}) \in B^\omega$, the probability space $(B^\omega, \mathcal{M}, \mu_{P_{VN}})$ induced by von Neumann normalisation is endowed with the measure $\mu_{P_{VN}}$. The measure $\mu_{P_{VN}}$ is defined on the sets xB^ω with $x \in B^*$ by

$$\mu_{P_{VN}}(xB^\omega) = \frac{\mu_P(VN^{-1}(x)B^\omega)}{\mu_P(VN^{-1}(B^{|x|})B^\omega)}.$$

By noting that $VN^{-1}(B^{|x|}) \subset VN^{-1}(B^*)$ it is clear that $\mu_{P_{VN}}$ satisfies the Kolmogorov axioms for a probability measure. While the set $VN^{-1}(B^{|x|})$ contains sequences for which normalisation produces a finite string, from Corollary 1 we know that the set of such sequences has measure zero, so the definition of $\mu_{P_{VN}}$ is a good model of the target probability space.

Scolium 1. *Let $\mathbf{x} \in B^\omega$ be random in $(B^\omega, \mathcal{M}, \mu_P)$. Then $VN(\mathbf{x}) \in B^\omega$ is also random in $(B^\omega, \mathcal{M}, \mu_{P_{VN}})$.*

Corollary 2. *If $\mathbf{x} \in B^\omega$ is random in $(B^\omega, \mathcal{M}, \mu_P)$ then $VN(\mathbf{x})$ is Borel normal in $(B^\omega, \mathcal{M}, \mu_{P_{VN}})$.*

Theorem 4. *The probability space $(B^\omega, \mathcal{M}, \mu_{P_{VN}})$ induced by von Neumann normalisation is the uniform distribution $(B^\omega, \mathcal{M}, \mu_{P_L})$, where μ_{P_L} is the Lebesgue measure.*

Proof. By Lemma 1 von Neumann normalisation is measure preserving, so for $x \in B^*$ we have $\mu_{P_{VN}}(xB^\omega) = \mu_P(VN^{-1}(x)B^\omega) = p_0^{|x|} p_1^{|x|} \sum_{d_i \in D^*} p_0^{\#_0(d_1 \dots d_{|x|})} p_1^{\#_1(d_1 \dots d_{|x|})}$. The key point is that this only depends on $|x|$ not x itself. By using the fact that for any n, $\sum_{x \in B^n} \mu_{P_{VN}}(xB^\omega) = 1$, we have $\mu_{P_{VN}}(xB^\omega) = 2^{-|x|}$, for all $x \in B^*$, and hence $\mu_{P_{VN}} = \mu_{P_L}$, the Lebesgue measure. □

Theorem 5. *The set $\{\mathbf{x} \in B^\omega \mid VN(\mathbf{x}) \in B^* \text{ or } VN(\mathbf{x}) \in B^\omega \text{ is computable}\}$ has measure zero with respect to the probability space $(B^\omega, \mathcal{M}, \mu_P)$.*

Both unpredictability and uniformity of distribution are independent symptoms of randomness, and it is important that any method to remove bias and ensure uniformity does not decrease unpredictability. Von Neumann's method preserves randomness and Borel normality, but fails to preserve incomputability in general. It is not known whether sequences of bits from a QRNG are random or even Borel normal with respect to the space $(B^\omega, \mathcal{M}, \mu_P)$, so it follows that such sequences may well lose unpredictability when normalised since only strong incomputability is known to be guaranteed. Fortunately, this 'damage' is limited in measure: it holds only with probability zero. However, it remains an open question to determine if strong computability is preserved. Even if it isn't, it may still be the case that the preservation of unpredictability can be guaranteed, but further theoretical characterisation of such sequences is needed to examine such issues.

5   Role of Probability Spaces for QRNGs

The treatment of QRNGs as entirely probabilistic devices is grounded purely on the probabilistic treatment of measurement in quantum mechanics which originated with Born's decision to 'give up determinism in the world of atoms' [6], a viewpoint which has become a core part of our understanding of quantum mechanics. This is formalised by the Born rule, but the probabilistic nature of *individual* measurement is nonetheless postulated and tells us nothing about *how* the probability arises. Along with the assumption of independence this allows us to predict the probability of *successive* events, as we have done.

No-go theorems such as the Kochen-Specker Theorem [17] tell us something stronger: if we assume non-contextuality (i.e. that the result of an observation is independent of the compatible observables are co-measured alongside it [4,12]) then there can, in general, be no pre-existing definite values prescribable to certain sets of measurement outcomes in dimension three or greater Hilbert space. In other words, the unpredictability is not due to ignorance of the system being measured; indeed, since there are in general no definite values associated with the measured observable it is surprising there is an outcome at all [28]. While this does not answer the question as to where the unpredictability arises from, it does tell us something stronger than the Born Rule does. In [9] it is shown that every infinite sequence produced by a QRNG is (strongly) incomputable. In particular, this implies that it is *impossible* for a QRNG to output a computable sequence. The set of computable numbers has measure zero with respect to the probability space of the QRNG, but the impossibility of producing such a sequence is much stronger than, although not in contradiction with, the probabilistic results.

In the finite case every string is, of course, obtainable, and we would expect the distribution to be that predicted by the probability space derived from the Born Rule. However, the infinite case has something to say here too. We can view any finite string produced by a QRNG as the initial segment of an infinite sequence the QRNG would produce if left to run indefinitely. For any infinite sequence produced by the QRNG, it is impossible to compute the value of *any* bit before it is measured [2]; in the finite case this means there is no way to provably compute the value of the next bit before it is measured. In light of value indefiniteness this is not unexpected, but nonetheless gives mathematical grounding to the postulated unpredictability of each individual measurement, as well as the independence of successive measurements—indeed we can rule out any computable causal link within the system which may give rise to the measurement outcome.

6   Conclusions

The analysis developed in this paper involves the probability spaces of the source and output of a QRNG and the effect von Neumann normalisation has on these spaces. In the 'ideal case', the von Neumann normalised output of an independent constantly biased QRNG is the probability space of the uniform distribution (un-biasing). This result is true for both for finite strings and for the infinite sequences produced by QRNGs (the QRNG runs indefinitely in the second case). We have also examined the effect von Neumann normalisation has on various properties of infinite sequences. In particular, Borel normality and (algorithmic) randomness are invariant under normalisation, but for ε-random sequences with $0 < \varepsilon < 1$, normalisation can both decrease or increase the randomness of the source. Further results of this form are necessary in order to be assured that normalisation techniques preserve the strong incomputability of bits produced by QRNGs. Finally, we reiterate that a successful application of von Neumann normalisation—in, fact, any un-biasing transformation—does exactly what it promises, *un-biasing*, one (among infinitely many) symptoms of randomness; it will not produce 'true' randomness.

Acknowledgments

We thank Karl Svozil and Marius Zimand for many discussions and suggestions, and Tania Roblot and the anonymous referees for comments which helped improve the paper.

References

1. Abbott, A.A., Calude, C.S.: Von Neumann normalisation of a quantum random number generator. Report CDMTCS-392, Centre for Discrete Mathematics and Theoretical Computer Science, University of Auckland, Auckland, New Zealand (2010)
2. Abbott, A.A., Calude, C.S., Svozil, K.: Unpublished work on the incomputability of quantum randomness (in preparation)
3. Abbott, A.A., Calude, C.S., Svozil, K.: A quantum random number generator certified by value indefiniteness. CDMTCS Research Report, 396 (2010)
4. Bell, J.S.: On the problem of hidden variables in quantum mechanics. Reviews of Modern Physics 38(3), 447–452 (1966)
5. Billingsley, P.: Probability and Measure. John Wiley & Sons, New York (1979)
6. Born, M.: Quantenmechanik der Stoßvorgänge. Zeitschrift für Physik 38, 803–837 (1926); English translation by Wheeler, J. A., Zurek, W.H.: Quantum Theory and Measurement, ch. I.2. Princeton University Press, Princeton (1983)
7. Calude, C.S.: Information and Randomness: An Algorithmic Perspective, 2nd edn. Springer, Berlin (2002)
8. Calude, C.S., Hertling, P., Jürgensen, H., Weihrauch, K.: Randomness on full shift spaces. Chaos, Solutions & Fractals 12(3), 491–503 (2001)
9. Calude, C.S., Svozil, K.: Quantum randomness and value indefiniteness. Advanced Science Letters 1, 165–168 (2008)
10. Downey, R., Hirschfeldt, D.: Algorithmic Randomness and Complexity. Theory and Applications of Computability. Springer, Heidelberg (2010)
11. Halmos, P.R.: Measure Theory. Springer, New York (1974)
12. Heywood, P., Redhead, M.L.G.: Nonlocality and the Kochen-Specker paradox. Foundations of Physics 13(5), 481–499 (1983)

13. id Quantique. Quantis—quantum random number generators (12/08/2009),
    http://idquantique.com/products/quantis.htm
14. Jauch, J.M.: Foundations of Quantum Mechanics. Addison-Wesley, Reading (1968)
15. Jennewein, T., Achleitner, U., Weihs, G., Weinfurter, H., Zeilinger, A.: A fast and
    compact quantum random number generator. Review of Scientific Instruments 71,
    1675–1680 (2000)
16. Kac, M.: Statistical Independence in Probability, Analysis and Number Theory.
    The Carus Mathematical Monographs. The Mathematical Association of America
    (1959)
17. Kochen, S., Specker, E.: The problem of hidden variables in quantum mechanics.
    Journal of Mathematics and Mechanics 17, 59–87 (1967); Reprinted in Specker, E.:
    Selecta. Brikhäuser, Basel (1990)
18. Kuipers, L., Niederreiter, H.: Uniform Distribution of Sequences. John Wiley &
    Sons, New York (1974)
19. Kwon, O., Cho, Y., Kim, Y.: Quantum random number generator using photon-
    number path entanglement. Applied Optics 48(9), 1774–1778 (2009)
20. Ma, H., Wang, S., Zhang, D., Change, J., Ji, L., Hou, Y., Wu, L.: A random-
    number generator based on quantum entangled photon pairs. Chinese Physics
    Letters 21(19), 1961–1964 (2004)
21. Martin-Löf, P.: The definition of random sequences. Information and Control 9(6),
    602–619 (1966)
22. Peres, Y.: Iterating von Neumann's procedure for extracting random bits. The
    Annals of Statistics 20(1), 590–597 (1992)
23. Pironio, S., Acín, A., Massar, S., de la Giroday, A.B., Matsukevich, D.N., Maunz,
    P., Olmchenk, S., Hayes, D., Luo, L., Manning, T.A., Monroe, C.: Random numbers
    certified by Bell's theorem. Nature 464(09008) (2010)
24. Schmidt, H.: Quantum-mechanical random-number generator. Journal of Applied
    Physics 41(2), 462–468 (1970)
25. Shen, Y., Tian, L., Zou, H.: Practical quantum random number generator based on
    measuring the shot noise of vacuum states. Physical Review A 81(063814) (2010)
26. Stefanov, A., Gisin, N., Guinnard, O., Guinnard, L., Zbinden, H.: Optical quantum
    random number generator. Journal of Modern Optics 47(4), 595–598 (2000)
27. Svozil, K.: The quantum coin toss – testing microphysical undecidability. Physics
    Letters A 143(9), 433–437 (1990)
28. Svozil, K.: Quantum information via state partitions and the context translation
    principle. Journal of Modern Optics 51, 811–819 (2004)
29. Svozil, K.: Three criteria for quantum random-number generators based on beam
    splitteres. Physical Review A 79(5), 054306 (2009)
30. Vadhan, S.: Pseudorandomness. Foundations and Trends in Theoretical Computer
    Science. Now publishers (to appear, 2011)
31. von Neumann, J.: Various techniques used in connection with random digits. Na-
    tional Bureau of Standards Applied Math Series 12, 36–38 (1951); In: Traub, A.H.
    (ed.) John von Neumann, Collected Works, pp. 768–770. MacMillan, New York
    (1963)

Robustness of Cellular Automata in the Light of Asynchronous Information Transmission

Olivier Bouré, Nazim Fatès, and Vincent Chevrier

Nancy Université – INRIA Nancy–Grand-Est – LORIA
{olivier.boure,nazim.fates,vincent.chevrier}@loria.fr

Abstract. Cellular automata are classically synchronous: all cells are simultaneously updated. However, it has been proved that perturbations in the updating scheme may induce qualitative changes of behaviours. This paper presents a new type of asynchronism, the β-synchronism, where the transmission of information between cells is disrupted randomly. After giving a formal definition, we experimentally study the behaviour of β-synchronous models. We observe that, although many effects are similar to those induced by the perturbation of the update, novel phenomena occur. We study the qualitative variation of behaviour triggered by continuous change of the disruption probability. In particular we observe that phase transitions appear, which belong to the directed percolation universality class.

Keywords: asynchronous cellular automata, discrete dynamical systems, robustness, phase transitions, directed percolation.

1 Introduction

Cellular automata are a parallel, spatially-extended, model of computation, which has been studied as an alternative to the sequential computing models, for instance Turing machines. By their very structure, they are well-suited for modelling natural phenomena and for the design of massively parallel algorithms. These mathematical objects have been mostly considered in the synchronous case, that is, when all their components are simultaneously updated. However, this hypothesis of perfect synchrony is somehow inadequate when modelling systems that are subject to noise or non-ideal information transmission, as often met in various natural systems or in asynchronous parallel computing devices. This is why authors tackled the question of whether a cellular automaton is robust to non-ideal updating, either without changing its local transition rule (*e.g.* [4]), or by adding adapted constructs (*e.g.* [8]).

The aim of this paper is to study the robustness of cellular automata by considering the possibility of disruptions in the transmission of information between cells. To this end, we describe the updating process of the system in the frame of a *cellular cycle*. This cycle consists of two steps: (a) the local computation and (b) the transmission of the updated state of a cell to its neighbours. This

C.S. Calude et al. (Eds.): UC 2011, LNCS 6714, pp. 52–63, 2011.

dichotomy induces two related types of asynchronism: the update-wise asynchronism, or α-*synchronism*, which disrupts step (a) and a novel type of asynchronism, the influence-wise asynchronism or β-*synchronism*, which disrupts step (b) and perturbs the interaction between a cell and its neighbours.

The α-synchronous updating is now a relatively well-studied perturbation, whose effect can either be "inoffensive" or trigger drastic qualitative changes, such as phase transitions [2,3,6]. Our goal is now to examine the effects of β-synchronism on simple cellular automata, in particular, in comparison with those of α-synchronism. We aim at extending the range of perturbations, in order to gain insight on how complex collective behaviour emerges from numerous simple local interactions.

On the formal side, this type of asynchronism leads us to extend the cell state space in order to distinguish the *eigenstate*, the actual state of the cell, and the *observable state*, the state of the cell which is observed by its neighbours. The corresponding definitions are introduced in Sec. 2 while experimental observations are presented in Sec. 3. We then analyze more particularly the occurrence of phase transitions in Sec. 4 and proceed to bring discussion in Sec. 5.

2   Asynchronous Cellular Automata

2.1   Cellular Automata

A synchronous cellular automaton is a discrete dynamical system defined by $A = \{\mathcal{L}, Q, \mathcal{N}, f\}$ where :

- $\mathcal{L} \subset \mathbb{Z}^d$ the array of the cellular space, where an element of $\mathcal{L}$ represents a *cell*.
- Q is a nonempty finite set of states.
- $\mathcal{N} \subset \mathcal{L}$ is a finite set of vectors called the *neighbourhood*, which associates to a cell the set of its neighbouring cells. $\mathcal{N}$ and $\mathcal{L}$ are such that for all $c \in \mathcal{L}$ and for all $n \in \mathcal{N}$, the *neighbour* $c + n$ is in $\mathcal{L}$.
- f is the *local transition rule*, which defines the next state of a cell according to the state of this cell and the ones of its neighbours.

A *configuration* x^t represents the state of the automaton at time t; it is defined as a function $x^t : \mathcal{L} \to Q$ which maps each cell to a state. Classically, cellular automata are synchronously updated, meaning that at each time the local transition rule is applied simultaneously of all cells. The *global transition function* is therefore defined as $x^{t+1} = F(x^t)$, so that, for $\mathcal{N} = \{n_1, ..., n_k\}$:

$$\forall c \in \mathcal{L}, \ x^{t+1}(c) = f\left(x^t(c), x^t(c + n_1), ..., x^t(c + n_k)\right).$$

Without loss of generality, we assume that the neighbourhood $\mathcal{N}$ does *not* contain the cell itself. This hypothesis is necessary to explicitly represent the flow of information between a cell and its neighbours. Note that this does not restrict the expressiveness of f since the current state of a cell $x^t(c)$ is always a parameter of f, possibly not taken into account in the transition calculus.

2.2 Asynchronism as a Disruption of Cell Activity

Cell cycle. The update of a cell can be represented by a *cell cycle*, which we decompose into two steps (see Fig. 1):

- the *state update* step, where a cell changes its state according to the local transition function.
- the *information transmission* step, where the cell transmits the updated state to its neighbours.

We give an example of the cell cycle for the different updating schemes on Fig. 2. The update-wise asynchronous updating in cellular automata, or α-*synchronism* [3], is defined as follows: at each time step, each cell is updated with a fixed probability α, or else left unchanged. We introduce a new type of asynchronism, the β-*synchronism*, where each cell is always updated but the transmission of the new state to the neighbourhood is realized with a fixed probability β. As a result, both perturbations consist in applying one of the two steps of the cell cycle with a probability defined as the *synchrony rate*. Please note that for the sake of simplicity, α and β denote both the type of asynchronism and the associated synchrony rates.

Formally, we introduce a *selection function* $\Delta_\alpha : \mathbb{N} \to \mathcal{P}(\mathcal{L})$ which returns for time t the subset of cells to be updated, where each cell has a probability α to be selected. Note that when $\alpha = 1$ the updating is fully synchronous and the system is deterministic. The global transition function becomes $\forall t \in \mathbb{N}, \forall c \in \mathcal{L}, \mathcal{N} = \{n_1, ..., n_k\}$:

$$f_{\Delta_\alpha}(x^t(c)) = \begin{cases} f(x^t(c), x^t(c+n_1), ..., x^t(c+n_k)) & \text{if } c \in \Delta_\alpha(t) \\ x^t(c) & \text{otherwise.} \end{cases}$$

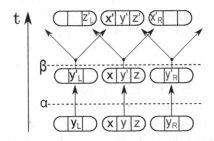

Fig. 1. Representation of a cell cycle, for a cell with two neighbours, denoted by indices L and R. A triplet represents a cell with its state (y) and the observable states of its neighbours (x for left cell and z for right). The prime sign indicates states that are updated. Solid arrows show the updates of states, otherwise states are conserved from the previous step. In α-synchronism, the local transition update applies with probability α and in β-synchronism, the information transmission is applied with probability β. The lower arrow is the local transition update, which performs $y' = f(x, y, z)$ when applied. The forking upper arrow is the information transmission, which performs $z'_L = y'$ and $x'_R = y'$ when applied.

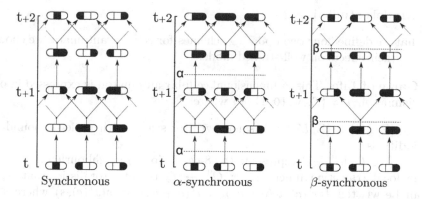

Fig. 2. Example of the time cycle for a 3-cell sample for synchronous (left), α-synchronous (middle) and β-synchronous (right) updating schemes (0 is denoted by a white space, 1 by black). The rule is ECA 50 (see Sec. 2.3) but is not important here.

β-synchronism. To define this new asynchronism, we need to extend the classical definition of cellular automata by taking into account the difference between the *eigenstate* of a cell, and the *observable state*.

Let us consider a cellular automaton $A = \{\mathcal{L}, Q, \mathcal{N}, f\}$. From A we derive a new cellular automaton $A' = \{\mathcal{L}, Q', \mathcal{N}, f'\}$ where :

- $Q' = Q^2$ is the new set of states.

  For a given configuration x^t, a cell state is denoted by $x^t = \begin{pmatrix} x_e^t \\ x_o^t \end{pmatrix}$ with $x_e^t : \mathcal{L} \to Q$ the eigenstate of the cell, and $x_o^t : \mathcal{L} \to Q$ the state of the cell observable by its neighbourhood.

- the local transition function f' is splitted into two parts to decompose its action: state update and information transmission.

We thus write $f' = f_t \circ f_u$, so that:

- $f_u : Q'^{k+1} \to Q'$ is the *update function*, which computes the new state of the cell based on its eigenstate and the observable state of the neighbours:

$$f_u \left(\begin{pmatrix} e \\ o \end{pmatrix}, \begin{pmatrix} \cdot \\ o_1 \end{pmatrix}, ..., \begin{pmatrix} \cdot \\ o_k \end{pmatrix} \right) = \begin{pmatrix} f(e, o_1, ..., o_k) \\ o \end{pmatrix}$$

- $f_t : Q' \to Q'$ is the *transmission function*, which replaces the observable state by the eigenstate:

$$f_t \left(\begin{pmatrix} e \\ o \end{pmatrix} \right) = \begin{pmatrix} e \\ e \end{pmatrix}.$$

We use a similar selection function Δ_β as introduced for α-synchronism, where cells are selected with probability β, and define the global transition function as, $\forall t \in \mathbb{N}$, $\forall c \in \mathcal{L}$, $\mathcal{N} = \{n_1, ..., n_k\}$:

$$f_{\Delta_\beta}(x^t(c)) = \begin{cases} f_t \circ f_u \left(x^t(c), x^t(c + n_1), ..., x^t(c + n_k) \right) & \text{if } c \in \Delta_\beta(t) \\ f_u \left(x^t(c), x^t(c + n_1), ..., x^t(c + n_k) \right) & \text{otherwise.} \end{cases}$$

2.3   Models studied

Now that we defined our two updating schemes for cellular automata, we choose to study their effects on well-studied models.

The Game of Life. This 2-dimensional cellular automaton is expressed in our formalism as $A_{GL} = \{\mathcal{L}, \mathcal{N}, \{0, 1\}, f\}$ where:

- $\mathcal{L} = \{\mathbb{Z}/L\mathbb{Z}\} \times \{\mathbb{Z}/L\mathbb{Z}\}$ is a square grid of size L with periodic boundary conditions.
- $\mathcal{N} = \{c \in \mathcal{L}, ||c|| = 1\}$ represents the 8-cell (Moore) neighbourhood.
- the local transition function $f : Q^9 \rightarrow Q$ is outer-totalistic, that is, it can be written $f(x^t(c), x^t(c + n_1), ..., x^t(c + n_8)) = \delta(x^t(c), s)$ where $s = \sum_{c' \in \mathcal{N}} x^t(c')$.

$$\text{If } x^t(c) = 0, \text{ then } \delta(x^t(c), s) = \begin{cases} 1 & \text{if } s = 3 \\ 0 & \text{otherwise.} \end{cases} \quad \textit{(Birth rule)}$$

$$\text{If } x^t(c) = 1, \text{ then } \delta(x^t(c), s) = \begin{cases} 1 & \text{if } s \in \{2, 3\} \\ 0 & \text{otherwise.} \end{cases} \quad \textit{(Survival rule)}$$

Elementary Cellular Automata (ECA). An *ECA* is a 1-D binary cellular automaton with nearest-cell neighbourhood, whose transition function is determined according to Wolfram's notation. In our formalism, ECAs are denoted by $A_E = \{\mathcal{L}, \mathcal{N}, \{0, 1\}, f\}$ where:

- $\mathcal{L} = \{\mathbb{Z}/L\mathbb{Z}\}$ is a 1-dimensional ring.
- $\mathcal{N} = \{-1, +1\}$, i.e. the 2-cell neighbourhood.
- the local transition function δ is determined by its code:
  $W = f(0, 0, 0).2^0 + f(0, 0, 1).2^1 + ... + f(1, 1, 1).2^7$.

3   Qualitative observations

We now observe qualitatively the effect of β-synchronism. We are in particular interested in knowing whether its effects will differ from those of α-synchronism. In the rest of this paper, initial conditions are constructed by assigning to each cell an equal value to the observable state and the eigenstate, and choosing this value randomly and uniformly in $\{0, 1\}$.

3.1   The Game of Life

Figure 3 shows sample configurations appearing for random initial conditions after a transient time of 1000, which was observed sufficient for reaching a steady state. We observe that the behaviour separates into two distinct phases for both

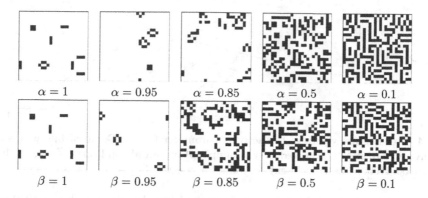

Fig. 3. Game of Life configurations for different values of α and β (0 is denoted by a white space, 1 by blue). These are obtained with the software FiatLux for identical initial states, for a transient time $t \approx 1000$. The first row shows the different behaviors obtained with different values of α-synchronism. The second row displays them for the same values of β-synchronism.

types of asynchronism: the system converges to a fixed point for values of the synchrony rate α or β higher than 0.9, whereas a *labyrinth-like pattern* appears for values lower than 0.9. To quantify this observation, we use the following macroscopic parameters:

- the *density* is the ratio of cells with state (or eigenstate) 1.
- the *activity* is the ratio of unstable cells. A cell is said *unstable* at time t if a synchronous update modifies its state (or eigenstate).

Note that the latter definition cannot be readily transposed for the β-synchronism. Indeed what does a synchronous update mean for a cell whose eigenstate and observable state are different? As neighbourhood knowledge is always absolute in synchronous cellular automata, we reckon that it should also be the cases for the estimation of activity. Therefore, using the notations defined in Fig. 1, we define a cell as instable if $y \neq f(y_L, y, y_R)$.

Figure 4 compares the two types of asynchronism through the activity parameter after a steady state has been reached. At first sight, the β-synchronous Game of Life seems to react in a similar way to its α-synchronous counterpart :

1. A singularity occurs for the activity parameter in the synchronous case $(\alpha, \beta = 1)$. This phenomenon has been explained for α-synchronism as the loss of stable periodic patterns when a noise is introduced no matter how small ($\alpha < 1$) [1]. The same explanation stands for β-synchronism. This means that the traditional construct to make the Game of Life Turing-universal, that is, by using periodical patterns such as gliders, no longer holds. Nevertheless, the question of the Turing-universality of the asynchronous Game of Life still stands.
2. The macroscopic behaviour confirms the phase separation for a critical value of the synchrony rate α_c (resp. β_c). For values $\alpha > \alpha_c$ (resp. $\beta > \beta_c$), the

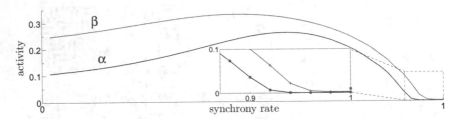

Fig. 4. Steady-state values of the activity parameter for the Game of Life with synchrony rates α and β, averaged over a sample of 50 initial conditions of size L=100, $t \in [10000, 11000]$. Inset: a closeup for values $\alpha, \beta \in]0.9, 1]$.

system converges to a stable fixed point of low density, which constitutes a *passive* phase. However, for values $\alpha < \alpha_c$ (resp. $\beta < \beta_c$), the system enters an *active* phase, characterized by the labyrinth pattern (though less regular in β- than in α-synchronism).

The comparison of both types of asynchronism reveals a similar reaction of the Game of Life, but this calls for a closer observation of the characteristics of the phase transition (see Sec. 4.1).

3.2 Elementary Cellular Automata

Through the search of reflexive and complementary symmetries, it is possible to narrow down the number of ECA to study from the 256 possible models to 88 inequivalent ECA. To quantify these observations on the 88 ECA, we use the following macroscopic parameters:

- the *density d*, as defined above.
- the 01 *block density* (00, respectively) is the ratio of successive cells with states 0 − 1 (resp., 0 − 0).

When comparing visually the plot profiles of the 88 minimal ECA under α- and β-synchronism[1], it can be observed that most rules display similar reactions to both types of asynchronism. However, some surprising divergences of plot profile appeared for lower synchrony rates, for example for ECA 6, 22, 72 and 200. For the sake of conciseness, we choose to focus on a single rule, namely ECA 50 and leave for future work the exhaustive study of the entire range of phenomena.

The case of ECA 50. A notable phenomenon appears for the β-synchronous ECA 50. The study of the density parameter (see Fig. 5a) shows a similar plot profile for both α- and β-synchronism, including a phase transition, which suggests that ECA 50 reacts in a similar way to these two perturbations.

However, a difference of behaviour is observed with the visual inspection of the evolution of the automaton over a few time steps (see Fig. 6). For instance, for the same synchrony rate 0.75, the patterns in β-synchronism appear much more

---

[1] For the complete results: http://www.loria.fr/~boure/results/ecaparambeta/

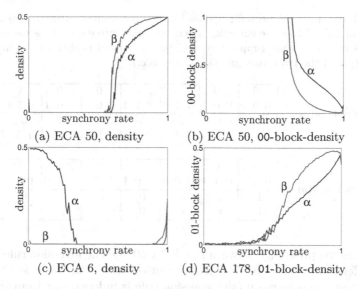

(a) ECA 50, density

(b) ECA 50, 00-block-density

(c) ECA 6, density

(d) ECA 178, 01-block-density

Fig. 5. Steady-state values of various observation parameters for different ECA. The grid size is $F = 1000$, transient time is $T_t = max(10000, 1000/\alpha)$ (β respectively) and averages are obtained for a sampling time $T_s = max(1000, 100/\alpha)$ (resp. β).

regular than α-synchronism. In particular, in the long run clusters of 0-states appear less frequently, and 111 patterns seem inexistent.

This can be explained for values of β close to 1 with the following observations:

- the *checkerboard regions* (*i.e.* alternated 0s and 1s) are robust to the β-synchronism, that is, the few anomalies (regions of 0s) that disturb the regularity of the pattern are quickly restored to the original pattern.
- the *pairs* (*i.e.* 00 and 11 pairs) follow a non-biased random walk, and annihilate when they meet.

As a consequence, if L is even, in the long run, the system tends to be covered by checkerboard patterns, which is a new property exclusive to β-synchronism.

This example illustrates the qualitative differences that may occur between α- and β-synchronous cellular automata. More specifically, we note that ECA 50 is more regular for β-synchronism than for α-synchronism. This regularity can

$\alpha=0.75$ $\beta=0.75$ $\alpha=0.95$ $\beta=0.95$ synchronous

Fig. 6. Configurations for ECA 50 under different updating schemes (0 is denoted by a white space, 1 by blue)

Table 1. Table of probabilities for updated states to become 1 after an β-synchronous transition of ECA 51 (inversion rule). Each column represents a possible state for a cell and its neighbourhood (upper box) and the associated probability to become a 1 for the output of the transition function (lower box).

Synchronized states	. 0 .	. 0 .	. 1 .	. 1 .	. 0 .	. 0 .	. 1 .	. 1 .
	0 0 0	0 0 1	0 1 0	0 1 1	1 0 0	1 0 1	1 1 0	1 1 1
	1	1	0	0	1	1	0	0
	β	β	$1-\beta$	$1-\beta$	β	β	$1-\beta$	$1-\beta$

Desynchronized states	. 0 .	. 0 .	. 1 .	. 1 .	. 0 .	. 0 .	. 1 .	. 1 .
	0 1 0	0 1 1	0 0 0	0 0 1	1 1 1	1 1 0	1 0 0	1 0 1
	1	1	0	0	1	1	0	0
	1	1	0	0	1	1	0	0

be "seen" as the proximity between rule ECA 50 and the inversion rule ECA 51, which differ only by one bit in their transition table. As the rule ECA 51 is insensitive to β-synchronism (the inversion rule is independent from the neighbours state), there also seems to exist a proximity in terms of global behaviour between ECA 50 and 51 for the β-synchronism. One may thus wonder what is the origin of such a radical difference between the two updating schemes. In the next paragraph, we endeavour to capture this difference in the framework of stochastic cellular automata.

Stochastic cellular automata. It is straightforward to describe α-synchronous ECA in terms of *elementary stochastic cellular automata*, that is, to define them with a function $f_\alpha : \{0,1\}^3 \rightarrow [0,1]$ which associates to each neighbourhood state (x, y, z) the probability to update to 1. Indeed, the local rule f_α is simply obtained as the barycentre of the transition function f and the identity rule Id with weights α and $1 - \alpha$, respectively.

By contrast, β-synchronism requires us to extend the state space in order to differentiate the eigenstate and the observable state. As a result, the number of inputs of the transition table is doubled to take into account whether the two states are *synchronized*, that is, if the eigenstate and the observable state are identical.

As said before, ECA 51 is insensitive to β-synchronism. However, this particular property of ECA 51 does not appear readily on its transition table (see Table 1). This makes the stochastic table all the more "cryptic", as the divergence from synchronism cannot be deduced from the reading of the table, unlike α-synchronism. This shows that the difference between the two types of asynchronism is non-trivial, and justifies the experimental approach adopted to study these systems.

4 Study of Phase Transitions

The occurrence of phase transitions is probably one of the most remarkable phenomena that arises in asynchronous cellular automata: there exists a non-trivial value of the synchrony rate, the critical threshold, which separates two distinct qualitative behaviours of the system. We are now interested in measuring quantitatively this phenomenon.

4.1   The Game of Life

As pointed out in Sec. 3.1, the activity parameter reveals the existence of two distinct phases in the α- and β-synchronous Game of Life. For the α-synchronism, the phase transition has been proved to be second-order [1], that is, if the macroscopic measures that describe the behaviour are continuous, their derivative curve is discontinuous for a critical value of the synchrony rate.

How to determine the critical synchrony rate α_c or β_c? A simple method to measure it consists in estimating the singularity point where the slope jumps from null to infinity, but for practical purposes, this method is rather imprecise as it introduces systematic biases. In order to reduce them, we follow a different protocol (see *e.g.* [5]):

1. We fix β, start from a random initial condition and let it evolve for a given number of steps.
2. We monitor the evolution of the order parameter for a long simulation time until we observe a sub-critical or super-critical behavior. In a log-log plot, a concave curve occurs for the passive phase (activity converges to zero) and a convex curve for the active phase (activity converges to a non-zero value). As we expect the order parameter to follow a power law $K.t^{-\delta}$ near criticality, the curve at critical value should appear as a straight line of slope $-\delta$.
3. We repeat the experiment with a value closer to the critical point until a satisfactory precision is reached (here 10^{-3}).

For the α-synchronous Game of Life, it has been measured that $\alpha_c = 0.9083$ and that the evolution of the order parameter of this value followed a power law in the form $K.t^{-\delta}$ with $\delta_{2D} = 0.451$ [3], which is the expected critical exponent for the directed percolation universality-class in 2 dimensions [7]. Figure 7 shows that the measures for β-synchronous updating. They also confirm the directed percolation hypothesis, with $\beta_c = 0.945$.

4.2   Elementary Cellular Automata

Among the 88 minimal ECAs, it has been observed that several ECA display second-order phase transition in their α-synchronous version. These rules were proved to belong to the directed percolation class [2], and have been divided into 3 distinct subclasses:

- rules 18, 26, 50, 58, 106 and 146 are the DP_{hi} class, for which the active phase of density (respectively the passive phase) occurs for $\alpha > \alpha_c$ (resp. $\alpha < \alpha_c$).
- rules 6, 38 and 134 are the DP_{low} class, where active and passive phases are inverted.

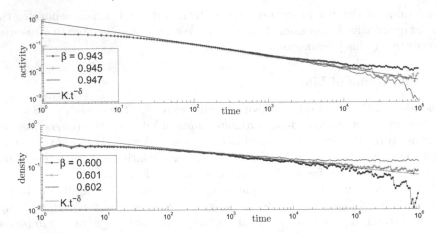

Fig. 7. Phase transition analysis for the Game of Life (top) and ECA 50 (bottom). The straight lines follow a power law $f(t) \sim t^{-\delta}$. For the Game of Life, measures are averaged over 25 samples of size 800×800. For the ECA 50, measures are averaged over 25 samples of size 20000. The straights line give the expected critical exponent for directed percolation: $\delta_{2D} = 0.451$ (top) and $\delta_{1D} = 0.1595$ (bottom).

- rule 178 is the sole element of the DP_2 class, where the density is stable in average but for which a phase transition appears for the 01-block-density.

Our observations show that these subclasses react differently to β-synchronism:

1. The rules of the DP_{hi} class have shown little behavioural change between the two types of asynchronism: the phase transition appearing for the density parameter in Fig. 5 is conserved.
   For α-synchronism, the same protocol was applied to ECA 50 [2], and displayed good evidence of a directed percolation phenomenon for $\alpha_c = 0.6282$. For β-synchronism (see Fig. 7), it appears the behaviour at critical synchrony rate $\beta_c = 0.601$ is in good agreement with a power law of critical exponent $\delta_{1D} = 0.1595$, the expected value for the directed percolation.
2. Surprisingly enough, no phase transition was observed for the three rules of the DP_{low} class in β-synchronism (see ECA 6 in Fig. 5), leaving a constant null-density convergent phase for any value of β.
3. Finally, the ECA 178 (DP_2 class) reproduced a similar plot profile for the 01-block-density (see ECA 178 in Fig. 5).

These first results show how rich the study of β-synchronism can be, and that although α- and β-synchronism are intuitively similar in their mechanisms, their effects may differ radically.

5   Discussion

This paper presented a formalism for a new type of asynchronous updating in cellular automata, the β-synchronism, based on the disruption of information transmission between cells. We compared this perturbation to α-synchronism and observed from a macroscopic point of view that α- and β-synchronism had similar effects. It was observed that β-synchronous updating also produces phase transitions, but for a smaller set of rules than α-synchronism. In particular, we remarked that there was no phase transition for the three ECA of the DP_{low} class but we have no explanation for this phenomenon so far.

By studying ECA 50 more closely, we could exhibit an example for which a macroscopic parameter behaviour was similar for both types of asynchronism, but for which novel lower-scale properties were observed.

This extension is a first step towards a unified view of asynchronism for cellular automata, based on the idea of cell cycles. There exist plenty of other ways the cellular activity can be perturbed. This raises the question of how the space of perturbations can be described.

Although this paper focused on cellular automata, the idea behind the β-synchronism can be transposed to other synchronous collective systems, such as lattice-gas cellular automata, neural networks or multi-agent systems. By extending the range of variations applied to the definition of the discrete dynamical systems, can we perfect our understanding of the robustness of natural systems? Can we gain some insight from their mechanisms to improve the design of spatially-extended computing models?

References

1. Blok, H.J., Bergersen, B.: Synchronous versus asynchronous updating in the "Game of Life". Physical Review E 59, 3876–3879 (1999)
2. Fatès, N.: Asynchronism induces second order phase transitions in elementary cellular automata. Journal of Cellular Automata 4(1), 21–38 (2009)
3. Fatès, N.: Does Life resist asynchrony? In: Adamatzky, A. (ed.) Game of Life Cellular Automata, pp. 257–274. Springer, Heidelberg (2010)
4. Fatès, N., Morvan, M.: An experimental study of robustness to asynchronism for elementary cellular automata. Complex Systems 16, 1–27 (2005)
5. Grassberger, P.: Synchronization of coupled systems with spatiotemporal chaos. Physical Review E 59(3), R2520 (1999)
6. Grilo, C., Correia, L.: Effects of asynchronism on evolutionary games. Journal of Theoretical Biology 269(1), 109–122 (2011)
7. Hinrichsen, H.: Nonequilibrium critical phenomena and phase transitions into absorbing states. Advances in Physics 49, 815–958 (2000)
8. Peper, F., Isokawa, T., Takada, Y., Matsui, N.: Self-timed cellular automata and their computational ability. Future Generation Computer Systems 18(7), 893–904 (2002)

Hidden Variables Simulating Quantum Contextuality Increasingly Violate the Holevo Bound

Adán Cabello[1] and Joost J. Joosten[2]

[1] Departamento de Física Aplicada II
Universidad de Sevilla, 41012 Sevilla, Spain
adan@us.es
www.adancabello.com
[2] Dept. Lògica, Història i Filosofia de la Ciència
Universitat de Barcelona, Montalegre 6, 08001 Barcelona, Spain
jjoosten@ub.edu
www.phil.uu.nl/~jjoosten/

Abstract. In this paper we approach some questions about quantum contextuality with tools from formal logic. In particular, we consider an experiment associated with the Peres-Mermin square. The language of all possible sequences of outcomes of the experiment is classified in the Chomsky hierarchy and seen to be a regular language.

We introduce a very abstract model of machine that simulates nature in a particular sense. A lower-bound on the number of memory states of such machines is proved if they were to simulate the experiment that corresponds to the Peres-Mermin square. Moreover, the proof of this lower bound is seen to scale to a certain generalization of the Peres-Mermin square. For this scaled experiment it is seen that the Holevo bound is violated and that the degree of violation increases uniformly.

1 Introduction

In this paper we will focus on an experiment that is associated to the famous Peres-Mermin square [8,9]. The experiment consists of a sequence of measurements performed consecutively on a two-qubit system. All the measurements are randomly chosen from a subset of those represented by two-fold tensor products of the Pauli matrices X, Y and Z, and the identity $\mathbb{I}$. The set $\mathcal{L}$ of all sequences of outcomes consistent with Quantum Mechanics is studied as a formal language.

In the theory of formal languages, the Chomsky hierarchy [4,10] defines a classification of languages according to their level of complexity. In Section 2,

C.S. Calude et al. (Eds.): UC 2011, LNCS 6714, pp. 64–76, 2011.
© Springer-Verlag Berlin Heidelberg 2011

this language $\mathcal{L}$ will be classified in the Chomsky hierarchy. It will be seen to live in the lower regions of the hierarchy. More concrete, it will be seen that the language is of Type 3, also called *regular*.

In Section 3, the question is addressed how much memory is needed to simulate Quantum Mechanics in experiments with sequential measurements. The question naturally arises: *how* are we allowed to simulate nature. We wish to refrain from technical implementation details of these simulations as much as possible. To this extent, we invoke the Church-Turing thesis that captures and mathematically defines the intuitive notion of what is computable at all by what mechanized and controlled means so-ever. This gives rise to our notion of MA-GAs: Memory-factored Abstract Generating Automata. We prove a lower bound for the amount of memory needed for MAGAs that simulate extensions of the experiment associated to the Peres-Mermin square and shall see that this bound directly and increasingly so violates the Holevo bound.

2   Language Defined by the Experiment

In this paper we will denote the Peres-Mermin square by the following matrix

$$\begin{pmatrix} A & B & C \\ a & b & c \\ \alpha & \beta & \gamma \end{pmatrix} \quad (\dagger),$$

where these variables can get assigned values in $\{1, -1\}$. The corresponding – classically impossible to satisfy– restriction is that the product of any row or column should be 1 except for C, c, γ which should multiply to -1.

Basically, contextuality amounts to the phenomenon that the outcome of a measurement on a system relates to and depends on other (compatible) measurements performed on that system.

In the next Subsection 2.1, we will first formally describe the possible outcomes of the experiment that corresponds to the Peres-Mermin square. We will describe this in almost tedious detail as we later need to formalize the corresponding language.

2.1   The Experiment

We will collect the nine observables of our experiment into an alphabet Σ which we denote by

$$\Sigma := \{A, B, C, a, b, c, \alpha, \beta, \gamma\}.$$

The experiment consists of arbitrarily many discrete consecutive measurements of these nine observables which can take values in the two-element set $\{-1, 1\}$. For reference we reiterate that $(\dagger)$ in this paper coincides with the well-studied Peres-Mermin square [8,9].

Definition 1 (Context; Compatible observables). *The rows and columns of matrix* (†) *are called* contexts. *Two observables within the same context are called* compatible *and two observables that do not share a common context are called* incompatible.

It is clear that each observable belongs to exactly two contexts. Likewise, each observable is compatible with 4 other observables and incompatible with 4 yet other observables. Now, let us define what it means for an observable to be *determined.*

Definition 2 (Determined observables; Value of a determined observable). *An observable becomes (or stays)* **determined** *if:*

(E1) Once we measure an observable, it becomes (or stays) determined and its value is the value that is measured, either 1 or −1. If the observable X that was measured was already determined, then the value of X that is measured anew must be the same as in the most recent measurement.

(E2) If two observables within one context are determined, and if the third observable in this context was not yet determined, this third value becomes determined too. Its corresponding value is such that the product of the three determined values in this context equals 1. The sole exception to this value assignment is the context $\{C, c, \gamma\}$ that should multiply to −1.

The notion of an observable being undetermined *is defined by the following clause:*

(D1) By default, an observable is **undetermined** *and only becomes determined in virtue of (E1) or (E2). An observable X that is determined remains determined* **if and only if** *all successive measurements are in one of the two contexts of X that is, all successive measurements are compatible with X. As soon as, according to this criterion, an observable is no longer determined we say that it has become* **undetermined.** *Undetermined observables stay undetermined until they become determined. Undetermined observables have no value assigned. Sometimes we will say that the value of an undetermined observable is* undefined.

A sequence of measurements is *consistent* with our experiment if its determined observables meet the restrictions above. In essence, part of our definition is of inductive nature. To see how this works, let us see this, by way of example, the sequence of measurements $[A = 1; B = 1; c = 1; \gamma = 1]$ is inconsistent with our experiment:

Measured observable	Measured value	Comments
A	1	The experiments starts so by default, all observables were undetermined (D1). After the measurement, by (E1) the observable A is determined and assigned the value 1.
B	1	As B is a new measurement, the observable becomes determined (E1) with value 1. The observable A remains determined by $(D1)$. Moreover, the observable C which is in the context $\{A, B, C\}$, now becomes determined in virtue of (E2) with value 1 too as the product $A \cdot B \cdot C$ should multiply to 1.
c	1	By (E1), c becomes determined with value 1 and the observables A and B become undetermined in virtue of $(D1)$. The observable C remains determined by (D1) with value 1 as the new measurement of c is in the context $\{C, c, \gamma\}$. Thus, in virtue of (E2), the observable γ becomes determined too. Its value must be -1 as $C \cdot c \cdot \gamma$ should multiply to -1.
γ	1	As we said before, in virtue of $(E2)$ the value of γ should be -1. Thus this is inconsistent with the measurement of 1 for γ.

Note that only the last measurement was inconsistent with our experiment.

Remark 1. If we fill the square (†) with any assignment of 1s and -1s, then the number of products of contexts that equal -1 will always be even.

Proof. By induction on the number of -1s. If all observables in (†) are set to one, then all contexts multiply to one thus yielding zero –an even number– of negative products. If we add one more -1 to (†), this -1 will occur in exactly two contexts thereby flipping the sign of the respective products of these two contexts.

2.2   The Formal Language

In this subsection we shall specify a formal language comprising exactly the possible strings of measurements of our above specified experiment. We mention where the complexity of this language resides in the Chomsky hierarchy: the regular languages. Let us first briefly introduce some terms and definitions from the theory of formal languages.

We shall call a collection of symbols an *alphabet* and commonly denote this by Σ. A *string* or *word* over an alphabet is any finite sequence of elements of Σ in whatever order. We call the sequence of length zero the *empty string*, and will denote the empty string/word by λ. We will denote the set of all strings over

Σ by Σ^* using the so-called Kleene-star. Thus, formally and without recurring to the notion of sequence, we can define Σ^*, the set of all finite strings over the alphabet Σ as follows.

$$\lambda \in \Sigma^*;$$
$$\sigma \in \Sigma^* \ \& \ s \in \Sigma \ \Rightarrow \ \sigma s \in \Sigma^*.$$

Instead of writing $\lambda\sigma$, we shall just write σ. It is clear that Σ^* is an inductive definition so that we also have an induction principle to prove or define properties over Σ^*. For example, we can now formally define what it means to *concatenate* – stick the one after the other– two strings: We define $\star$ to be the binary operation on Σ^* by $\sigma \star \lambda = \sigma$ and $\sigma \star (\tau s) = (\sigma \star \tau)s$. Any subset of Σ^* is called a *language* over Σ.

The study of formal languages concerns, among others, which kind of grammars define which kind of languages, and by what kind of machines these languages are recognized. In the current paper we only need to provide a formal definition of so-called *regular* languages. We do this by employing regular grammars. Basically such a grammar is a set of rules that tells you how strings in the language can be generated.

Definition 3 (Regular Grammar). *A regular grammar over an alphabet Σ consist of a set $\mathcal{G}$ of generating symbols together with a set of rules. In this paper we shall refer to generating symbols by using a line over the symbols. The generating symbols always contain the special start-symbol $\overline{S}$. Rules are of the form*

$$\overline{X} \to \lambda \quad or$$
$$\overline{X} \to s\overline{Y},$$

where $\overline{X}, \overline{Y} \in \mathcal{G}$ and $s \in \Sigma$. The only restriction on the rules is, that there must be at least one rule where the left-hand side is $\overline{S}$.

Informally, we state that a *derivation* in a grammar is given by repeatedly applying possible rules starting with $\overline{S}$, where the rules can be applied within a context. Thus, for example, when we apply the rule $\overline{A} \to a\overline{B}$ in the context σA, we obtain $\sigma a\overline{B}$. A more detailed example of a derivation is given immediately after Definition 4. We say that a string σ over Σ is derivable within a certain grammar if there is a derivation resulting in σ. The *language defined by a grammar* is the set of derivable strings over Σ^*. A language is called *regular* if it is definable by a regular grammar.

We are now ready to give a definition of a regular language that, as we shall see, exactly captures the outcomes of our experiment. To this end, we must resort to a richer language than just Σ as Σ only comprises the observables and says nothing over the outcomes. So, we shall consider a language where, for example, $\tilde{A}$ will stand for, "A was measured with value -1", and A will stand for, "A was measured with value 1". We will denote this alphabet by $\tilde{\Sigma}$. We will use the words *compatible* and *incompatible* in a similar fashion for $\tilde{\Sigma}$ as we did for our observables in Σ. Thus, for example, we say that both B and $\tilde{B}$ are compatible with C. The only difference will be that $\tilde{A}$ is not compatible with A whereas A is.

Definition 4 (A grammar for $\mathcal{L}$). *The language $\mathcal{L}$ will be a language over the alphabet $\tilde{\Sigma} := \{A, \tilde{A}, B, \tilde{B}, \ldots, \beta, \tilde{\beta}, \gamma, \tilde{\gamma}\}$, where the intended reading of A will be that the observable A was measured to be 1, and $\tilde{A}$ will stand for measuring -1, etc.*

Before we specify the grammar that will generate $\mathcal{L}$, we first need some notational conventions. In the sequel, U, V, X, Y and Z will stand for possible elements of our alphabet. If X and Y are compatible symbols, we will denote by $Z(XY)$ the unique symbol that is determined in (E2) by X and Y. Thus, for example, $Z(A\tilde{B}) = \tilde{C}$ and $Z(C\gamma) = \tilde{c}$.

The generating symbols of the grammar will be denoted by a string with a line over it. Note that, for example, $\overline{XY}$ is regarded as one single generating symbol. The intended reading of such a string is that the two symbols are different and compatible, the last symbol is the one that can be generated next, and the remainder of the string codifies the relevant history. As usual we will denote the initial generating symbol by $\overline{S}$. Let $\mathcal{L}$ be the formal language generated by the following grammar.

$$
\begin{aligned}
\overline{S} &\longrightarrow \lambda \\
\overline{S} &\longrightarrow \overline{X} && \textit{for any symbol } X \in \Sigma \\[6pt]
\overline{X} &\longrightarrow X \\
\overline{X} &\longrightarrow X\overline{X} \\
\overline{X} &\longrightarrow X\overline{Z} && \textit{for } Z \textit{ incompatible with } X \\
\overline{X} &\longrightarrow X\overline{XY} && \textit{for } Y \textit{ compatible with } X \textit{ (but not equal)} \\[6pt]
\overline{XY} &\longrightarrow Y \\
\overline{XY} &\longrightarrow Y\overline{XY} \\
\overline{XY} &\longrightarrow Y\overline{YX} \\
\overline{XY} &\longrightarrow Y\overline{YZ} && \textit{for } Z \textit{ compatible with } Y \textit{ (not equal),} \\
&&& \textit{but not with } X \\[6pt]
\overline{XY} &\longrightarrow Y\overline{YZ(XY)} \\
\overline{XY} &\longrightarrow Y\overline{Z(XY)U} && \textit{for } U \textit{ compatible with } Z(XY) \textit{ (but not equal)} \\
&&& \textit{but not compatible with } X \textit{ or } Y
\end{aligned}
$$

We emphasize that the conditions on the right are not part of the rules. Rather, they indicate how many rules of this type are included in the grammar. For example, $\overline{S} \longrightarrow \overline{X}$ for any symbol $X \in \Sigma$ is our short-hand notation for nine rules of this kind.

Let us give an example of how this grammar works to the effect that $ABc\tilde{\gamma}$ is in our language. Recall our reading convention that says that A stands for measuring $A = 1$, B for measuring $B = 1$, c for $c = 1$, and $\tilde{\gamma}$ for measuring $\gamma = -1$. Here goes a derivation of the string $ABc\tilde{\gamma}$:

String derived	Instantiation of rule	General rule applied
$\overline{A}$	$\overline{S} \longrightarrow \overline{A}$	By the rule $\overline{S} \longrightarrow \overline{X}$ with $X = A$
$A\overline{AB}$	$\overline{A} \longrightarrow A\overline{AB}$	By the rule $\overline{X} \longrightarrow X\overline{XY}$ (X and Y compatible) with $X = A$, $Y = B$. Note that A and B are indeed compatible.
$AB\overline{Cc}$	$\overline{AB} \longrightarrow B\overline{Cc}$	By the rule $\overline{XY} \longrightarrow Y\overline{Z(XY)U}$ for U compatible with $Z(XY)$ (but not equal), but not compatible with X or Y, where $X = A$, $Y = B$, $Z(XY) = C$ and $U = c$.
$ABc\overline{c\tilde{\gamma}}$	$\overline{Cc} \longrightarrow c\overline{c\tilde{\gamma}}$	By the rule $\overline{XY} \longrightarrow Y\overline{YZ(XY)}$ with $X = C$, $Y = c$ and $Z = \tilde{\gamma}$.
$ABc\overline{\tilde{\gamma}}$	$\overline{c\tilde{\gamma}} \longrightarrow \tilde{\gamma}$	By the rule $\overline{XY} \longrightarrow Y$ with $X = c$ and $Y = \tilde{\gamma}$.

In a previous example in Subsection 2.1, we showed that the string $ABC\gamma \in \tilde{\Sigma}$ is not consistent with the experiment. It is not hard to prove that in our grammar there is no derivation of this string either.

We note and observe that the grammar has various desirable properties. As such, the grammar is monotone[1], where each generated string contains at most one generating symbol. Moreover, it is easily seen that once a string generated by the grammar contains a composite generating symbol (like $\overline{XY}$), each subsequently generated string will also contain a composite generating symbol if it contains any generating symbols at all. Indeed, the grammar is very simple.

Theorem 1. *The language $\mathcal{L}$ as defined in Definition 4 coincides with the set of consistent measurements as defined in Definition 2.*

Proof. We must show that, on the one hand any string in $\mathcal{L}$ is consistent with the experiment, and on the other hand, any sequence of measurements consistent with the experiment is derivable in $\mathcal{L}$.

The first implication is proven by an induction on the length of the sequence of measurements. It is easy to see that any possible extension of a sequence of measurements by a new measurement is covered by one of the rules in the grammar.

A proof of the second implication proceeds by a simple induction on the length of a derivation in the grammar.

Note that by the mere syntactic properties of the definition of $\mathcal{L}$ we see that $\mathcal{L}$ is indeed a regular language.

Corollary 1. *The set of consistent sequences of measurements of the Peres-Mermin experiment is a regular language.*

---

[1] That is, the length of each subsequent string in a derivation is at least as long as the length of the previous one.

Regular languages are of Type 3 in the Chomsky hierarchy. As such we have access to a corpus of existing theory. In particular, there exists a method to determine the minimal amount of states in a Deterministic Finite State Automata that will accept $\mathcal{L}$. Moreover, we also have access to the following proposition [5].

Proposition 1. *Let $\mathcal{L}$ be a regular language. Then, there exist polynomials $p_1, \ldots, p_k$ and constants $\lambda_1, \ldots, \lambda_k$ such that the number of strings of length n in $\mathcal{L}$ is given by*

$$p_1(n)\lambda_1^n + \ldots + p_k(n)\lambda_k^n.$$

As we have seen already, now that we have access to a smooth inductive definition of $\mathcal{L}$ and thus of the set of possible measurements according to the experiment described in Definition 2, various properties are readily proved using induction on the length of a derivation in $\mathcal{L}$.

Definition 5. *We say that a string $\sigma \in \tilde{\Sigma}^*$ **determines** some observable $s \in \Sigma$ with value v, whenever the sequence of measurements corresponding to σ determines s as defined in Definition 2 with value v. We say that a string $\sigma \in \tilde{\Sigma}^*$ determines some context c, if σ determines each observable in c. We say that two strings $\sigma, \sigma' \in \tilde{\Sigma}^*$ **agree** on $s \in \Sigma$, whenever either both do not determine s or both determine s with the same value.*

With this definition at hand, we explicitly re-state some observations that were already used in the proof of Theorem 1.

Lemma 1. *If some $\sigma \in \tilde{\Sigma}^*$ determines a context, then any extension/continuation $\sigma \star \tau$ of σ also defines a context.*

This lemma does not scale to systems of more qubits. The following lemma does.

Lemma 2. *Each $\sigma \in \tilde{\Sigma}^*$ determines at most one context.*

3   Hidden Variables and the Holevo Bound

Holevo [6] showed that the maximum information carrying capacity of a qubit is one bit. Therefore, a machine which simulates qubits but has a density of memory (in bits per qubit) larger than one violates the Holevo bound. In this section we show that a very broad class of machines that simulate the Peres-Mermin square violate the Holevo bound.

In [7] deterministic automata are presented that generate a subsets of $\mathcal{L}$. In that paper lower bounds on the amount of states of these automata are presented. Of course, as $\mathcal{L}$ inhibits a genuine amount of non-determinism any deterministic automata will generate only a proper subset of $\mathcal{L}$ but never the whole set $\mathcal{L}$ itself.

In the next subsection we shall introduce the notion of an Memory-factoring Abstract Generating Automata (MAGA) for $\mathcal{L}$ and prove a lower bound on the number of states any MAGA should have if it were to generate $\mathcal{L}$. In a sense, a MAGA for $\mathcal{L}$ will generate all of $\mathcal{L}$.

3.1 Memory-Factoring Abstract Generating Automata

In this project we are not interested in the details of (abstract) machine 'hardware'. Thus, in our definition of a MAGA we will try to abstract away from the implementation details of language generating automata. We will do this by invoking the notion of *computability*. By the Church-Turing thesis (see, a.o. [10]) any sufficiently strong and mechanizable model of computation can generate the same set of languages, or equivalently, solve the same same set of problems. Thus, instead of fixing one particular model of computation and speak of computability therein, we may just as well directly speak of computable outright leaving the exact details of the model unspecified.

Basically, a MAGA $\mathcal{M}$ for $\mathcal{L}$ is an abstract machine that will predict the outcome of a measurement of some observable $s \in \Sigma$ in the experiment as described in Definition 2 given that a sequence of measurements $\sigma \in \tilde{\Sigma}^*$ has already been done. If according to the experiment s is determined by σ with value $v \in \{1, -1\}$, then $\mathcal{M}$ should output v. If the observable s is not determined by σ then $\mathcal{M}$ should output r indicating that the experiment can randomly output a -1 or a 1.

The only requirement that we impose on a MAGA is that its calculation in a sense *factors through* a set of memory states[2] in the sense that before outputting the final value, the outcome of the calculation is in whatever way reflected in the internal memory of the machine. Let us now formulate the formal definition of a MAGA for $\mathcal{L}$.

Definition 6 (MAGA). *A Memory-factoring Abstract Generating Automata (MAGA) for $\mathcal{L}$ is a quadruple $\langle M, M_0, M_1, S \rangle$ with*

1. *S is (finitely or infinitely) countable set of memory states;*
2. *all of M, M_0 and M_1 are computable functions such that*
   - *(a) $M = M_1 \circ M_0$;*
   - *(b) $M_0 : \mathcal{L} \times \Sigma \rightarrow S \times \Sigma$,*
     where[3] $\Pi_2 \circ M_0 = \mathbb{I}$;
   - *(c) and $M_1 : S \times \Sigma \rightarrow \{1, -1, r\}$,*

such that

$$M(\sigma, s) = \quad 1 \quad \text{if } \sigma \text{ determines } s \text{ with value} \quad 1;$$
$$M(\sigma, s) = -1 \quad \text{``} \qquad\qquad\qquad\qquad\qquad\qquad \text{''} -1;$$
$$M(\sigma, s) = \quad r \quad \text{if } \sigma \text{ does not determine } s.$$

In our definition, we have that $M_0 : \tilde{\Sigma}^* \times \Sigma \rightarrow S \times \Sigma$, where M_0 does nothing at all on the second coordinate, that is, on the Σ part. We have decided

---

[2] A memory state is something entirely different from, and hence is not to be confused with, a quantum state.

[3] Here Π_2 is the so-called projection function that projects on the second coordinate: $\Pi_2(\langle x, y \rangle) = y$. Basically, $\Pi_2 \circ M_0 = \mathbb{I}$ just says that M_0 only tells us which state is defined by a sequence $\sigma \in \tilde{\Sigma}^*$.

to nevertheless take the second coordinate along so that we can easily compose M_0 and M_1 to obtain M. This is just a technical detail. The important issue is that the computation *factors* through the memory. That is, essentially we have that $M : \mathcal{L} \times \Sigma \xrightarrow{M_0} S \xrightarrow{M_1} \{1, -1, r\}$ where 'M_1 borrows some extra information on the Σ-part of the original input'.

Theorem 2. *The class of MAGA-computable functions is the full class of computable functions.*

Proof. It is easy to see that if we have infinite memory, we can conceive any Turing Machine μ with the required input-output specifications as a MAGA where M_0 is just the identity, S is coded by the tape input, and M_1 is the function computed by μ. Thus, the class of MAGA-computable functions is indeed the full class of computable functions.

3.2   A Lower Bound for the Peres-Mermin Square

With the formal definition at hand we can now state and prove the main theorem for the Peres-Mermin square. In the proof we will use the so-called Pigeon Hole Principle (PHP). The PHP basically says that there is no injection of a finite set into a proper subset of that set. Actually we will only use a specific case of that which can be rephrased as, if we stuck $n + 1$ many pigeons in n many holes, then there will be at least one hole that contains at least two pigeons.

Theorem 3. *Any MAGA for $\mathcal{L}$ contains at least 24 states. That is, if $\langle M, M_0, M_1, S \rangle$ is such a MAGA, then $|S| \geq 24$.*

Proof. We shall actually use a slightly modified version of a MAGA to prove our theorem. In the unmodified MAGA, the function M_0 tells us what state is attained on what sequence of measurements $\sigma \in \tilde{\Sigma}^*$. In the modified MAGA we will only require that M_0 will tell us in what state the machine is whenever σ determines a full context.

To express this formally we define

$$\tilde{\Sigma}^+ := \{\sigma \in \tilde{\Sigma}^* \mid \sigma \text{ determines a full context of observables}\}.$$

Thus, instead of requiring that M_0 maps from $\tilde{\Sigma}^* \times \Sigma$ to $S \times \Sigma$, we will require that M_0 maps from $\tilde{\Sigma}^+ \times \Sigma$ to $S \times \Sigma$. Clearly, if we have a lower bound for any MAGA $\mathcal{M}$ with this restriction on M_0, we automatically have the same lower bound for any MAGA $\mathcal{M}'$ outright. This is so as any MAGA $\mathcal{M}'$ trivially defines a restricted MAGA $\mathcal{M}$ by just restricting the domain of M_0 to $\tilde{\Sigma}^+ \times \Sigma$.

To continue our proof, let $s_1, \ldots, s_{24}$ enumerate all possible combinations $\langle c, v_1, v_2 \rangle$ of contexts and the first two[4] values of the first two observables of that

---

[4] For horizontal contexts we will enumerate from left to right and for vertical contexts we will enumerate from top to bottom. Thus, for example, the first two observables of the context $\langle C, c, \gamma \rangle$ are C and c.

context c. Note that there are indeed $6 \times 2 \times 2 = 24$ many such combinations. We define a map

$$C : \quad \tilde{\Sigma}^{+} \rightarrow \{s_1, \ldots, s_{24}\}$$

in the canonical way, mapping an element $\sigma \in \tilde{\Sigma}^{+}$ to that s_i that corresponds to the triple consisting of the context that is determined by σ followed by the first two values of the the first two observables of that context. By Lemmas 1 and 2 the function C is well-defined.

Now, let $\sigma_1, \ldots, \sigma_{24}$ be representatives in $\tilde{\Sigma}^{+}$ of $s_1, \ldots, s_{24}$ such that $C(\sigma_i) = s_i$. For a contradiction, let us assume that there exists some restricted MAGA $\langle M, M_0, M_1, S \rangle$ for $\mathcal{L}$ with $|S| \leq 23$. By the Pigeon Hole Principle, we can choose for this MAGA some σ_i and some different σ_j such that[5]

$$M_0(\sigma_i) = M_0(\sigma_j).$$

We now use the following claim that shall be proved below. Recall from Definition 5 what it means for two sequences to agree on some variable.

Claim. If $\sigma_k \neq \sigma_l$ then there is some $s \in \Sigma$ such that σ_k and σ_l disagree on s.

Once we know this claim to hold it is easy to conclude the proof. Consider any $s \in \Sigma$ on which σ_i and σ_j disagree. By the definition of M we should have that $M(\sigma_i, s) \neq M(\sigma_j, s)$. However as $M_0(\sigma_i, s) = M_0(\sigma_j, s)$ we see that $M_1 \circ M_0(\sigma_i, s) = M_1 \circ M_0(\sigma_j, s)$. But $M_1 \circ M_0 = M$, which contradicts $M(\sigma_i, s) \neq M(\sigma_j, s)$. We conclude that M can not have 23 or less states.

Thus to finalize our proof we prove the claim. Let $C(\sigma_k) = \langle c^k, v_0^k, v_1^k \rangle \neq \langle c^l, v_0^l, v_1^l \rangle = C(\sigma_l)$. If $c^k \neq c^l$ then c^k contains at least two observables on which σ_k and σ_l agree as each of these σ's only determine observables in their respective contexts.

In case $c^k = c^l$, then one of v_0^k, v_1^k differs from the corresponding one in v_0^l, v_1^l giving rise to a disagreement between σ_k and σ_l.

This concludes the proof of the claim and thereby of Theorem 3.

Remark 2. Note that the proof of Theorem 3 nowhere invokes the notion of computability therefore proving actually something stronger.

One can easily see that for $\mathcal{L}^{+}$ the obtained lower bound is actually sharp in the sense that there is a MAGA with 24 memory states for $\mathcal{L}^{+}$. However, is seems that for $\mathcal{L}$ this is not the case.

3.3   Scaling

We first note that the proof of Theorem 3 is very amenable to generalizations:

Remark 3. The proof of Theorem 3 easily generalizes under some rather weak conditions giving rise to lower bounds of #contexts $\times\, 2^{(\#\text{degrees of freedom in one context})}$.

---

[5] Par abus de langage we will write $M_0(\sigma_i)$ as short for $\Pi_2(M_0(\sigma_i, s))$.

The Peres-Mermin square (†) that corresponds to the two-qubit system can be generalized in various ways [1,2]. One particular generalization for n qubits gives rise to a system where each context consists of exactly d elements with $d = 2^n$ [2]. Moreover, there are $c := \prod_{k=1}^{m}(2^k + 1)$ different many such contexts. Each context is now determined by a particular selection of n of its elements. We shall denote the corresponding languages by $\mathcal{L}_n$. Thus, what we have called $\mathcal{L}$ so far in this paper, would correspond to $\mathcal{L}_2$.

Theorem 4. *Any MAGA for $\mathcal{L}_n$ contains at least $2^n \cdot \prod_{k=1}^{n}(2^k + 1)$ many different memory states.*

Proof. Basically this is just by plugging in the details of the languages $\mathcal{L}_n$ into remark 3. Let us very briefly note some differences with the proof of Theorem 3. The main difference is that the set $\mathcal{L}_2^+$ is nice: once a string is in there, any extension is as well. However, this does not impose us to, again define a restricted MAGA by restricting the domain of M_0 to $\mathcal{L}_n^+$. Again, we consider the $n+1$-tuples consisting of a context with some values for the[6] n observables that determine this context and choose some correspondence between these tuples and some representing sequences $\sigma_i \in \tilde{\sigma}_n^*$. Clearly, these $\sigma_i \in \mathcal{L}_n^+$. As the claim obviously holds also in the general setting, the assumption that the memory states $s_1, \ldots, s_{2^n \cdot \prod_{k=1}^{n}(2^k+1)-1}$ suffice yields together with the PHP to a contradiction as before.

As was done in [3] and in [7], we can consider the information density d_n for the corresponding languages defined as the number of classical bits of memory needed to simulate a qubit:

$$d_n := \frac{\log_2(|S_n|)}{n}. \qquad (+)$$

If we apply the lower bound for S_n –the number of memory states for a MAGA for $\mathcal{L}_n$– from Theorem 4 to (+) we obtain

$$
\begin{aligned}
S_n &\geq 2^n \cdot \prod_{k=1}^{n}(2^k + 1) \\
&\geq 2^n \cdot \prod_{k=1}^{n}(2^k) \\
&\geq 2^n \cdot (2^{\sum_{k=1}^{n} k}) \\
&\geq 2^n \cdot (2^{\frac{n(n+1)}{2}})
\end{aligned}
$$

whence d_n is approximated (from below) in the limit by $\frac{\log_2(2^n \cdot (2^{\frac{n(n+1)}{2}}))}{n} = \frac{n+3}{2} \sim \frac{n}{2}$. Thus, the information density in this generalization of the Peres-Mermin square grows linear in the number of qubits. However, as observed before, any density more than 1 implies a violation of the Holevo bound.

---

[6] For each context, we fix some n observables that determine that context as these are not uniquely defined.

References

1. Cabello, A.: Proposed test of macroscopic quantum contextuality, Phys. Rev. A (82), 032110 (2010)
2. Cabello, A., et al.: State-independent quantum contextuality for n qubits, arXiv:1102.....
3. Cabello, A.: The role of bounded memory in the foundations of quantum mechanics. Found. Phys., published online (September 16, 2010), doi: 10.1007/s10701-010-9507-2
4. Chomsky, N.: Three models for the description of language. IRE Transactions on Information Theory (2), 113–124 (1956)
5. Flajolet, P., Sedgewick, R.: Analytic Combinatorics. Cambridge University Press, New York (2009) ISBN 978-0-521-89806-5
6. Holevo, A.S.: Some estimates of the information transmitted by a quantum communication channel. Probl. Inf. Trans (9), 177 (1973)
7. Kleinmann, M., Gühne, O., Portillo, J.R., Larsson, J.-Å., Cabello, A.: Memory cost of quantum contextuality. arXiv:1007.3650
8. Mermin, N.D.: Simple unified form for the major no-hidden-variables theorems. Phys. Rev. Lett (65), 3373 (1990)
9. Peres, A.: Incompatible results of quantum measurements. Phys. Lett. A (151), 170 (1990)
10. Sipser, M.: Introduction to the Theory of Computation. PWS Publishing (1997) ISBN 0-534-94728-X

Can Anything from Noether's Theorem Be Salvaged for Discrete Dynamical Systems?

Silvio Capobianco[1,*] and Tommaso Toffoli[2]

[1] Inst. of Cybernetics at TUT, Tallinn, Estonia
silvio@cs.ioc.ee
[2] ECE Dept., Boston Univ., Boston, MA
tt@bu.edu

Abstract. The dynamics of a physical system is linked to its phase-space geometry by Noether's theorem, which holds under standard hypotheses including continuity. Does an analogous theorem hold for discrete systems? As a testbed, we take the Ising spin model with both ferromagnetic and antiferromagnetic bonds. We show that—and why—energy not only acts as a generator of the dynamics for this family of systems, but is also conserved when the dynamics is time-invariant.

Keywords: analytical mechanics of cellular automata, second-order dynamics, energy conservation, energy as generator of the dynamics, Noether's theorem in the discrete.

1 Introduction

In the last three hundred years, analytical mechanics has turned up a wealth of profound concepts and powerful methods for dealing with the dynamical systems of *physics*. Can these results—of which Noether's Theorem is one of the best known—be applied to the dynamics of *information systems*, which today make up such a large part of our culture, and to what extent?

Here one may anticipate two difficulties. On one hand, the above results take advantage of specific aspects of physics and may not be valid outside that territory. On the other, the methods of analytical dynamics are intimately tied to the calculus of the *continuum*, and may fail when applied to *discrete* systems such as automata and networks.

In sum, is analytical dynamics just a "bag of tricks" for doing continuum physics, or does it provide templates for tools useful in a more general context? Does the insight that it offers transcend the distinction between physical and man-made systems (cf. [12]) and that between discrete and continuous ones?

The prototypical dynamical system that we all know and love, namely, *physics*, displays a striking property. Its explicit evolution law has the form of a

* This research was supported by the European Regional Development Fund (ERDF) through the Estonian Centre of Excellence in Computer Science (EXCS), and by the Estonian Science Foundation under grant #7520.

C.S. Calude et al. (Eds.): UC 2011, LNCS 6714, pp. 77–88, 2011.

lookup table that to each possible *state* of the system—say, a $2n$-tuple of real variables (or a *vector*) consisting of n positions and n momenta (where n is the number of degrees of freedom)—associates another $2n$-tuple—its "next state." For physical systems, it turns out that this table can be "compressed" into one that to each state associates a *single real variable*—the *energy* of that state. The first table explicitly expresses the dynamics as a *vectorial* function, while the second implicitly expresses the dynamics through a *scalar* function. This "compression" is *nondestructive*, that is, the full table can be reconstructed from the compressed table. Both the $2n$-to-1 encoding and the converse 1-to-$2n$ decoding are achieved by simple, standardized algorithms (in analogy with "zipping" and "unzipping" a file). For this reason, the *energy function*, that is, the above energy lookup table, is said to be the *generator* of the dynamics, in the sense that plugging this function into a generic kinematic scheme (such as Hamilton's canonical equations) allows one to explicitly reconstruct a specific dynamics.

The usual way to derive the above property of energy relies on the fact that a physical system (both in classical and quantum mechanics) is imagined to evolve *continuously* in time. However, in order to have this property, being time-continuous is not enough—for this, the dynamics must have much additional structure, namely, *symplectic*, in classical mechanics, and *unitary*, in quantum.

As we generalize the concept of dynamical system, in particular to systems that, like automata of all kinds, have *discrete* state variables and evolve *discretely* in time, the question arises whether an energy-like quantity can be retained at all, and, if so, under what conditions. The first question we'll ask here is whether a cellular automaton can have something that can rightfully be called "energy," what conditions it would have to satisfy for that, and what it would look like. We'll then ask whether and to what extent the method proposed by Noether for deriving *conserved quantities* from the *symmetries* of the dynamics of a continuous systems may be adapted to discrete systems.

2   Energy Conservation

We've all heard of the principle of *conservation of energy*: In an isolated physical system, no matter what transformations may take place within it, there is a quantity, called *energy*, that remains conserved. Our belief in this principle is so strong that, whenever a violation of energy conservation is observed, our first reflex is not to question the principle, but to look more carefully into the experimental evidence and ask whether something may be escaping us. To defend the reputation of energy as a bona fide conserved quantity against new incriminating evidence, now and then physicists are even willing to stretch energy's very definition (see Feynman's peerless parable[5]).

How far shall we be able to retain a viable concept of energy as our conception of "dynamical system" keeps widening? Can energy be conserved in a *cellular automaton*? (cf. [4]) Conserved or not, should an energy be expected to be found *at all* in the latter? Is the idea of "energy" meaningful for a *Turing machine* or a *finite automaton*? In sum, what features does a dynamical system need to exhibit in order to possess a recognizable generalization of energy? *Why* should energy be conserved?

3   Noether's Theorem

Nearly a century ago, Emmy Noether formulated a principle of great generality and beauty, commonly known as *Noether's Theorem*[8,13], which we shall state for a moment just in headline form, ignoring the fine print: "*To every symmetry of a dynamical system there corresponds a conserved quantity.*" According to this principle, the conservation of energy (a *state variable*) would be the consequence of the constancy of a system's dynamical laws, that is, of the invariance in time of the system's *structural parameters*.

Noether's theorem plays a unifying role in our understanding of mechanics; according to [13], it "has become a fundamental tool of modern theoretical physics." Since it can take advantage of many kinds of symmetries, it is very productive: for instance, it predicts that, if a system's laws are rotationally invariant, then, just because of that structural invariance, there will be a specific state variable that is conserved, namely, *angular momentum*.

However, when we get down to reading Noether's theorem's fine print, we discover a number of restrictive provisos. A fuller version of the contract would spell "To every *one-parameter, continuous group* of symmetries of a *Lagrangian* dynamical system there corresponds a *scalar, real-valued* conserved quantity." Thus, a *discrete* group of symmetries, such as the invariance of Conway's "game of LIFE" under quarter-turn rotations of the orthogonal grid, won't do (bye-bye angular momentum?). Similarly, even though the dynamics of LIFE is the same at every step, and thus *time-invariant*, it is not *invertible* (and thus only forms a semigroup, not a group), and in any event the game advances in discrete steps, instead of along a one-dimensional continuum (bye-bye energy?).

4   The Ising Playground

The best way to identify the issues at play is to examine a specific class of models, and then map the conceptual question of our investigation to concrete features of a model. For our purposes, we shall choose a standard cellular-automaton realization of the *Ising spin model* (cf. [2,3,9]). This consists of a regular array of elementary magnets, or *spins*, that can assume one of two orientations—'up' and 'down'.

Spins interact with one another. In the Ising model stylization of physics, long-range forces are ignored and short-range ones represented by the interaction of a spin with only its immediate neighbors: its four first neighbors, in a 2D orthogonal array. In the simplest case, these four *bonds* (or "couplings") are all of a *ferromagnetic* nature, that is, they behave like rubber bands that are relaxed when the spins at the two ends of a band point in the same direction ("parallel"), and stretched, when in the opposite direction ("anti-parallel").

Different kinds of dynamics may be used for the Ising model; in order to have a unifying criterion for prescribing a dynamics, one starts with assigning one unit of a notional "potential energy" to stressed bonds, and zero to unstressed ones. The class of dynamics we choose here, the so-called *microcanonical Ising model*, are strictly deterministic and invertible: on a given step, a spin will *flip* (that is, reverse its orientation) *if and only if* doing so will leave the sum of the potential energies of the four surrounding bonds *unchanged*. Note that, in the ferromagnetic case, this happens when two of the neighbors are parallel to the given spin and the other two anti-parallel. In this case, in fact, we start with two stressed bonds. After the flipping of a given spin, its parallel neighbors will have turned anti-parallel and vice versa, so that the two stretched bonds relax at the same time as the two relaxed ones stretch, and consequently the overall potential energy stored in these bonds is two units both *before* and *after* the step. In all other cases, since no flip occurs, potential energy is conserved trivially.

A final stipulation. The above rule would become inconsistent if one attempted to update two adjacent spins *at the same time*. In fact, if *both* spins are instructed to flip, the assumption on the part of either spin that the shared bond's stress status will thereupon be complemented *fails* (as its stress status will remain unchanged), and energy is no longer conserved. A standard solution is to treat the array of spins as a checkerboard consisting of two intermeshed subarrays called *even* and *odd* (the red and black squares of the board) according to the parity of $x + y$, where x and y are the spatial coordinate of a site. A system state s consists of an *ordered pair* $\langle q_A, q_B \rangle$ of *configurations* of different parity (i.e., one even and one odd, in either order). Together, an even and an odd configuration fill up the whole array; however, their order in the pair is relevant, that is, state $\langle q_A, q_B \rangle$ is distinct from $\langle q_B, q_A \rangle$. Depending on the context, it may be convenient to call the two configurations of a pair either the "past" and the "present" or, alternatively, the "present" and the "future" (the latter abbreviated, when subscript, as "pres" and "futr").

The Ising dynamics is a second-order recurrence relation of the form

$$q_{x,y}^{t-1} \oplus f(q_{x-1,y}^t, q_{x+1,y}^t, q_{x,y-1}^t, q_{x,y+1}^t) = q_{x,y}^{t+1}. \tag{1}$$

This is a *conditional permutation*: a spin $q_{x,y}$ in the *past* (time $t - 1$) is mapped into a spin $q_{x,y}$ in the *future* (time $t + 1$) by XORing it with an *enable* bit—the output of the binary function f when given as arguments the *present* values (time t) of the first-neighbor spins $q_{x-1,y}, q_{x+1,y}, q_{x,y-1}, q_{x,y+1}$, which belong to the complementary subarray and thus affect the updating of $q_{x,y}$ but are not affected it. Depending on the value of the enable bit, this updating either leaves a spin unchanged or complements it—both invertible operations. The above "shifting-frame" mechanism—of the form $\langle q_{\text{past}}, q_{\text{pres}} \rangle \mapsto \langle q_{\text{pres}}, q_{\text{futr}} \rangle$, where the right item q_{pres} of the old state shifts to the left unchanged, pushing out the left item q_{past}, and a new item q_{futr}, computed according to (1), shifts in from the right—is called in numerical analysis *leapfrog updating*.

The macroscopic behavior of this kind of model is trivial in one dimension; conceptually productive in two dimensions, as it provides by very simple means an insight into the physics of phase transitions; and increasingly rich and challenging in more dimensions.

5   Why Call It Energy?

Setting physical motivation aside, we have in the Ising spin model an abstract symbolic dynamics consisting of a two-dimensional array over the binary alphabet $\{0, 1\}$ (for what was 'down' and 'up'; it will still be convenient to call "spin" the state variable at each site). We distinguish between even and odd sites according to the parity of the sum $x + y$ of their coordinates. We update even and odd sites on alternating steps according to the rule (stated the rule in a way that obviously generalizes to any number of dimensions) that

"A spin u at $\langle x, y \rangle$ flips (i.e., complements) *if and only if* its first neighbors (those at $\langle x \pm 1, y \rangle$ and $\langle x, y \pm 1 \rangle$) are *evenly divided* between state u and its complement $\bar{u}$."

Note that we had no use for an "energy." However, if as an afterthought we choose to call *energy* a distributed quantity whose value, for each *pair* of adjacent first-neighbor spins, is 0 if these two spins are equal and 1 if different, then it follows that energy is *conserved* by the above dynamics. Also note that this energy is but the *length of the boundary* between 'up' and 'down' domains. While the *magnetization*—number of spins up minus number of spins down—may change with time, that length, and thus the energy, remains constant. Is this use of the term "energy" just physics nostalgia—like calling a British colony "New Hampshire"? Or is there something more to it?

For most people, energy is a fungible resource of the form $E = T + U$, that is, stored either in the inertia of a moving object ("kinetic energy" T) or in the state of compression of an elastic medium such as a spring ("potential energy" U). Such a concrete presentation of energy will do for many practical applications.

But if you look at what energy is *really* supposed to be in physics, you'll discover an abstract quantity that is totally unconcerned with the *nature* of the objects and materials that make up a dynamical system, and instead totally absorbed with the topological structure of the system's *trajectories*. In fact, the total *energy* of a system may be defined as

1. A *real-valued function* of the system's *state*,
2. that is *additive*,   (2)
3. and is a *generator of the dynamics*.

Incidentally, there is no mention of "energy conservation" in all this.

NOTE.   We wish we could give a precise reference for the above definition. In fact, here we rely on disparate loose hints culled and integrated with some effort from the literature. Physicists are supposed to know what energy is, and they don't need—perhaps have good reasons not to *want*—a final definition (cf. the Feynman parable mentioned in §2). So, not only college textbooks, but even such beloved conceptually-minded references as Arnold[1], Goldstein[6], and Lanczos[7] proceed by giving *examples* of energy in increasingly elaborate contexts, and then bringing one's attention to increasingly more abstract *properties* of energy. Definitions—hardly ever.

Let's verify that the "energy" of the Ising model agrees with our abstract definition. In order to do so, we'll be forced to re-examine the definition and pore over the fine print that by tacit agreement stands behind such terms as "state," "additive," and "generator."

5.1   Additivity

As we've seen, what we proposed to call Ising "energy" of a state is simply the *number of stretched bonds* in that state. Incidentally, since the Ising model consist of an *infinite* array of sites, the state of the system consists of an infinite array of spins, and virtually any state will have *infinite* energy. For a function to be *additive*, here the physicist means that (a) it is meaningful to partition the system into separate subsystems such that each of these subsytems has its own well-defined state, and the global state is the composition, or Cartesian product, of the states of these subsystems; (b) the function is defined on the state of each of these subsystems, and (c) its value for the whole system is the sum of its values for the individual subsystems.

If a system could be partitioned into subsystems or "blocks" that are totally independent of one another, each one having a well-defined energy independent of the state of the other blocks, additivity would be trivial to achieve—just *define* the energy of the whole system as the sum of the energy of its subsystems—but by the same token would be a *vacuous* concept. In general, however, there will be some residual coupling between blocks, and even if one knew the precise state of a block one would be able to assign an energy to it only to within an upper bound and a lower bound, which reflect the lack of knowledge of the block's environment's state. Fortunately, in many cases, when interactions are *local* and the blocks are appropriately chosen, this uncertainty is roughly proportional to the length of a block's *boundary*—where it interfaces with its environment, and thus the *relative* uncertainty is bounded above by the boundary/volume ratio for a block, decreases with block size, and vanishes in the limit of arbitrarily large blocks. This is what "additivity" means to the physicist, and it is consistent with the approach taken by topological dynamics.

In the Ising model, sites interact with one another through an unbroken mesh of couplings; there are no strictly independent subsystems. However, the stipulation made at the end of §4, whereby the odd mesh is kept constant when the even mesh is updated (and vice versa)—and thus effectively acts as a parameter rather than a state-variable—enforces a discipline by which energy variations corresponding to local variations of a configuration can be determined *exactly*. We shall see this at the end of §5.2 and use it to full advantage in §7.

5.2   Generator of the Dynamics

A *generator of the dynamics* is a function of the system's state through the knowledge of which one can reconstruct the system's dynamics in an explicit, vectorial form (cf. §1)—of course only *up to an isomorphism*. It is understood that this reconstruction should be achievable by an algorithm given once-and-for-all, independent of the specific system (otherwise, one could "cheat" and hide information about the system in the algorithm itself—big deal!). Suppose the function is a "Hamiltonian" that for each state $\langle q, p \rangle = \langle q_1, \ldots, q_n, p_1, \ldots, p_n \rangle$ yields its energy $H(q, p)$. Our goal is to determine where any state $\langle q, p \rangle$ will go.

For a continuous dynamics as we imagine in the present example one would want the instantaneous "direction and rate" at which the state progresses, that is, the $2n$ derivatives $\langle \dot{q} = \frac{dq}{dt}, \dot{p} = \frac{dp}{dt} \rangle$, In other words, we want to arrive at *vectorial* mapping, of the form

$$\langle q, p \rangle \mapsto \langle \dot{q}, \dot{p} \rangle, \tag{3}$$

even though mere evaluation of the function at $\langle q, p \rangle$ will perform the mapping $\langle q, p \rangle \mapsto H(q, p)$ and thus only give us a a scalar. However, if we performed *repeated* samplings of H in the vicinity of $\langle q, p \rangle$ (after all, we have the entire function H and we can apply it as many time as we want) so as to get a "sense of direction," or, equivalently, in the continuum case, if we could get the *derivatives* of H at that point, we would get more information associated with that point than its mere energy $H(q, p)$. In Hamiltonian mechanics, mapping (3) is provided by the standard *canonical equations*

$$\dot{q} = \frac{\partial H}{\partial p}; \quad \dot{p} = -\frac{\partial H}{\partial q}. \tag{4}$$

Note that, since q and p are actually abbreviations for n-tuples, here we have $2n$ equations yielding a *vector* consisting of n pairs $\langle \dot{q}_i, \dot{p}_i \rangle$, $(i = 1, \ldots, n)$.

It will presently become obvious that in the Ising model the energy function *does* generate the dynamics—and by means of a very simple recipe. In a *discrete* Hamiltonian dynamics, a state is no longer a "position/momentum" pair $\langle q, p \rangle$ as in the continuous case, but an ordered *pair of configurations* $\langle q_0, q_1 \rangle$, as explained at the end of §4. The rules of the game are that one may propose *any* such state, *as many times as one wishes*, and each time get as an answer the energy $H(q_{\text{past}}, q_{\text{pres}})$ for that state. Isn't this number, in the range $\{0, 1, 2, \ldots, \infty\}$ (and almost always ∞) much too little to determine the infinite q_{futr} configuration? The answer lies in the principle of *virtual displacements*. That is, given q_{past} and q_{pres}, one may propose *any conceivable candidate* for q_{futr}, ask for the energy of the resulting new state $\langle q_{\text{pres}}, q_{\text{futr}} \rangle$, compare it with the old energy, and on the basis of this comparison decide whether to accept or reject the candidate. Just as in the "hot, hot, cold, cold" game, in the game of "battleships"—or, for that matter, in biological evolution—one is never explicitly shown the target. Instead, one must describe an object, and will only be told ("yes," "no") whether that is target, or possibly given an intermediate scalar value ("hot", "warm", etc.) that indicates *how far* one's proposal is from the target. In principle, one can arrive at the solution by going through all possible candidates. The point of the game is to speed up the search by dynamically arranging the proposals so that the feedback will give one a "sense of direction."

In the Ising case, we want to find a q_{futr} such that the energy difference $\Delta E = H(q_{\text{pres}}, q_{\text{futr}}) - H(q_{\text{past}}, q_{\text{pres}})$ is zero not only globally but also *locally*, that is, at any scale down to the radius of the first-neighbor neighborhood (the range of interspin "interactions"). Forget about getting finite energies by taking finite configurations by blocking (§5.1); just take the 32 *infinite* configurations of the form

$$
\begin{array}{c}
\vdots\ \vdots\ \ \vdots\ \vdots\ \vdots \\
\cdots\ 0\ 0\ 0\ 0\ 0\ \cdots \\
\cdots\ 0\ 0\ n\ 0\ 0\ \cdots \\
\cdots\ 0\ w\ c\ e\ 0\ \cdots \\
\cdots\ 0\ 0\ s\ 0\ 0\ \cdots \\
\cdots\ 0\ 0\ 0\ 0\ 0\ \cdots \\
\vdots\ \vdots\ \ \vdots\ \vdots\ \vdots
\end{array}
\tag{5}
$$

with c denoting a "center cell" in the past and n, s, w, e denoting its four neighbors in the present. This infinite state will have a finite energy ranging from 0 through 4. Let's make a table of energy for the 32 possible collective values of those five sites,

c nswe	H f c'		c nswe	H f c'		c nswe	H f c'		c nswe	H f c'	
0 0000	0 0 0		0 1000	1 0 0		1 0000	4 0 1		1 1000	3 0 1	
0 0001	1 0 0		0 1001	2 1 1		1 0001	3 0 1		1 1001	2 1 0	
0 0010	1 0 0		0 1010	2 1 1		1 0010	3 0 1		1 1010	2 1 0	
0 0011	2 1 1		0 1011	3 0 0		1 0011	2 1 0		1 1011	1 0 1	
0 0100	1 0 0		0 1100	2 1 1		1 0100	3 0 1		1 1100	2 1 0	
0 0101	2 1 1		0 1101	3 0 0		1 0101	2 1 0		1 1101	1 0 1	
0 0110	2 1 1		0 1110	3 0 0		1 0110	2 1 0		1 1110	1 0 1	
0 0111	3 0 0		0 1111	4 0 0		1 0111	1 0 1		1 1111	0 0 1	

$$\tag{6}$$

where H is the "Hamiltonian energy" of the Ising model for the entire global state (5), f the "conditioning" function that appears in (1), and c', as we shall see in a moment, the effective future value of the center site. As we recall from §4, the Ising dynamics is "flip if and only if that leaves the energy of the neighborhood unchanged." Therefore, for any state of the form (5), we may propose as candidates for the new state either (a) the same state, i.e., with $c' = c$, or (b) the one where c has flipped, i.e., where $c' = \bar{c}$. By using table (6) *two* times, i.e., for the two c' candidates c and $\bar{c}$, we see that the energy is (obviously) unchanged in the first case, and *may* have changed in the second case. If the energy is the *same* in both cases, then the *actual* value for c' (that is, the one prescribed by the Ising dynamics) will be the default value c; otherwise, it will be $\bar{c}$. This is the value given in column c' of (6).

Finally, we must show that this new value c', given the old value of the entire neighborhood $w\overset{n}{\underset{s}{c}}e$, is independent of all the others sites of the global state— those assigned 0 in global state(5). In fact, by the definition, in the Ising model, of energy as a *sum over bonds* (or adjacent-spin pairs), the energy spanning, say, the "extended patch" $\overset{u}{\underset{s}{w}}\overset{nar}{c}e$ is the sum of the energies spanning the two elementary patches $w\overset{n}{\underset{s}{c}}e$ and $n\overset{u}{\underset{e}{a}}r$; in other words, energy is *strictly additive*, and the *local dynamics* table (6) can be applied separately to every elementary patch.

Thus, the "energy" of the Ising model satisfies the abstract, "model-independent" definition of *energy* given in (2). ... Or *does* it?

6 Raising the Kinematics Bar

Indeed the Ising energy (a) is a real-valued function of the system's state; (b) it is additive, since the alternating updating scheme guarantees that the accounting of energy can be done independently on every elementary patch; and (c) a simple rule allows us to reconstruct the dynamics by sampling the energy function in the (topological) neighborhood of a state.

However, point (c) is in this case *vacuous*, since here we are dealing with a *single* system—the Ising model—and there is no indication of how one would go about generating the dynamics of *other* systems using other energy functions. Consider a telegraph channel with a single input key, pressing which will invariably transmit the text of the *Kamasutra*. In spite of the length and the intrinsic interest of this text, the capacity of the channel is obviously $\log 1 = 0$ bits. In a similar way, the information contents of H_{ising} *as a function*, for the sake of specifying a dynamics, is zero, because it is a single "key": the Ising dynamics is the only dynamics we have in mind, it is known, and we might as well have stored a copy of it at the channel's receiving end. What one needs is a generic function scheme—a look-up table "blank form," as it were—which, when filled with specific contents and processed by an algorithm given once and for all, will allow one to describe any one of an agreed upon class of dynamical systems.

This issue is addressed by physicists by distinguishing between "dynamics" and "kinematics," where the latter term is used in a somewhat specialized sense. That is, a *kinematics* is a framework—a set of rules—that characterizes a whole class of dynamical systems of a similar type or "texture," while a *dynamics* is any one of the systems sharing that kinematics. Until we place the Ising model within a larger class of models, and show the existence of an energy-function-valued *variable*—whose values will be specific energy functions each of which will in turn allow a fixed algorithm to generate the different dynamics of that class—until then we will not have a bona fide *generator of the dynamics*.

There is a natural and a very simple generalization of the Ising model (used, specifically, in the modeling of *spin glasses*) that yields a class of *sixteen* dynamics. We'll show that for that class the energy function plays a nonvacuous role as a generator of the dynamics.

In §4, we called *ferromagnetic* a bond between adjacent spins that is unstressed when the two spins are parallel and stressed when antiparallel ("type 0" bond). We now introduce a bond of "type 1," or *antiferromagnetic*, which is stressed when the spins are *parallel*, and unstressed otherwise. With bonds of two kinds thus available, and four bonds surrounding a spin, there are $2^4 = 16$ possible *structural environments* by which a spin can find itself surrounded. We shall retain our definition of *Ising energy* as the sum of the energy of the bonds, recalling only that the stress $u \overset{b}{\longleftrightarrow} v$ of a bond of type b between a pair of parallel spins u and v will $u \oplus b \oplus v$, rather than just $u \oplus v$ as in the original Ising model. As part of our kinematics—not as a peculiarity of an individual dynamics—we shall retain the Ising prescription that a spin will flip if and only if such a move leaves the energy unchanged for a given assignment of the four surrounding bond types. However, within this kinematics, we shall for the moment restrict our attention to systems whose structural parameters (those determining a dynamics, and in our case consisting of the assignment of bond types) be *time- and space-invariant*, that is, for each specific dynamics of our kinematics, the behavior of a spin shall not depend on *where* the spin is located and on *what time it is*, but only on the values of its neighbors.

With the latter restriction, we are left with as many *global* assignments of structural parameters as there are choices of *local* assignments—that is, the 16 distinct assignments of four bond types around a spin. For each of these structures we must tabulate, by running over all possible states, the corresponding dynamics (a function from old global state to new global state) and the corresponding energy function (a function from global state to an integer). Observe that two distinct structures may well end up yielding the same dynamics or the same energy function. Our goal will be achieved if (a) no two dynamics are specified by the *same* energy function (in other words, the energy function *distinguishes* between dynamics), and (b) every dynamics has *at least one* energy function that specifies it. Since, by construction, energy is additive down to the scale of individual four-bond "patches," the local "variational principle" employed in §5.2 will then allow us not only to indirectly denote, but to explicitly *generate* the corresponding vectorial-form (that is, state-to-state) dynamics.

This is indeed the case. Since we are presenting here an original approach, it is important not to deceive ourselves on delicate aspects of an argument. For this reason, as a "sanity check" we have explicitly tabulated dynamics and energy function for all 16 elements of our generalized Ising kinematics on a finite, toroidally wrapped-around model. (We used just a 4×4 torus; though small, this size is large enough to rule out spurious degeneracies. Full scripts, output data, and statistics are available at www.ioc.ee/~silvio/nrg/.) A formal proof is subsumed within the properties of the numerical integration scheme of Fig. 1, which constitute a proof in the more general context of space- and time-dependent dynamics. In fact, in the next section we shall remove the constraint that the structure—and the attendant dynamics—be spacetime uniform.

7   Space- and Time-Dependent Dynamics

Energy in these generalized Ising models is a function of the global state—a function ultimately specified by the structure parameters of the specific model. If the parameters are allowed to change from one step to the next, then this function is in turn a function *of time*. In fact, the type of the bond joining two given spins will depend on "what time it is" at the moment of performing a transition $\langle q_A, q_B \rangle \to \langle q_B, q_C \rangle$. The "new state" $\langle q_B, q_C \rangle$ at the end of this transition reappears as the "old state" at the beginning of the following transition, $\langle q_B, q_C \rangle \to \langle q_C, q_D \rangle$, but now it is a *different time*, and the energy of the *same state* (of spins) has to be re-evaluated accordingly to the new assignment of structural parameters.

This arrangement of things, which is essentially a numerical integration scheme, is graphically illustrated by Fig. 1. There, the integers along the time axis represent discrete time instants, those along x or y, discrete spatial positions. A spin is represented by a thick vertical line, interrupted by a conditional-permutation gate ($\oplus$) every time that it is the spin's turn to be updated. A system state is an ordered pair of configurations associated with two consecutive integer times. This state may conveniently be labeled with the "half-time" in between; that is, one may denote by $\frac{1}{2}$ the state consisting of the configurations at 0 and 1, and so forth. Steps of the dynamics are nominally associated with integer values of time. So, for instance, state $1\frac{1}{2}$ is the "new state" for step 1 but the "old state" for step 2. Horizontal arcs of unit length (thin lines) represent interspin bonds as at the time of a step; letters may be used to specify the nature of a bond; so r denotes the nature of the bond between spins 1 and 2 at step 1, while s denotes the nature of the same bond at step 2.

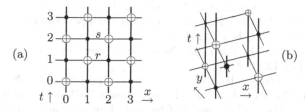

Fig. 1. (a) Spacetime spin grid (y not shown). (b) Detail of full grid, with both x and y.

The value of a bond remains steady during spin updating (at integer times) but may change between updatings (at half-integer times). On this occasion, the energy of a state is accordingly re-evaluated. Thus, even if energy is conserved during spin updating, in a time-dependent system the energy may still change when the *bonds* are altered.

Thus, it will generally happen that energy changes with time. With the present *full* class of dynamics (i.e., one which allows arbitrary spacetime variation of the structural parameters) of the given generalized Ising kinematics, we arrive at the conclusion that energy is *necessarily conserved if* the spacetime structure is time-invariant (because the transitions themselves are by construction energy conserving, and with time-invariant bond types no re-evaluation of the energy need occur). On the other hand, when the spacetime structure is *not* time invariant, *at least some* of the dynamics are not energy-conserving. We can thus rightfully say that, in the present context, "Ising energy" *as a whole* (not the energy of a specific Ising dynamics) is the quantity that is conserved *because of* the time-invariance of the dynamics.

For this class of dynamics, then, in spite of the discreteness of the dynamics, a non-frivolous transliteration of Noether's theorem applies with full force, yielding energy conservation from time invariance.

8   Conclusions

We've shown that certain aspects of Noether's theorem apply to dynamical systems beyond ordinary analytical mechanics, and *why they do*. We were able to shed the requirement that the dynamics be *continuous*; still, some form of *second-order* discipline seems to remain essential for a symmetry to give rise to a conservation law. This is just the beginning of what promises to be a productive line of research.

References

1. Arnold, V.: Mathematical Methods of Classical Mechanics, 2nd edn., corr. 4th printing. Springer, Heidelberg (2010)
2. Bach, T.: Methodology and implementation of a software architecture for cellular and lattice-gas automata programming. PhD thesis, Boston University 2007; http://ww.pm1.bu.edu/tt/Bach07thesis.pdf
3. Bennett, C.H., Margolus, N., Toffoli, T.: Bond-energy variables for Ising spin-glass dynamics. Phys. Rev. B 37, 2254 (1988)
4. Boykett, T., Kari, J., Taati, S.: Conservation Laws in Rectangular CA. J. Cell Autom. 3, 115–122 (2008)
5. Feynman, R., Leighton, R., Sands, M.: Conservation of Energy, ch. 4. The Feynman Lectures on Physics, vol. 1, pp. 4-1–4-8. Addison-Wellesly, Reading (1963)
6. Goldstein, H., Poole, C., Safko, J.: Classical Mechanics, 3rd edn. Addison-Wesley, Reading (2001)
7. Lanczos, C.: The Variational Principles of Mechanics, 4th edn. Dover, New York (1986)
8. Noether, E.: Invariante Variationsprobleme. Nachr. D. König. Gesellsch. D. Wiss. Zu Göttingen, Math-phys. Klasse 1918, 235–257 (English translation at arxiv.org/abs/physics/0503066v1)
9. Pomeau, Y.: Invariant in cellular automata. J. Phys. A 17, L415–L418 (1984)
10. Toffoli, T.: Action, or the fungibility of computation. In: Hey, A. (ed.) Feynman and Computation, pp. 349–392. Perseus, Cambridge (1999)
11. Toffoli, T., Capobianco, S., Mentrasti, P.: A new inversion scheme, or how to turn second-order cellular automata into lattice gases. Theor. Comp. Sci. 325, 329–344 (2004)
12. Tyagi, A.: A principle of least computational action. In: Workshop on Physics and Computation, pp. 262–266. IEEE Computer Society Press, Los Alamitos (1994)
13. Wikipedia. Noether's theorem (sampled on January 19, 2011), http://www.en.wikipedia.org/wiki/Noether's_theorem

On Normal Forms for Networks
of Evolutionary Processors

Jürgen Dassow, Florin Manea*, and Bianca Truthe

Otto-von-Guericke-Universität Magdeburg, Fakultät für Informatik
PSF 4120, D-39016 Magdeburg, Germany
{dassow,manea,truthe}@iws.cs.uni-magdeburg.de

Abstract. In this paper we show that some aspects of networks of evolutionary processors can be normalized or simplified without loosing generative power. More precisely, we show that one can use very small finite automata for the control of the communication. We first prove that the networks with evolutionary processors remain computationally complete if one restricts the control automata to have only one state, but underlying graphs of the networks have no fixed structure and the rules are applied in three different modes. Moreover, we show that networks where the rules are applied arbitrary, and all the automata for control have one state, cannot generate all recursively enumerable languages. Finally, we show that one can generate all recursively enumerable languages by complete networks, where the rules are applied arbitrary, but the automata for control have at most two states.

1 Introduction

Motivated by a series of basic computing paradigms for parallel and distributed symbolic processing ([1,2,3]), E. Csuhaj-Varjú and A. Salomaa defined in [4] networks of language processors as a formal languages generating model. Such a network can be viewed as a graph whose nodes contain sets of productions and, at any moment of time, a language is associated with a node. In a derivation step, any node derives from its language all possible words as its new language. In a communication step, any node sends those words that satisfy a series of filtering conditions, given as a regular language, to other nodes and any node takes those words sent by the other nodes that satisfy an input condition also given by a regular language. The language generated by a network of language processors consists of all (terminal) words which occur in the languages associated with a given node.

Having a biological motivation, in [5] one considers a computing model inspired by the evolution of cell populations, which might model some properties of evolving cell communities at the syntactical level. Cells are represented by

---

* Also at: *Faculty of Mathematics and Computer Science, University of Bucharest, Str. Academiei 14, RO-010014 Bucharest, Romania* (flmanea@fmi.unibuc.ro). The work of Florin Manea is supported by the *Alexander von Humboldt Foundation*.

C.S. Calude et al. (Eds.): UC 2011, LNCS 6714, pp. 89–100, 2011.

words which describe their DNA sequences. Informally, at any moment of time, the evolutionary system is described by a collection of words, a representation of a collection of cells. Cells belong to species, and their community evolves according to mutations and division, which are defined formally as operations on words. Only those cells which are represented by a word in a given set of words, called the genotype space of the species, are accepted as surviving (correct) ones. This feature parallels with the natural process of evolution. Similar ideas may be met in other bio-inspired models, such as *tissue-like membrane systems* ([6]), or models from distributed computing area, like *parallel communicating grammar systems* ([7]).

In this context, in [8], networks of evolutionary processors (NEPs for short) were defined. More precisely, one considers that in each node of such a network there exists a processor which is able to perform very simple operations, namely point mutations in a DNA sequence (insertion, deletion or substitution of a pair of nucleotides). More generally, each node may be viewed as a cell having a genetic information encoded in DNA sequences, which may evolve by local evolutionary events, namely point mutations; moreover, each node is specialized just for one of these evolutionary operations. Furthermore, the data in each node are organized in the form of multisets of words, each word appearing in an arbitrarily large number of copies, and all the copies are processed as follows: if at least one rule can be applied to a word w, we obtain all the words that are derived from the word w by applying exactly one of the possible rules at exactly one feasible position in the word w. Clearly, this computational process is not exactly an evolutionary process in the Darwinian sense, but the rewriting operations considered might be viewed as mutations and the filtering process might be viewed as a selection process. Recombination is missing but it was asserted that evolutionary and functional relationships between genes can be captured by taking only local mutations into consideration ([9]). The computation of a NEP is conducted just as in the case of networks of language processors: initially, the nodes contain some finite sets of words, and further, these words are processed according to the rules in each node, and then, they are communicated to the other nodes, as permitted by some filtering condition associated with the nodes, and so on; the language generated by a NEP consists of all the words which appear in a given node, called the output node. Results on NEPs, seen as formal languages generating devices, can be found, e. g., in [10,8,11,12,13]. In the seminal paper [8], it was shown that such networks are computationally complete, i. e., they are able to generate all recursively enumerable languages; however, in the constructions, they use different types of underlying graphs and relatively large automata for the control of the communication.

In this paper, we want to show that some aspects of evolutionary networks can be normalized or simplified without loosing generative power. Especially, we are interested in a use of very small finite automata for the control of the communication. We first prove that the networks with evolutionary processors remain computationally complete if one restricts the control automata to have only one state. However, the resulting underlying graphs have no fixed structure and the rules are applied in three different modes.

In the remaining part of the paper, we show that one can generate all recursively enumerable languages by networks, where the rules are applied in only one mode (any rule can be applied to any position), and the rule sets are singletons or the underlying graph is a complete graph; however, the automata for control are a little larger, they have at most two states.

Due to space limits, we sometimes give only ideas and sketches of proofs or omit the proofs. All proofs in full length are given in the technical report [14].

2   Definitions

We assume that the reader is familiar with the basic concepts of formal language theory (see e.g. [15]). We here only recall some notations used in the paper.

By V^* we denote the set of all words (strings) over an alphabet V (including the empty word λ). For the number of elements of a set A, we write $\mathrm{card}(A)$. The minimal alphabet of a word w and a language L is denoted by $\mathrm{alph}(w)$ and $\mathrm{alph}(L)$, respectively.

In the proofs we shall often add new letters of an alphabet U to a given alphabet V. In all these situations, we assume that $V \cap U = \emptyset$.

A phrase structure grammar is specified as a quadruple $G = (N, T, P, S)$ where N is a set of nonterminals, T is a set of terminals, P is a finite set of productions which are written as $\alpha \to \beta$ with $\alpha \in (N \cup T)^+ \setminus T^*$ and $\beta \in (N \cup T)^*$, and $S \in N$ is the axiom.

A grammar $G = (N, T, P, S)$ is in Geffert normal form ([16]) if the set of nonterminals only consists of the axiom S and three additional letters A, B, C and all rules in P have the form $ABC \to \lambda$ or $S \to v$ with $v \in (N \cup T)^*$.

By REG and RE we denote the families of regular and recursively enumerable languages, respectively. For $i \in \mathbb{N}$, we designate the family of all languages L that can be accepted by a deterministic finite automaton with at most i states working over the alphabet $\mathrm{alph}(L)$ by MIN_i.

We call a production $\alpha \to \beta$ a
  - substitution if $|\alpha| = |\beta| = 1$,
  - deletion if $|\alpha| = 1$ and $\beta = \lambda$.

We introduce insertion as a counterpart of deletion and write it as a rule $\lambda \to a$, where a is a letter.

Besides the usual context-free rewriting we also consider derivations where the rules are applied to the left or right end of the word. Formally, for a substitution, deletion or insertion rule $p : \alpha \to \beta$ and words v and w, we define
  - $v \Longrightarrow_{*,p} w$ by $v = x\alpha y$, $w = x\beta y$ for some $x, y \in V^*$,
  - $v \Longrightarrow_{l,p} w$ by $v = \alpha y$, $w = \beta y$ for some $y \in V^*$,
  - $v \Longrightarrow_{r,p} w$ by $v = x\alpha$, $w = x\beta$ for some $x \in V^*$.

The indices are omitted whenever they can be easily determined from the context.

We now introduce the basic concept of this paper, the networks of evolutionary processors with regular filters.

Definition 1. (i) *A network of evolutionary processors (of size n) with filters in the class X is a tuple*

$$\mathcal{N} = (V, N_1, N_2, \ldots, N_n, E, j)$$

where

- *V is a finite alphabet,*
- *for $1 \leq i \leq n$, $N_i = (M_i, A_i, I_i, O_i, \alpha_i)$ where*
  - *M_i is a set of evolutionary rules of a certain type, i. e.,*
    $M_i \subseteq \{ a \rightarrow b \mid a, b \in V \}$ or $M_i \subseteq \{ a \rightarrow \lambda \mid a \in V \}$, or
    $M_i \subseteq \{ \lambda \rightarrow b \mid b \in V \}$,
  - *A_i is a finite subset of V^*,*
  - *I_i and O_i are regular languages, included in V^*, from the class X,*
  - *$\alpha_i \in \{*, l, r\}$ indicates the way the rules from M_i are applied: arbitrary in the word ($*$), at the left (l) or right (r) end of the word,*
- *E is a subset of $\{1, 2, \ldots, n\} \times \{1, 2, \ldots, n\}$, and*
- *j is a natural number such that $1 \leq j \leq n$.*

(ii) *A configuration C of $\mathcal{N}$ is an n-tuple $C = (C(1), C(2), \ldots, C(n))$ where $C(i)$ is a subset of V^* for $1 \leq i \leq n$.*

(iii) *Let $C = (C(1), C(2), \ldots, C(n))$ and $C' = (C'(1), C'(2), \ldots, C'(n))$ be two configurations of $\mathcal{N}$. We say that C derives C' in one*

- *evolutionary step (written as $C \Longrightarrow C'$) if, for $1 \leq i \leq n$, $C'(i)$ consists of all words $w \in C(i)$ to which no rule of M_i is applicable and of all words w for which there are a word $v \in C(i)$ and a rule $p \in M_i$ such that $v \Longrightarrow_{\alpha_i, p} w$ holds,*
- *communication step (written as $C \vdash C'$) if, for $1 \leq i \leq n$,*

$$C'(i) = (C(i) \setminus O_i) \cup \bigcup_{(k,i) \in E} (C(k) \cap O_k \cap I_i).$$

The computation of $\mathcal{N}$ is a sequence of configurations

$$C_t = (C_t(1), C_t(2), \ldots, C_t(n)), \quad t \geq 0,$$

such that

- *$C_0 = (A_1, A_2, \ldots, A_n)$,*
- *C_{2t} derives C_{2t+1} in an evolutionary step: $C_{2t} \Longrightarrow C_{2t+1}$ ($t \geq 0$),*
- *C_{2t+1} derives C_{2t+2} in a communication step: $C_{2t+1} \vdash C_{2t+2}$ ($t \geq 0$).*

(iv) *The language $L(\mathcal{N})$ generated by $\mathcal{N}$ is defined as*

$$L(\mathcal{N}) = \bigcup_{t \geq 0} C_t(j)$$

where the sequence of configurations $C_t = (C_t(1), C_t(2), \ldots, C_t(n))$, $t \geq 0$, is a computation of $\mathcal{N}$.

Intuitively, a network with evolutionary processors is a graph consisting of some, say n, nodes $N_1, N_2, \ldots, N_n$ (called processors) and the set of edges given by E such that there is a directed edge from N_k to N_i if and only if $(k, i) \in E$. Any processor N_i consists of a set of evolutionary rules M_i, a set of words A_i, an input filter I_i and an output filter O_i. We say that N_i is a substitution, deletion, or insertion node if $M_i \subseteq \{ a \to b \mid a, b \in V \}$ or $M_i \subseteq \{ a \to \lambda \mid a \in V \}$ or $M_i \subseteq \{ \lambda \to b \mid b \in V \}$, respectively. The input filter I_i and the output filter O_i control the words which are allowed to enter and to leave the node, respectively. With any node N_i and any time moment $t \geq 0$, we associate a set $C_t(i)$ of words (the words contained in the node at time t). Initially, N_i contains the words of A_i. In an evolutionary step, we derive from $C_t(i)$ all words applying rules from the set M_i. In a communication step, any processor N_i sends out all words $C_t(i) \cap O_i$ (which pass the output filter) to all processors to which a directed edge exists (only the words from $C_t(i) \setminus O_i$ remain in the set associated with N_i) and, moreover, it receives from any processor N_k such that there is an edge from N_k to N_i all words sent by N_k and passing the input filter I_i of N_i, i.e., the processor N_i gets in addition all words of $C_t(k) \cap O_k \cap I_i$. We start with an evolutionary step and then communication steps and evolutionary steps are alternately performed. The language consists of all words which are in the node N_j (also called the output node, j is chosen in advance) at some moment.

We say that a NEP $\mathcal{N} = (V, N_1, N_2, \ldots, N_n, E, j)$ is in weak normal form if, for $1 \leq i \leq n$, the working mode of N_i is $*$, and we say it is in normal form, if it is in weak normal form and $E = \{ (i, j) \mid 1 \leq i \leq n, \ 1 \leq i \leq n, \ i \neq j \}$ holds (i.e., it has a complete underlying graph).

For a family $X \subseteq REG$, we denote the family of languages generated by networks of evolutionary processors (in weak normal form and normal form, respectively), where all filters are of type X by $\mathcal{E}(X)$ ($\mathcal{E}_*(X)$ and $\mathcal{E}_N(X)$, respectively).

The following results are known (see, e.g., [8], [17, Theorem 4.3]).

Theorem 1. *i)* $\mathcal{E}_N(REG) = RE$.
ii) $\mathcal{E}_*(MIN_2) = RE$.

3 Simplifying the Filters

We start by showing that Theorem 1 ii) is optimal in that sense that one state automata are not sufficient.

Lemma 1. *The regular language* $L = \{ wb \mid w \in \{a, b\}^* \}$ *is not contained in the class* $\mathcal{E}_*(MIN_1)$.

Proof. Suppose $L \in \mathcal{E}_*(MIN_1)$. Then there is a NEP $\mathcal{N}$ which has only filters that belong to the class MIN_1 (i.e., the filters are of the form Q^* for some alphabet Q or are the empty set) and which generates the language L. Since L is infinite and networks with only substitution and deletion nodes generate finite languages, $\mathcal{N}$ contains an inserting processor. The number of as is un-bounded (for each natural number n, there is a word $w \in L$ with more than n occurrences of a). Hence, there are a natural number $s \geq 0$ and letters $x_0, x_1, \ldots, x_s$ with $x_s = a$ such that the network contains the rules $\lambda \to x_0$ and $x_i \to x_{i+1}$ for $0 \leq i \leq s-1$ and there is a word $w_1 a w_2 \in L$ which is derived from a word $v_1 v_2$ by applying these rules (possibly not only these rules), starting with the insertion of x_0 between v_1 and v_2. Instead, x_0 could also be inserted at the end of $v_1 v_2$. All words derived from $v_1 x_0 v_2$ are letter equivalent to those derived from $v_1 v_2 x_0$. Thus, if a word derived from $v_1 x_0 v_2$ can pass a filter then also a word that is derived from $v_1 v_2 x_0$ in the same manner can pass that filter. Hence, in the same way how $w_1 a w_2$ is derived and communicated to the output node, also the word $w_1 w_2 a$ is derived and communicated to the output node. But $w_1 w_2 a \notin L$. Thus, the language L cannot be generated by a NEP where the filters belong to MIN_1. This implies $L \notin \mathcal{E}_*(MIN_1)$.   □

However, if we also allow the other two modes of derivation, then we can im-prove the bound given in Theorem 1 ii), i. e. we can prove that every recursively enumerable language can be generated by a network of evolutionary processors where each filter is accepted by a deterministic finite automaton with one state only.

Theorem 2. $\mathcal{E}(MIN_1) = RE$.

Proof. Let L be a recursively enumerable language and G a grammar in Geffert normal form generating the language $L(G) = L$ where the set of nonterminals is $N = \{ S, A, B, C \}$ with the axiom S, the set of terminal symbols is T, and the set of rules is P with the rules being $S \to v$ with $v \in (N \cup T)^*$ or $ABC \to \lambda$. We construct a NEP $\mathcal{N}$ that simulates the derivation process in G and, hence, generates the same language. The idea is to rotate the sentential form until the subword which has to be replaced is a suffix and then to delete and insert (if necessary) at the right end of the word.

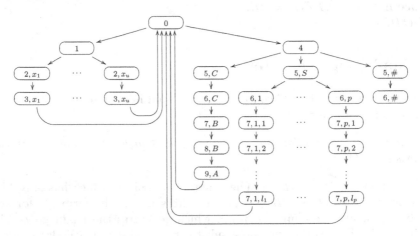

Fig. 1. NEP for a Geffert normal form

Let $V = N \cup T$. Let $\# \notin V$ be a symbol that marks the actual end of a word during the rotation and $U = V \cup \{\#\}$. Let $u = \text{card}(U)$ be the number of the elements of U, $x_1, x_2, \ldots, x_u$ be the elements of U, and $U' = \{ x' \mid x \in U \}$. Furthermore, let p be the number of the rules of the form $S \to v$ and let these rules be $P_i = S \to a_{i,1} a_{i,2} \cdots a_{i,l_i}$ with $1 \le i \le p$, $l_i \ge 0$, and $a_{i,j} \in V$ for $1 \le j \le l_i$.

We construct subnetworks for rotating a word and for simulating the application of a rule. The structure of the network can be seen in Figure 1. We only give the initial set and/or the derivation mode if they are different from $\emptyset$ and $*$, respectively.

The node N_0 is defined by $A_0 = \{ S\# \}$, $M_0 = \emptyset$, $I_0 = U^*$ and $O_0 = U^*$. The work starts with the first sentential form of G where the end is marked by $\#$. This node distributes the words to the nodes N_1 where the rotation will start and N_4 where the simulation of a rule will start.

The rotation of a letter x_i $(1 \le i \le u)$ is performed by the nodes N_1, N_{2,x_i}, and N_{3,x_i} defined as:

$$I_1 = U^*, \quad M_1 = \{ x \to x' \mid x \in U \}, \quad \alpha_1 = l, \quad O_1 = (U' \cup U)^*,$$
$$I_{2,x_i} = (\{x_i'\} \cup U)^*, \quad M_{2,x_i} = \{ x_i' \to \lambda \}, \quad O_{2,x_i} = U^*,$$
$$I_{3,x_i} = U^*, \quad M_{3,x_i} = \{ \lambda \to x_i \}, \quad \alpha_{3,x_i} = r, \quad O_{3,x_i} = U^*.$$

The simulation of a rule starts in the node N_4. It is defined by $I_4 = U^*$, $M_4 = \{ x \to x' \mid x \in U \}$, $\alpha_4 = r$, and $O_4 = (U' \cup U)^*$. In this node, the right most symbol will be changed to the primed version. If the last symbol was C and the rule $ABC \to \lambda$ exists in P then this rule can be started to be simulated (maybe C is not preceded by AB then the simulation does not work which will be noticed later). If the last symbol was S then one of the rules P_i $(1 \le i \le p)$ can be simulated. If the last symbol was $\#$ then this end marker should be removed to obtain a real sentential form of G (which is the end of the simulation of G). In all other cases, we do not need the word anymore.

We now construct subnetworks for these three cases.

If the rule $ABC \to \lambda$ exists in P then we define the following nodes: $N_{5,C}$, $N_{7,B}$, and $N_{9,A}$ for checking that a C, a B, or an A is marked and deleting it, as well as $N_{6,C}$ and $N_{8,B}$ for marking the rightmost symbol:

$$I_{5,C} = (\{C'\} \cup U)^*, \quad M_{5,C} = \{ C' \to \lambda \}, \quad O_{5,C} = U^*;$$
$$I_{6,C} = U^*, \quad M_{6,C} = \{ x \to x' \mid x \in U \}, \quad \alpha_{6,C} = r, \quad O_{6,C} = (U' \cup U)^*,$$
$$I_{7,B} = (\{B'\} \cup U)^*, \quad M_{7,B} = \{ B' \to \lambda \}, \quad O_{7,B} = U^*,$$
$$I_{8,B} = U^*, \quad M_{8,B} = \{ x \to x' \mid x \in U \}, \quad \alpha_{8,B} = r, \quad O_{8,B} = (U' \cup U)^*,$$
$$I_{9,A} = (\{A'\} \cup U)^*, \quad M_{9,A} = \{ A' \to \lambda \}, \quad O_{9,A} = U^*.$$

For the rules of the form $S \to v$, we define the node $N_{5,S}$ by $I_{5,S} = (\{S'\} \cup U)^*$, $M_{5,S} = \{ S' \to S_i \mid 1 \le i \le p \}$, and $O_{5,S} = (U \cup \{ S_i \mid 1 \le i \le p \})^*$. This node chooses a rule P_i that is simulated afterwards. The word obtained is sent to all nodes $N_{6,j}$ with $1 \leq j \leq p$ but it is accepted only by node $N_{6,i}$ which corresponds to the rule selected.

For each rule P_i with $1 \leq i \leq p$, we define the following nodes: $N_{6,i}$ defined by $I_{6,i} = (\{S_i\} \cup U)^*$, $M_{6,i} = \{ S_i \to \lambda \}$, and $O_{6,i} = U^*$ deletes the left hand side of the rule under consideration. If $P_i = S \to \lambda$ then this word is sent back to N_0, otherwise the rule is $S \to a_{i,1} a_{i,2} \cdots a_{i,l_i}$ for a natural number $l_i \geq 1$ and symbols $a_{i,j} \in V$ for $1 \leq j \leq l_i$. These symbols are appended to the word one by one by the nodes $N_{7,i,j}$ with $I_{7,i,j} = U^*$, $M_{7,i,j} = \{ \lambda \to a_{i,j} \}$, $\alpha_{7,i,j} = r$, and $O_{7,i,j} = U^*$ for $j \in \{1, \ldots, l_i\}$.

For the case that the last symbol in N_4 is the end marker #, we define two nodes $N_{5,\#}$ and $N_{6,\#}$ by $I_{5,\#} = (\{\#'\} \cup U)^*$, $M_{5,\#} = \{ \#' \to \lambda \}$, $O_{5,\#} = U^*$, $I_{6,\#} = T^*$, $M_{6,\#} = \emptyset$, $O_{6,\#} = T^*$. In $N_{5,\#}$, the end marker is deleted. The word obtained is a sentential form of G and it is sent to node $N_{6,\#}$ which serves as the output node and accepts the word only if it is a terminal word.

The NEP $\mathcal{N}$ is defined as $\mathcal{N} = (X, N_0', N_1', \ldots, N_t', E, j)$ with the working alphabet $X = U \cup U' \cup \{ S_i \mid 1 \leq i \leq p \}$, an enumeration $N_0', N_1', \ldots, N_t'$ of the nodes defined above, the set E of edges given in Figure 1, and the output node $N_j' = N_{6,\#}$. From the explanations given along with the definitions of the nodes and edges, it follows that the network $\mathcal{N}$ generates the same language as the grammar G. Moreover, all filters belong to MIN_1.

Thus $RE \subseteq \mathcal{E}(MIN_1)$. The converse inclusion follows from Theorem 1 i), so we have $RE = \mathcal{E}(MIN_1)$.   □

We note that in [18] the bound 2 was given for the number of states accepting the filter languages, and that this bound cannot be improved. However, in that paper, the automaton accepting a given language L has the input alphabet V of the network and not the input alphabet $alph(L)$ as it is done in this paper. The transformation from $alph(L)$ to V requires often/mostly one additional state. Moreover, Theorem 2 does not follow from the constructions in [18].

4   Transformations into (Weak) Normal Form

The results in this section show how we can transform a given network into a network in weak normal form, where in addition the set of rules of all nodes are singletons, and into a network in normal form.

Lemma 2. *Let $\mathcal{N}$ be a network of evolutionary processors with filters in the class MIN_k and $k' = \max\{k, 2\}$. Then a network $\mathcal{N}' = (U, N_1', N_2', \ldots, N_m', E', j')$ of evolutionary processors can be algorithmically constructed such that $L(\mathcal{N}') = L(\mathcal{N})$, $N_j' = (\emptyset, \emptyset, I', O', *)$, no edge is leaving N_j', and all nodes N_i', $1 \leq i \leq m$, have input and output filters from the class $MIN_{k'}$.*

This lemma can be proven analogously to the corresponding lemma in [17].

After normalizing the output node, we continue with normalizing the remaining nodes.

Lemma 3. *Let $\mathcal{N} = (V, N_1, N_2, \ldots, N_n, E, j)$ be a network of n evolutionary processors with filters from the class MIN_k and $k' = \max\{k, 3\}$. Then a network $\mathcal{N}' = (U, N_1', N_2', \ldots, N_m', E', j')$ in weak normal form can be algorithmically constructed such that $L(\mathcal{N}) = L(\mathcal{N}')$ and all nodes N_i', $1 \le i \le m$, have input and output filters in the class $MIN_{k'}$.*

Proof. We can assume without loosing generality that the output node of $\mathcal{N}$ has no rules and axioms. We will show how we can construct, for each node x of the network, a subnetwork $s(x)$ that simulates its computation; altogether, these subnetworks will form the new network $\mathcal{N}'$. We denote by $nodes(s(x))$ the nodes of the subnetwork $s(x)$ and distinguish two nodes of this subnetwork, namely $i(s(x))$ and $o(s(x))$ that facilitate the communication with the subnetworks constructed for the other nodes of the original network (these two nodes can be seen as the entrance node and exit node of the subnetwork). We denote by $edges(s(x))$ the set of edges of the subnetwork. Note that all the processors of the new network work in arbitrary mode. Finally, let $V_\# = \{ \#_a \mid a \in V \}$ be a set of symbols with $V \cap V_\# = \emptyset$, and $U = V \cup V_\# \cup \{\#\}$.

First we approach the simple cases: nodes with no rules, nodes that have only substitutions of the type $a \to a$ with $a \in V$, and nodes where rules can be applied anywhere in the words. If x is such a node we will simply copy it into the second network with the application mode $*$.

Now, let us consider the case of left-insertion nodes. Assume that the processor x is (M, A, I, O, l). Let $V_1 = \{ a \in V \mid \lambda \to a \in M \}$. For this node, we define the following subnetwork $s(x)$:

- $nodes(s(x)) = \{x_0, x_1, x_2, x_3\}$,
  $edges(s(x)) = \{(x_0, x_1), (x_1, x_2), (x_1, x_3), (x_2, x_1)\}$;
- $i(s(x)) = x_0$ and $o(s(x)) = x_3$;
- $x_0 = (\{\lambda \to \#\}, A, I, (V \cup \{\#\})^*, *)$;
- $x_1 = (\{ \# \to a \mid a \in V_1 \}, \emptyset, \{\#\}(V \cup \{\#\})^*, V^*, *)$;
- $x_2 = (\{\lambda \to \#\}, \emptyset, U^* \setminus O, (V_2 \cup \{\#\})^*, *)$ where $V_2 = \mathrm{alph}(O)$;
- $x_3 = (\emptyset, \emptyset, O, V^*, *)$.

All filters of this subnetwork belong to MIN_k or to MIN_3 (where exactly 3 states are needed for accepting the input filter of node x_1).

A word enters the node x if and only if it enters the node x_0. In x_0, the intermediate symbol $\#$ is inserted. Then the word moves to x_1 where $\#$ must be found at the beginning of the word such that the substitution by some $a \in V_1$ simulates the insertion of a in x. If the word obtained in x_1 passes the output filter O, then the word enters the exit node x_3. If it does not pass the output filter because it contains a symbol which is not in V_2, then the word is trapped in node x forever and is also blocked by the subnetwork $s(x)$ in the node x_2. In the remaining case, the word is further processed by x and may finally leave the node which is simulated by the cooperation of the nodes x_1 and x_2.

We now discuss the case of left-deletion nodes. Assume that the processor x is (M, A, I, O, l). Let $V_1 = \{ a \in V \mid a \to \lambda \in M \}$ and $V_2 = \mathrm{alph}(O)$. For this node, we define the following subnetwork $s(x)$:

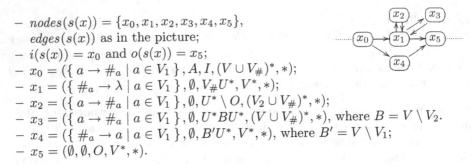

- $nodes(s(x)) = \{x_0, x_1, x_2, x_3, x_4, x_5\}$,
  $edges(s(x))$ as in the picture;
- $i(s(x)) = x_0$ and $o(s(x)) = x_5$;
- $x_0 = (\{\, a \to \#_a \mid a \in V_1 \,\}, A, I, (V \cup V_\#)^*, *)$;
- $x_1 = (\{\, \#_a \to \lambda \mid a \in V_1 \,\}, \emptyset, V_\# U^*, V^*, *)$;
- $x_2 = (\{\, a \to \#_a \mid a \in V_1 \,\}, \emptyset, U^* \setminus O, (V_2 \cup V_\#)^*, *)$;
- $x_3 = (\{\, a \to \#_a \mid a \in V_1 \,\}, \emptyset, U^* B U^*, (V \cup V_\#)^*, *)$, where $B = V \setminus V_2$.
- $x_4 = (\{\, \#_a \to a \mid a \in V_1 \,\}, \emptyset, B' U^*, V^*, *)$, where $B' = V \setminus V_1$;
- $x_5 = (\emptyset, \emptyset, O, V^*, *)$.

All filters of this subnetwork belong to MIN_k or to MIN_3.

In node x_0, a symbol is marked for deletion. The word enters node x_1 if the leftmost symbol was marked, and this symbol is deleted now, or it enters node x_4, where the marking is undone, if no rule of x can be applied to the leftmost symbol of the word. In both cases, the word enters then the exit node if it can pass the output filter O of x. If the word cannot pass the output filter and no rule could be applied then, in the original network, the word is trapped in node x while in $s(x)$ is eventually trapped in the node x_2. If a rule was applied but the word does not pass the output filter, then the word remains in node x and is further processed. This is simulated by the cooperation of the node x_1 with x_2 (which takes a word if it does not belong to O and returns it to x_1 if after a new deletion it will have only letters from V_2) and x_3 (which takes a word if it contains a letter not in V_2).

For the case of left-substitution nodes, we only change the set V_1 of all left hand side symbols to $V_1 = \{\, a \in V \mid \exists c \in V : a \to c \in M \,\}$ and the rule set of x_1 such that $x_1 = (\{\, \#_a \to c \mid a \to c \in M \,\}, \emptyset, V_\#(V \cup V_\#)^*, V^*, *)$.

The cases of right-operations are left to the reader. The main difference is that where we used to check for the leftmost symbol by some filter XY^* we now have to check for the rightmost symbol by the filter $Y^* X$. In this case, all the filters are in MIN_k or MIN_2.

Finally, let us return now to the given network $\mathcal{N} = (V, N_1, \ldots, N_n, E, j)$. We define the network $\mathcal{N}' = (U, N_1', \ldots, N_m', E', j')$, where

- $N_1', \ldots, N_m'$ is an enumeration of the nodes in the set $\bigcup_{i=1}^n nodes(s(N_i))$;
- the output node of the network is the only node from the subnetwork $s(N_j)$ (which contains only a copy of N_j, since the original node had no rules);
- the set E' of edges is

$$\{\, (o(s(N_i)), i(s(N_k))) \mid (i,k) \in E \,\} \cup \bigcup_{i \in \{1, \ldots, n\}} edges(s(N_i));$$

- the rules, axioms and filters of the nodes are defined as above.

From the remarks made when we explained how each subnetwork works it is clear that the network $\mathcal{N}$ generates exactly the same language as $\mathcal{N}$. □

Obviously, the application of Lemma 3 to the networks constructed in Theorem 2 gives only the bound 3 for the number of states whereas Theorem 1 ii) gives the better bound 2. Nevertheless, Lemma 3 is of interest since we have to consider the form in which a recursively enumerable language is given; Lemma 3 requires a description by a network with evolutionary processors, whereas Theorem 1 ii) uses a description by a grammar in Kuroda normal form.

Now we normalize the number of rules present in each node.

Lemma 4. *Let* $\mathcal{N} = (V, N_1, N_2, \ldots, N_n, E, j)$ *be a network of* n *evolutionary processors in weak normal form and with filters from the class* MIN_k, *and let* $k' = \max\{k, 2\}$. *Then a network* $\mathcal{N}' = (V, N_1', N_2', \ldots, N_m', E', j')$ *in weak normal form can be algorithmically constructed such that* $L(\mathcal{N}) = L(\mathcal{N}')$ *and all nodes* $N_i' = (M_i', A_i', I_i', O_i', *)$, $1 \leq i \leq m$, *have a singleton set* M_i' *and input and output filters from the class* $MIN_{k'}$.

Proof. According to Lemma 2, we can assume that the output node of $\mathcal{N}$ has the form $N_j = (\emptyset, \emptyset, I_j, O_j, *)$. Those nodes in $\mathcal{N}$ that have at most one rule can be used for $\mathcal{N}'$ without change. The other nodes can be simulated by a network in which there is a node for each rule that takes a word if and only if the rule can be applied and a node for the case that no rule can be applied. Similarly to the method used in the proof of Lemma 3, we need three more nodes that simulate the original output filter of each node. □

Once we know how to transform the NEPs in order to have only processors containing at most one rule and where the rules can be applied at an arbitrary position, we can focus on normalizing the topology of the network, i.e. to get networks in normal form.

Lemma 5. *Let* $\mathcal{N} = (V, N_1, N_2, \ldots, N_n, E, j)$ *be a network of* n *evolutionary processors in weak normal form such that, for* $1 \leq i \leq n$, $N_i = (M_i, A_i, I_i, O_i, *)$ *satisfies* $\text{card}(M_i) \leq 1$ *and* $N_j = (\emptyset, \emptyset, I_j, O_j, *)$ *holds. Also, let* $k' = \max\{k, 2\}$. *Then a network* $\mathcal{N}' = (V', N_1', N_2', \ldots, N_m', E', j')$ *in normal form can be algorithmically constructed such that* $L(\mathcal{N}) = L(\mathcal{N}')$ *and, all filters of* $N_i', 1 \leq i \leq m$, *are from the class* $MIN_{k'}$.

By reasons of space the proof is omitted.

The Lemmas 3, 4, and 5 yield the following result.

Theorem 3. *For a network of evolutionary processors* $\mathcal{N}$ *with filters from the class* MIN_k, *a network* $\mathcal{N}'$ *in normal form can be algorithmically constructed such that* $L(\mathcal{N}) = L(\mathcal{N}')$ *and all filters of* $\mathcal{N}'$ *are from the class* $MIN_{k'}$, *where* $k' = \max\{k, 3\}$.

Let L be a recursively enumerable language. By Theorem 1, $L = L(\mathcal{N})$ for a network of evolutionary processors in weak normal form, where all filters are in the class MIN_2. If we now apply the Lemmas 4 and 5, then we obtain a network in normal form where all filters are in MIN_2. This result can be reformulated as follows.

Theorem 4. $\mathcal{E}_N(MIN_2) = RE$

This result shows that the bound for the number of states to accept the filter languages is not increased if we go from networks in weak normal form to networks in normal form.

Since $\mathcal{E}_N(MIN_1)$ is properly contained in RE (by Lemma 1), the result in Theorem 4 is optimal.

References

1. Hillis, W.D.: The Connection Machine. MIT Press, Cambridge (1986)
2. Errico, L.D., Jesshope, C.: Towards a New Architecture for Symbolic Processing. In: Proc. of AIICSR 1994, pp. 31–40. World Scientific Publishing Co., Inc., River Edge (1994)
3. Fahlman, S.E., Hinton, G.E., Sejnowski, T.J.: Massively Parallel Architectures for AI: NETL, Thistle, and Boltzmann Machines. In: Proc. AAAI 1983, pp. 109–113 (1983)
4. Csuhaj-Varjú, E., Salomaa, A.: Networks of Parallel Language Processors. In: Păun, G., Salomaa, A. (eds.) New Trends in Formal Languages. LNCS, vol. 1218, pp. 299–318. Springer, Heidelberg (1997)
5. Csuhaj-Varjú, E., Mitrana, V.: Evolutionary Systems: A Language Generating Device Inspired by Evolving Communities of Cells. Acta Informatica 36(11), 913–926 (2000)
6. Păun, G.: Computing with Membranes. J. Comput. Syst. Sci. 61(1), 108–143 (2000)
7. Păun, G., Sântean, L.: Parallel Communicating Grammar Systems: The Regular Case. Annals of University of Bucharest, Ser. Matematica-Informatica 38, 55–63 (1989)
8. Castellanos, J., Martín-Vide, C., Mitrana, V., Sempere, J.M.: Networks of evolutionary processors. Acta Informatica 39(6-7), 517–529 (2003)
9. Sankoff, D., Leduc, G., Antoine, N., Paquin, B., Lang, F., Cedergren, R.: Gene Order Comparisons for Phylogenetic Inference: Evolution of the Mitochondrial Genome. Proc. of the National Academy of Sciences of the United States of America 89(14), 6575–6579 (1992)
10. Martín-Vide, C., Mitrana, V.: Networks of evolutionary processors: Results and perspectives. In: Molecular Computational Models: Unconventional Approaches, pp. 78–114 (2005)
11. Alhazov, A., Dassow, J., Martín-Vide, C., Rogozhin, Y., Truthe, B.: On networks of evolutionary processors with nodes of two types. Fundamenta Informaticae 91, 1–15 (2009)
12. Alhazov, A., Csuhaj-Varjú, E., Martín-Vide, C., Rogozhin, Y.: On the size of computationally complete hybrid networks of evolutionary processors. Theoretical Computer Science 410, 3188–3197 (2009)
13. Castellanos, J., Martín-Vide, C., Mitrana, V., Sempere, J.M.: Solving NP-Complete Problems With Networks of Evolutionary Processors. In: Mira, J., Prieto, A.G. (eds.) IWANN 2001. LNCS, vol. 2084, pp. 621–628. Springer, Heidelberg (2001)
14. Dassow, J., Manea, F., Truthe, B.: A Normal Form for Networks of Evolutionary Processors. Technical report, Otto-von-Guericke-Universität Magdeburg, Fakultät für Informatik (2010),
    http://theo.cs.uni-magdeburg.de/pubs/preprints/pp-afl-2010-02.pdf
15. Rozenberg, G., Salomaa, A.: Handbook of Formal Languages. Springer, Heidelberg (1997)
16. Geffert, V.: Normal forms for phrase-structure grammars. RAIRO – Theoretical Informatics and Applications 25, 473–496 (1991)
17. Dassow, J., Manea, F., Truthe, B.: Networks of Evolutionary Processors with Subregular Filters. Technical report, Otto-von-Guericke-Universität Magdeburg, Fakultät für Informatik (2011),
    http://theo.cs.uni-magdeburg.de/pubs/preprints/pp-afl-2011-01.pdf
18. Dassow, J., Truthe, B.: On networks of evolutionary processors with filters accepted by two-state-automata (submitted)

Geometrical Accumulations and Computably Enumerable Real Numbers*

Jérôme Durand-Lose

LIFO, Université d'Orléans,
B.P. 6759, F-45067 ORLÉANS Cedex 2

Abstract. Abstract geometrical computation involves drawing colored line segments (traces of signals) according to rules: signals with similar color are parallel and when they intersect, they are replaced according to their colors. Time and space are continuous and accumulations can be devised to unlimitedly accelerate a computation and provide, in a finite duration, exact analog values as limits.

In the present paper, we show that starting with rational numbers for coordinates and speeds, the time of any accumulation is a *c.e.* (*computably enumerable*) real number and moreover, there is a signal machine and an initial configuration that accumulates at any *c.e.* time. Similarly, we show that the spatial positions of accumulations are exactly the *d-c.e.* (*difference of computably enumerable*) numbers. Moreover, there is a signal machine that can accumulate at any *c.e.* time or *d-c.e.* position.

Keywords: Abstract geometrical computations; Computable analysis; Geometrical accumulations; *c.e.* and *d-c.e.* real numbers; Signal machine.

1 Introduction

Starting from a few aligned points, lines are initiated. When they intersect, they end and new line segments start. Each line is given a color and lines with the same color should be parallel. The new line segments are colored according to the colors of the removed ones.

What can kind of figure can one build with finitely many colors? Is this system computing in some way?

Such a system computes. It does so in the understandings of both *Turing* computability, the original *Blum, Shub and Smale* model [Blum et al., 1989, Durand-Lose, 2007, 2008a] and *Computable analysis* [Weihrauch, 2000, Durand-Lose, 2009b, 2010b]. The so-called *Black-hole model* of computation can be embedded too [Etesi and Németi, 2002, Hogarth, 2004, Lloyd and Ng, 2004, Andréka et al., 2009, Durand-Lose, 2006a, 2009a].

Given that the underlying space and time are Euclidean, thus continuous, can there be any accumulation? What can be said about them?

Geometrical accumulation is a common phenomenon (as in Fig. 1, time is always elapsing upward). With a shift and a rescaling, it could happen anywhere. It is the key to embedding analog computing as well as the Black-hole model.

---

* http://www.univ-orleans.fr/lifo/Members/Jerome.Durand-Lose,
  Jerome.Durand-Lose@univ-orleans.fr

C.S. Calude et al. (Eds.): UC 2011, LNCS 6714, pp. 101–112, 2011.

In the present paper, we show that if the system is based on rational numbers then the temporal and spatial coordinates of any isolated accumulation belong to some particular sets of real numbers. The times are exactly the *computably enumerable numbers* (*c.e.* numbers for short): the limits of (converging) increasing computable sequences of rational numbers. The spatial positions are exactly the differences of two such numbers (*d-c.e.* numbers). Usual *computable* numbers (limits of effectively converging computable sequences of rational) are a strict subset of *c.e.* numbers which is a strict subset of *d-c.e.* numbers [Zheng, 2006].

The geometric system described above is a *signal machine* in the context of *abstract geometrical computations*. It is inspired by a continuous time and space counterpart of cellular automata [Durand-Lose, 2008b] and related to the approaches of Jacopini and Sontacchi [1990], Takeuti [2005] and Hagiya [2005]. It is also an idealization of collision computing [Adamatzky, 2002, Adamatzky and Durand-Lose, 2010].

A *signal machine* gathers the definition of meta-signals (colors, like zig and right in Fig. 1(a)) and collision rules (like {zig, right} → {zag, right}). An instance of a meta-signal is a dimensionless point called a *signal*. Each signal moves uniformly, its speed only depends on the associated meta-signal. The traces of signals on the space-time diagrams form line segments and as soon as they correspond to the same meta-signal, they are parallel. When signals meet, they are removed and replaced by new signals. The emitted signals only depend on the nature of colliding ones.

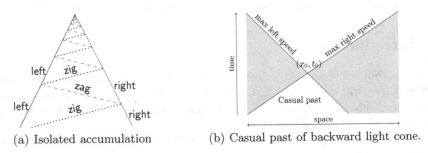

(a) Isolated accumulation   (b) Casual past of backward light cone.

Fig. 1. Example of a space-time diagram and light cone

One key feature of AGC is that space and time are continuous. This has been used to do fractal parallelism [Duchier et al., 2010]. Moreover, Zeno effects can be implemented to generate unbounded acceleration; in particular to allow infinitely many discrete transitions during a finite duration. This has been used to decide the halting problem and implementing the Black-hole model [Durand-Lose, 2009a]. It has also been used to carry out exact analog computations [Durand-Lose, 2008a, 2009b].

This is achieved with rational signal machines: speeds as well as initial positions are rational numbers. Since the positions of collisions are defined by linear equations in rational numbers, the collisions all happen at rational positions.

This is important since rational numbers can be handled exactly in classical discrete computability.

One early question in the field was whether, starting from a rational signal machine, accumulation could lead to an irrational coordinate. An accumulation at $\sqrt{2}$ was provided in Durand-Lose [2007]. The question was then to characterize all the possible accumulation points. Please note that forecasting accumulation for a rational signal machine is as undecidable as the strict partiality of a computable function (Σ_2^0-complete in the arithmetical hierarchy [Durand-Lose, 2006b]).

In the present article, we are interested in *isolated accumulations*: in the space-time diagram, sufficiently close to it, there is no accumulation point and nothing except in the casual past as in Fig. 1(a). The accumulation on Fig. 2(a) is not isolated because of infinitely many left signals on the right. The ones on Fig. 2(b) form a cantor set and the ones of Fig. 2(c) are on a curve (right upper limit) and are almost all accumulations of signals away from any collision.

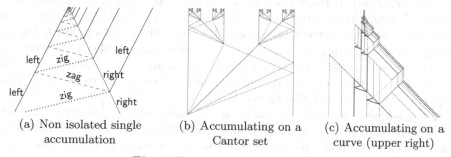

(a) Non isolated single accumulation

(b) Accumulating on a Cantor set

(c) Accumulating on a curve (upper right)

Fig. 2. Non-isolated accumulations

The time of an accumulation is *c.e.* because the system can be simulated exactly on a computer and the time is the above limit of a Zeno phenomenon. For the spatial position, the difference is exhibited by slanting the space-time diagram to exhibit an increasing sequence, the drifting correction provide the negative *c.e.* term.

In Durand-Lose [2009b], we use a two level construction scheme where: the inner level simulates a TM that output orders to the outer structure. The outer structure undergoes a shrinking process—generating the accumulation—driven by the received orders. We use specially designed inner and outer structures to provide (rational) signal machines and initial configurations that accumulates at any *c.e.* time (resp. *d-c.e.* spatial position).

This paper goes beyond Durand-Lose [2010a] which ends up having a major flaw: the accumulation positions do not need to be computable.

Definitions are gathered in Sect. 2. Section 3 shows that the temporal (resp. spatial) coordinate of isolated accumulations is always *c.e.* (resp. *d-c.e.*). Section 4 presents the two level structure. Section 5 provides the control layer. Sections 6 and 7 present outer structures to accumulate respectively at a *c.e.* time and at a *d-c.e.* spatial position. Section 8 concludes the paper.

2   Definitions

2.1   Abstract Geometrical Computation

A *signal machine* collects the definitions of available meta-signals, their speed and the collision rules. For example, the machine to generate Fig. 1(a) is composed of the following meta-signals (with speed): left ($\frac{1}{2}$), zig (4), zag (−4), and right (-$\frac{1}{2}$). Two collision rules are defined:

$$\{\mathsf{left}, \mathsf{zag}\} \longrightarrow \{\mathsf{left}, \mathsf{zig}\} \quad \text{and} \quad \{\mathsf{zig}, \mathsf{right}\} \longrightarrow \{\mathsf{zag}, \mathsf{right}\} \ .$$

It might happen that exactly three (or more) meta-signals meet. In such a case, collisions rules involving three (or more) meta-signals are used. There can be any number of meta-signals in the range of a collision rule, as long as they speeds differ.

A *configuration* is a function from the real line (space) into the set of meta-signals and collision rules plus two extra values: $\oslash$ (for nothing there) and $\circledast$ (for accumulation). If there is a signal of speed s at x, then, unless there is a collision before, after a duration Δt, its position is $x + s.\Delta t$. At a collision, all incoming signals are immediately replaced by outgoing signals in the following configurations according to collision rules. Moreover, any signal must be spatially isolated —nothing else arbitrarily closed—, locations with $\oslash$ value must form an open set and the accumulation points of non $\oslash$ locations must be $\circledast$. (This is a spatial, static, accumulation like on Fig. 2(c).)

A *space-time diagram* is the collection of consecutive configurations which form a two dimensional picture. It must also verify that any accumulation point of collisions in the picture is $\circledast$. (This is a dynamical accumulation like on Fig. 1(a).)

Considering the definition of light cone as on Fig. 1(b), an accumulation at (x_0, t_0) is *isolated* if, sufficiently close to (x_0, t_0):
 – there is nothing but $\oslash$ out of the casual past, and
 – there are infinitely many signals and collisions but no accumulation in the casual past.
It is a purely dynamical and local accumulation.

A signal machine is *rational* if all the speeds are rational (numbers) and only rational positions are allowed for signals in the initial configuration. Since the position of collisions are solutions of systems of rational linear equations, they are rational. In any space-time diagram of a rational signal machine, as long as there is no accumulation, the coordinates of all collisions are rational.

The dynamics is uniform in both space and time. Space and time are continuous; there is no absolute scale. So that if the initial configuration is shifted or scaled so is the whole space-time diagram.

2.2   *c.e.* and *d-c.e.* Real Numbers

A computable sequence is defined by a Turing machine that on input n output the nth term of the sequence.

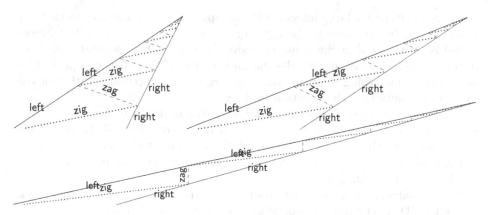

Fig. 3. Examples of drifts by 1, 2 and 4

Definition 1 (*c.e.* and *d-c.e.* numbers). *A real number is* c.e. *(computably enumerable) if there is an increasing computable sequence of rational numbers that converges to it.*

A real number is d-c.e. *(difference of computably enumerable) if it is the difference of two* c.e. *numbers.*

The *c.e.* numbers are closed by rational addition and positive rational multiplication but they not closed under subtraction. On the other side *d-c.e.* numbers form a closed field [Ambos-Spies et al., 2000] and are also characterized by:

Theorem 1 (Ambos-Spies et al. [2000]). *A real number is* d-c.e. *iff there is a computable sequence* (x_n) *that weakly effectively converges to it in the sense that the sum* $\sum_{n \in \mathbb{N}} |x_{n+1} - x_n|$ *converges.*

3 Only (*d-*)*c.e.* Coordinates

Let us consider any (rational) isolated accumulation at (x_0, t_0). The configurations is "clipped" sufficiently closed to the accumulation so that there is nothing out of the casual past. It is rational and finite.

From a (rational) signal machine and a (finite) configuration, it is easy to build a Turing machine that treats the collisions and updates the configuration forever (and indeed this has been programmed in java). Each time a collision is treated, let it output the (rational) time. This sequence is increasing and converges to the time of the accumulation.

Lemma 1. *The time of any (rational) isolated accumulation is* c.e.

A space-time diagram can be slanted by adding the same "drift" to all signals. This is done by increasing all the speeds by the same amount. For example, starting from Fig. 1(a), by adding 1, 2 and 4 to all the speeds, the diagrams of Fig. 3 are generated.

With a sufficiently large integer drift, all speed become positive, so that the configuration has to move to the right, or at least, the positions of its leftmost signal is. Consider a modification of above Turing machine so that each time it treats a collision, it outputs the spatial position of the leftmost signal. This produces an increasing sequence that converges to the spatial position of the drifted accumulation. This position, y_0 is *c.e.*

While the times remains unchanged, the spatial positions of signals and collisions (and hence accumulation) are moved by $d.t$ where d is the drift and t is the time. To correct the drift for the accumulation, $d.t_0$ has to be removed where t_0 is the time of the accumulation. Since t_0 is *c.e.* and d is an integer, $d.t_0$ is *c.e.* So that $x_0 = y_0 - d.t_0$ is *d-c.e.*

With a sufficiently large negative drift, a decreasing converging sequence is generated. This generates the opposite of a *c.e.* (a *co-c.e.*) real number.

Lemma 2. *The spatial position of any (rational) isolated accumulation is* d-c.e. *The coordinates of an isolated accumulation can be expressed as* $(y - d.t_0, t_0)$ *or* $(d.t_0 - y', t_0)$ *where* y, y' *and* t_0 *are c.e. and* d *is an integer.*

4   Controlled Shrinking Structure

This section presents a general scheme based on a two level structure. The outer one handles the shrinking of the whole structure according to the messages received from the inner one which acts as a control.

The shrinking process is not detailed here (it is in Durand-Lose [2010b]). It works by steps. Each step ensures that the structure and everything that is embedded inside it is scaled down by a constant factor. Three such steps are displayed in Fig. 4(c). This is repeated forever generating an accumulation.

4.1   Control/Inner Structure

The inner structure simulates a Turing machine that outputs orders to the outer structure. Each order is formed by signals sent on the left. Outputting is blocking: the control has to receive some acknowledgement signal to resume and send the next order.

The computation is embedded inside a shrinking structure to ensure a bounded delay between outputs. The shirking process is also blocked after the output and resume on relaying the acknowledgement. This is done to ensure that this structure does not generate an accumulation.

Simulating a Turing machine with a signal machine is only exemplified by figures 4(a) and 4(b) where $\overline{11}$ is output. The cells of the tape are encoded by motionless signals (vertical lines) displayed in a geometrical sequence so that it works in a bounded space. A more detailed construction can be found in [Durand-Lose, 2010b].

The output leaves on the left, unaffected by the inner shrinking. The latter is stopped by the presence of wait. The output is collected and processed by the outer structure.

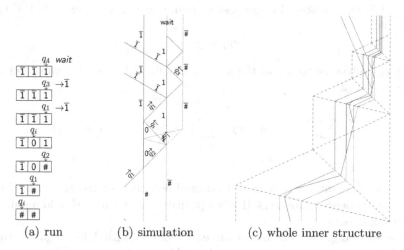

(a) run   (b) simulation   (c) whole inner structure

Fig. 4. Blocking inner structure including a Turing machine

4.2   Outer Structure

The outer structure waits a fixed time before collecting and processing the next output. So that the inner structure has a limited activation time for outputting. Since it has also a shrinking structure, the Turing machine has an unlimited number of iterations ahead of it —this is a form of unbounded acceleration to ensure the output comes in due time. (Special case is taken so that the inner structure have space as if it would never halt.)

When an order is issued, the inner structure is blocked with all its signal parallel. Parallel signals are easy to move preserving their relative position as illustrated on Fig. 4(c) where 0-speed signals amounting for the tape cells go from one triangle to the next. Each time their distances are scaled by one half (this is the shrinking scheme). This is more exemplified in the next sections where special constructions are provided.

This structure provides the isolated accumulation but it moves or waits at each step so that to make the accumulation happen at some position according to the control.

5   Controls for *d-c.e.*

Let x be any *d-c.e.* number. There is a Turing machine that generates a sequence x_n such that this sequence converges to x and $\sum_{n\in\mathbb{N}} |x_{n+1} - x_n|$ converges. If x is *c.e.*, it is also requested that the sequence is increasing.

Let α be any positive rational number. Let us define the sequences:

$$y_0 = \left\lfloor \frac{1}{\alpha} x_0 \right\rfloor \qquad\qquad y_{n+1} = \left\lfloor \frac{2^{n+1}}{\alpha}(x_{n+1} - x_n + e_n) \right\rfloor \qquad (1)$$

$$e_0 = x_0 - \alpha . y_0 \qquad\qquad e_{n+1} = x_{n+1} - x_n + e_n - \frac{\alpha}{2^{n+1}} y_{n+1} \qquad (2)$$

where $\lfloor u \rfloor$ is the greatest integer less or equal to u ($\lfloor u \rfloor \leq u < \lfloor u \rfloor + 1$). It follows that

$$|e_n| \leq \frac{\alpha}{2^n} . \tag{3}$$

So that e_n converges to 0 and the sequence defined below, z_n, converges to x because, using (2),

$$z_n = \sum_{i=0}^{n} \frac{\alpha}{2^i} y_i = x_0 - e_0 + \sum_{i=1}^{n} (-e_i + x_i - x_{i-1} + e_{i-1})$$
$$= x_n - e_n .$$

Since x_n is a computable sequence of rational numbers, so are y_n and e_n. Moreover y_n is a sequence of integers. If x is a positive $c.e.$, then y_n should a sequence of natural numbers.

In the following, the control output y_n in unary (with $\bar{1}$ for negative values and with 1 for positive values) then waits. This loop is repeated forever.

If x is $c.e.$, then the computable sequence is increasing, so that only non-negative values are output.

6   Accumulating at a *c.e.* Time

On the nth iteration, the outer structure receives y_n (an unary encoded natural number). It waits y_n time a delay and then shrinks the whole configuration by one half and wakes up the inner control.

If y_n is zero, the outer structure just shrinks as in Fig. 5(a). Figure 5(b) illustrates a unit delay: the bottom signal that crosses the configuration left to right encounters the unique 1 output. It collects it and goes forth and back to the left to start again. If the delay is more important, like in Fig. 5(c), the other 1's are stored (vertical black line) and each time one is collected and processed.

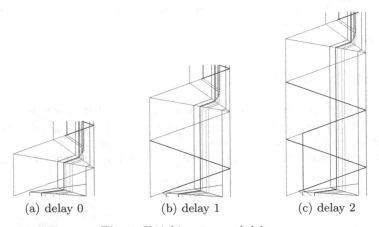

(a) delay 0   (b) delay 1   (c) delay 2

Fig. 5. Shrinking step and delays

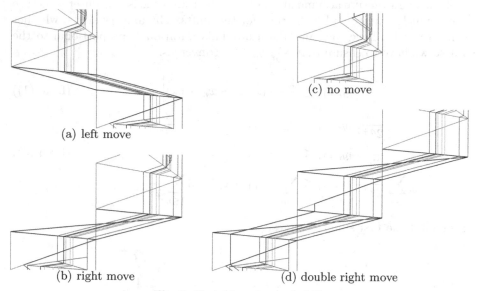

Fig. 6. Shrinking step and shifts

The inner structure does not produce any accumulation since it freezes each time it output anything (including for the empty word) which it does infinitely often.

Since the configuration is shrunk by one half each time, the unit delay sequence is geometrical with one half factor. The sum of delays is z_n.

The outer shrinking process alone provides a term to the final accumulation time. This term, g, is the sum of a geometrical sequence of factor one half, which is rational. It is easy to scale down the initial configuration so that g is less than t. Since *c.e.* are stable by rational addition, $t - g$ is *c.e.* and the control should output the sequence corresponding to it and α (as given by the scale of the outer structure).

Figure 7(a) illustrates a longer run with a machine that always outputs 11.

7 Accumulating at a *d-c.e.* Spatial Position

The whole structure shifts left or right according to the inner control. Figure 6(c) illustrates the case where nothing is output (it is similar to the Fig. 5(a)). For each $\overline{1}$ output the whole configuration is shifted by its width on the left. This is done as on Fig. 6(a): all signals are drifting on the left with identical speed (*i.e.*, they are parallel). Two fast parallel signals ensure that the distance is the same; they form a parallelogram.

For the right shift in Fig. 6(b), the parallelogram is incomplete, but the second diagonal is here. As before if the movement is of more that one unit, all other units are preserved and treated one after the other as depicted in Fig. 6(d).

A larger run is displayed in Fig. 7(b).

To accumulate on a given *d-c.e.* number, the right extremity of the structure should be set at coordinate 0; then α can be chosen to be 1 by scaling.

The whole structure accumulates because the total time is: the outer structure time—which is finite— plus the time for the shifts. The unitary shifts, whether on left or right, have the same duration. This duration is proportional to the shift so we have to ensure that $\sum_n |y_n| 2^{-n}$ converges.

$$|y_{n+1}| \leq \left| \frac{2^{n+1}}{\alpha}(x_{n+1} - x_n + e_n) \right| + 1 \qquad \text{(from (1))}$$

$$\frac{\alpha}{2^{n+1}} |y_{n+1}| \leq |x_{n+1} - x_n| + |e_n| + \frac{\alpha}{2^{n+1}}$$

$$\frac{\alpha}{2^{n+1}} |y_{n+1}| \leq |x_{n+1} - x_n| + 3\frac{\alpha}{2^{n+1}} \qquad \text{(from (3))}$$

$$\sum_{1 \leq n} \frac{\alpha}{2^{n+1}} |y_{n+1}| \leq \sum_{1 \leq n} |x_{n+1} - x_n| + 3 \sum_{1 \leq n} \frac{\alpha}{2^{n+1}}$$

From Th. 1 the first sum converges.

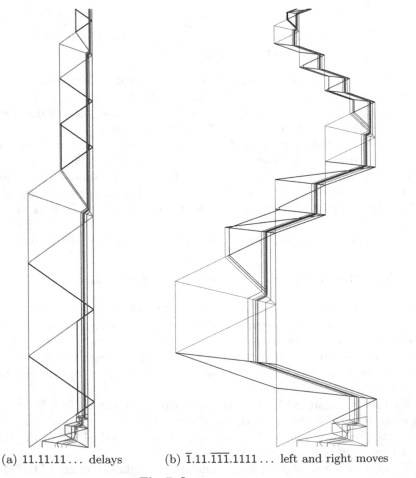

(a) 11.11.11... delays   (b) $\overline{1}.11.\overline{111}.1111\ldots$ left and right moves

Fig. 7. Longer runs

8 Conclusion

By considering a universal Turing machine in the control and union of signal machines (one for space and one for time), comes:

Theorem 2. *There is a rational signal machine that can generate isolated accumulation at any c.e. time or d-c.e. spatial position depending on the initial configuration.*

Following the restriction on the coordinated expressed in Lem. 2, we conjecture that there can be an isolated accumulation at any such coordinates, *i.e.*, time and spatial position simultaneously.

When spatial dimension 2 and above is addressed, it seems that each spatial coordinate can be traded independently and the same result holds with a similar relation with time.

Acknowledgement

The author would like to thank Laurent Bienvenu for inspiring discussions on *d-c.e.* numbers.

References

Adamatzky, A. (ed.): Collision based computing. Springer, London (2002)

Adamatzky, A., Durand-Lose, J.: Collision computing. In: Corne, D. (ed.) Handbook of Natural Computing: Theory, Experiments, and Applications, Part II. Springer, Heidelberg (2010)

Ambos-Spies, K., Weihrauch, K., Zheng, X.: Weakly computable real numbers. J. Complexity 16(4), 676–690 (2000), doi:10.1006/jcom.2000.0561

Andréka, H., Németi, I., Németi, P.: General relativistic hypercomputing and foundation of mathematics. Nat. Comput. 8(3), 499–516 (2009)

Blum, L., Shub, M., Smale, S.: On a theory of computation and complexity over the real numbers: NP-completeness, recursive functions and universal machines. Bull. Amer. Math. Soc. 21(1), 1–46 (1989)

Duchier, D., Durand-Lose, J., Senot, M.: Fractal parallelism: Solving SAT in bounded space and time. In: Cheong, O., Chwa, K.-Y., Park, K. (eds.) ISAAC 2010. LNCS, vol. 6506, pp. 279–290. Springer, Heidelberg (2010), doi:10.1007/978-3-642-17517-6_26

Durand-Lose, J.: Abstract geometrical computation 1: Embedding black hole computations with rational numbers. Fund. Inf. 74(4), 491–510 (2006a)

Durand-Lose, J.: Forecasting black holes in abstract geometrical computation is highly unpredictable. In: Cai, J.-Y., Cooper, S.B., Li, A. (eds.) TAMC 2006. LNCS, vol. 3959, pp. 644–653. Springer, Heidelberg (2006b), doi:10.1007/11750321_61

Durand-Lose, J.: Abstract geometrical computation and the linear blum, shub and smale model. In: Cooper, S.B., Löwe, B., Sorbi, A. (eds.) CiE 2007. LNCS, vol. 4497, pp. 238–247. Springer, Heidelberg (2007), doi:10.1007/978-3-540-73001-9_25

Durand-Lose, J.: Abstract geometrical computation with accumulations: Beyond the Blum, Shub and Smale model. In: Beckmann, A., Dimitracopoulos, C., Löwe, B. (eds.) Logic and Theory of Algorithms, 4th Conf. Computability in Europe (CiE 2008) (abstracts and extended abstracts of unpublished papers), pp. 107–116. University of Athens, Athens (2008a)

Durand-Lose, J.: The signal point of view: from cellular automata to signal machines. In: Durand, B. (ed.) Journées Automates Cellulaires (JAC 2008), pp. 238–249 (2008b)

Durand-Lose, J.: Abstract geometrical computation 3: Black holes for classical and analog computing. Nat. Comput. 8(3), 455–472 (2009a), doi:10.1007/s11047-009-9117-0

Durand-Lose, J.: Abstract geometrical computation and computable analysis. In: Costa, J.F., Dershowitz, N. (eds.) UC 2009. LNCS, vol. 5715, pp. 158–167. Springer, Heidelberg (2009b), doi:10.1007/978-3-642-03745-0_20

Durand-Lose, J.: The coordinates of isolated accumulations [includes] computable real numbers. In: Ferreira, F., Guerra, H., Mayordomo, E., Rasga, J. (eds.) Programs, Proofs, Processes, 6th Int. Conf. Computability in Europe (CiE 2010) (abstracts and extended abstracts of unpublished papers), pp. 158–167. CMATI, U. Azores (2010a)

Durand-Lose, J.: Abstract geometrical computation 5: embedding computable analysis. Nat. Comput (2010b); Special issue on Unconv. Comp. 2009, doi:10.1007/s11047-010-9229-6

Etesi, G., Németi, I.: Non-Turing computations via Malament-Hogarth space-times. Int. J. Theor. Phys. 41(2), 341–370 (2002), http://www.gr-qc/0104023

Hagiya, M.: Discrete state transition systems on continuous space-time: A theoretical model for amorphous computing. In: Calude, C., Dinneen, M.J., Paun, G., Pérez-Jiménez, M.J., Rozenberg, G. (eds.) UC 2005. LNCS, vol. 3699, pp. 117–129. Springer, Heidelberg (2005)

Hogarth, M.L.: Deciding arithmetic using SAD computers. Brit. J. Philos. Sci. 55, 681–691 (2004)

Jacopini, G., Sontacchi, G.: Reversible parallel computation: an evolving space-model. Theoret. Comp. Sci. 73(1), 1–46 (1990)

Lloyd, S., Ng, Y.J.: Black hole computers. Scientific American 291(5), 31–39 (2004)

Takeuti, I.: Transition systems over continuous time-space. Electr. Notes Theor. Comput. Sci. 120, 173–186 (2005)

Weihrauch, K.: Introduction to computable analysis. Texts in Theoretical computer science. Springer, Berlin (2000)

Zheng, X.: A computability theory of real numbers. In: Beckmann, A., Berger, U., Löwe, B., Tucker, J.V. (eds.) CiE 2006. LNCS, vol. 3988, pp. 584–594. Springer, Heidelberg (2006) ISBN 3-540-35466-2, doi:10.1007/11780342_60

Heterotic Computing

Viv Kendon[1], Angelika Sebald[2], Susan Stepney[3],
Matthias Bechmann[2], Peter Hines[3], and Robert C. Wagner[1]

[1] School of Physics and Astronomy, University of Leeds, UK
[2] Department of Chemistry, University of York, UK
[3] Department of Computer Science, University of York, UK

Abstract. Non-classical computation has tended to consider only single computational models: neural, analog, quantum, etc. However, combined computational models can both have more computational power, and more natural programming approaches, than such 'pure' models alone. Here we outline a proposed new approach, which we term *heterotic computing*[1]. We discuss how this might be incorporated in an accessible refinement-based *computational framework* for combining diverse computational models, and describe a range of physical exemplars (combinations of classical discrete, quantum discrete, classical analog, and quantum analog) that could be used to demonstrate the capability.

1 Introduction

Classical computation is epitomised by the Turing machine paradigm. We are concerned with more diverse models of computation, in particular determined by the physical properties of the system used as a computer [27].

Given that we have different basic types of computers, not necessarily Turing universal, it is natural to ask how to *compose* them into hybrid – "heterotic" – computers, and to ask about the computational power of the composition. Thus, we need a framework that not only allows different models of computation to be compared and contrasted, but also allows us to compose different models and determine the resulting computational power.

The structure of the paper is as follows. §2 describes two existing heterotic computers that are used to motivate our approach. §3 outlines the heterotic framework, introducing the proposed computational architecture, and outlining a semantic bases and refinement approach. §4 describes how to progress the classical nuclear magnetic resonance (NMR) computer within this framework. §5 describes an approach that challenges the proposed framework. §6 concludes.

2 Motivating Examples

The computational power of a given physical system is determined by:

1. How much of what type of data can be encoded in the system. Eg: identical objects that can be placed in different collections can be used to encode natural numbers in a unary representation, and to do simple arithmetic by

---

[1] *Heterotic*, from the Greek *heterosis*, a term in genetics meaning 'hybrid vigour'.

C.S. Calude et al. (Eds.): UC 2011, LNCS 6714, pp. 113–124, 2011.

concatenation; they cannot encode negative numbers, as there is no sign bit, or zero if only non-empty collections can exist. (See also work by Blakey [8]).

2. What operations are available to manipulate the system. Eg: quantum computing, following DiVincenzo's checklist [15], first identifies a physical system that can represent a qubit, then identifies a set of operations sufficiently rich to provide universal quantum computation. A CNOT gate combined with single qubit rotations through $\pi/8$ is universal, while the gates in the Clifford group (Pauli operators plus Hadamard and CNOT) are not universal.

3. What operations are available to decode the end result. Eg: in quantum computing, the measurements available to obtain the result can reveal only a limited amount of information, making the final decoding step highly nontrivial. Additionally, they are not just observations, but can form an essential part of the computation.

Classically, only point 2 is analysed in depth, but in wider models, points 1 and 3 are also crucial. In our heterotic approach, interaction between the computational layers requires consideration of 1 and 3 within the framework.

Two motivating examples that demonstrate a layered approach to heterotic computation are described here. The first motivating example from quantum computation is the cluster state, or one-way, quantum computer. The second example is classical computation in NMR [24], where the instruments controlling the NMR form part (but not all) of the computation. We discuss these further here, to illustrate the underlying heterotic principles in action.

2.1   Cluster State Computing

Anders and Browne [2] noticed that the classical computation required to control and feed forward information in a quantum cluster state computer is a crucial part of the computational power. Cluster state measurement without feed-forward is (efficiently) classically simulable, as is (trivially) the classical part of the computation. However, the combination of the two is equivalent to the quantum circuit model, which is not (efficiently) classically simulable. Hence the combination of layers is in a more powerful computational class than either layer alone. The base and control layer are shown in figure 1a.

2.2   Classical NMR Computing

NMR experiments provide both a classical and a quantum computing layer, intimately intermixed with discrete and continuous parameters at several levels. The classical layer typically comprises the spectrometer, radio-frequency (r.f.) pulses, the detector, and the macroscopic (nuclear) magnetisation of a sample in the magnetic field, as described by density matrix formalism. The quantum layer is provided by the (individual) nuclear spin systems, with their discrete states being very well described by quantum mechanics. NMR experiments thus involve both discrete (spin states) and continuous (e.g. r.f. pulses) parameters. See figure 1b.

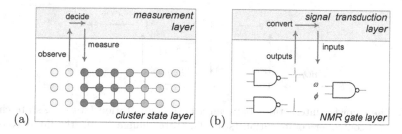

Fig. 1. (a) Cluster state computer. The base layer is a cluster state. The control layer performs measurements on the base layer, thereby changing its state; the control layer uses the observed results to decide what measurement to perform next. (b) Classical NMR computer [24]. The base layer gates are implemented as NMR experiments: inputs are frequencies ω and phase delays ϕ; outputs are the integrated output signal. The control layer performs "signal transduction": taking the integrated output, interpreting it as a 0 or 1, and converting that to the appropriate physical input signal.

Most importantly, NMR experiments give us an element of choice: we can opt for a purely classical behaviour by using only samples made up from isolated (non-coupled) nuclear spins, the dynamics of which are fully described by a classical vector model (for background, see [24]), while choosing a system with coupled nuclear spins provides a mixed classical/quantum combination. Moreover, we can easily switch between the two regimes within a single experiment (e.g. de- and recoupling [16]).

Prior work on computation using NMR mostly deals with implementations of *quantum* computations, predominantly based on solution-state NMR experiments [19], with some examples exploiting solid-state NMR [13]. Prior work by ourselves has proved the suitability of NMR for classical (binary) logic gates [24], based on a range of different solution-state NMR experiments. We implemented a half-adder constructed from binary NAND gates in an NMR system with a classical control performing 'signal transduction', transforming the physical outputs of one gate to the physical inputs of the next [24]. Further, we have produced a preliminary classification of the experimental NMR parameters for implementing classical logic gates. More recently, we have extended our work to take advantage of the inherently continuous nature of the NMR parameter space of non-coupled spin species [7] by implementing continuous gates, so the combined system performs an analog computation.

The classical and quantum computational layers have thus both been demonstrated experimentally in NMR, and the rich parameter space (together with the very well developed theory of NMR and superb degrees of experimental control) provide the source of the computational power of NMR implementations. However, the extent to which the control layer contributes to the computational power of quantum or classical NMR computing has yet to be analysed.

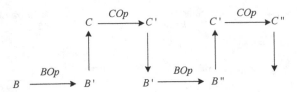

Fig. 2. The stepwise interactions between a base computation (state B, state change BOp) and a controller computation (state C, state change COp): the input to one is the output from the other

3   Heterotic Framework

3.1   Computational Framework

Our basic heterotic model is two coupled computers, one 'guiding' or controlling the other. So the base machine does a step, the guiding machine looks at its output, and tells it what to do next, and so on. The pattern of computation and communication alternates between the two (figure 2). In this basic model, the state of one layer does not change during the computation by the other (for example, the control layer remains in state C' as the base layer evolves from B' to B''). In a physical implementation this state might continue to evolve, yet if its subsequent computation depends only on its input (either it is essentially 'reset' to the previous state, or the input fully determines what happens next), it still fits the basic model. This case holds for both our motivating examples above.

We discuss a possible semantic framework to describe the hybrid coupling of the two machine types, and a refinement/retrenchment approach to support a program development approach. Together, these would provide the tools to determine the combined computational power of the heterotic computer.

3.2   Semantic Framework

The ultimate goal is a form of *refinement calculus for heterotic computers*, suitable for use by the working programmer. However, producing such a framework first requires theoretical input. In particular, we seek a suitable form of semantics on which the refinement calculus is based. Although such models exist for the individual systems described here, the theoretical challenge is to give a formal description of how such systems may interact in non-trivial ways.

Classical analog computation has been modelled in several ways, from the traditional approaches based on differential equations, to interval-based analyses relying on domain theory. Similarly, quantum computation has many models, including the stabiliser formalism, purely category-theoretic approaches, the circuit model, and density matrices particularly suitable for quantum/probabilistic hybrid systems. Classical probabilistic computation can be modelled via categories of stochastic relations, and non-determinism frequently requires categories of relations, or constructions based on the power set functor.

These approaches provide background for the development of a semantic framework. Due to the wide range of heterotic computing systems under consideration, we aim for an abstract categorical semantics, and seek concrete instantiations where appropriate.

Given two dissimilar systems A and B, and models of each of these in distinct categories $\mathcal{C}_A$ and $\mathcal{C}_B$, we require a formal setting in which both the joint system, and the non-trivial interactions between systems A and B may be modelled. A common approach to modelling a joint system is via the straightforward product category $\mathcal{C}_A \times \mathcal{C}_B$. However, for our purposes, this is entirely unsuitable: the real object of study should instead be the non-trivial interactions between the subsystems. Although *ad hoc* extensions or quotients of the straightforward product category may have some utility, we take a more structural approach, based on the theory of adjunctions [23].

In categorical models of logic and computation, the notion of a monoidal closed category is often fundamental: monoidal closure has a strongly computational interpretation as β-reduction (in Cartesian closed categories) or cut-elimination in logical systems [20], and even compositionality in models of Turing machines [18]. Formally, a monoidal closed category $\mathcal{C}$, has two functors, the *monoidal tensor* $\otimes : \mathcal{C} \times \mathcal{C} \to \mathcal{C}$, and the *internal hom* $[\_ \to \_] : \mathcal{C}^{op} \times \mathcal{C} \to \mathcal{C}$ satisfying various conditions laid out in [23]. In particular, these two functors are related by the following condition

$$\mathcal{C}(A \otimes B, C) \cong \mathcal{C}(B, [A \to C]) \tag{1}$$

This is a canonical example of an adjunction. Further, in the very special case where the system is *untyped* (so all objects of $\mathcal{C}$ are isomorphic), we expect to recover models of *universal computation* (e.g. the C-monoids of [20] or the untyped compact closure of [17]).

For our purposes, monoidal closure, in either its typed or untyped form, is too strong — it is clearly describing the situation where the computation is carried out in a single system. Thus, we take the notion of an adjunction between two functors as primitive, without the additional baggage imposed by categorical closure. The notion of an adjunction is simply a categorification of the concept of a galois connection, thus two functors $\Gamma : \mathcal{C}_A \to \mathcal{C}_B$ and $\Delta : \mathcal{C}_B \to \mathcal{C}_A$. form an adjoint pair when $\mathcal{C}_A(\Gamma(X), Y) \cong \mathcal{C}_B(X, \Delta(Y))$, for all $X \in Ob(\mathcal{C}_A)$, $Y \in Ob(\mathcal{C}_B)$. The duality provided by such an adjunction allows us to model the mutual update of system A by system B and system B by system A, without requiring that system B is fully able to simulate the behaviour of system A, or vice versa. We are thus able to capture the sometimes hidden symmetry within such interactions.

A concrete example of this notion of adjunction (via its characterisation as unit/co-unit maps in a 2-category setting), used to describe creation of quantum systems from classical data, and measurement of quantum systems (resulting in classical information), can be found in the categorical semantics approach of Abramsky and Coecke [1]. It thus appears that category theory provides

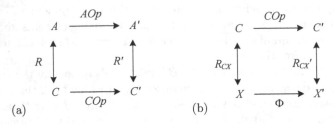

Fig. 3. (a) A simulation, used to prove refinement; (b) Physical and computational layer relationship

ready-made abstract conditions suitable for describing the mutual update of distinct systems in heterotic computing, along with real concrete examples of how this works in certain settings.

3.3 Refinement Framework

Given some suitable semantic framework, such as the one outlined above, it is necessary to cast it in a form suitable for enabling the working programmer to analyse and develop novel heterotic systems in (relatively) familiar ways. We suggest that a classical refinement framework is more appropriate than, say, a process algebra approach, since this is more accessible and familiar to the working programmer.

Introduction. State-and-operation refinement is the classical computational approach to program development. It takes an abstract, possibly non-determin-istic, specification of a state A evolving under a sequence of operations AOp, and 'refines' it (reducing non-determinism, changing data types) into a more concrete implementation with state C and operations COp, with the abstract state A 'retrieved' from the concrete state C through the retrieve relation R (figure 3a). We have the refinement correctness requirement (ignoring non-determinism here for simplicity) that the diagram 'commute' (we get the same value for C' either way round):

$$R'(AOp(A)) = COp(R(A)) \tag{2}$$

Usually the process of refinement stops at a computational level suitably con-crete to allow implementation, such as a mid-level programming language. It can in principle be carried further. Here we need to consider it all the way down to the physical implementation, since we are interested in non-classical execution models. So we continue refining from C down to the physical level, with a state X, that evolves under the laws of physics, Φ. The physical state variables in X are again 'retrieved' through relation R_{CX} as computational state variables in C (figure 3b). [2]

---

[2] Refinement 'reduces non-determinism' until we reach a completely deterministic im-plementation. It is interesting to interpret this in the case of quantum computation, where the implementation is intrinsically non-deterministic. We classically think of the resolution of non-determinism being under the control of the programmer.

Note that the induced computation COp depends on both the physical system Φ and the viewing interpretation R_{CX}.

We would like this diagram to 'commute' (to get the same value for X' either way round), but there will be errors (measurement, noise)[3]. So we can at best require the inexact commutation

$$R_{CX}'(COp(C)) = \Phi(R_{CX}(C)) \pm \epsilon \qquad (3)$$

Analog refinement as Retrenchment? Retrenchment [3,4,5,6] is a form of inexact refinement. It allows deviations from exact refinements by use of various forms of 'concedes' clauses; analysis of the retrenchment concessions provides insight into the way an implementation deviates from a pure refinement. In particular, retrenchment has been applied to developing discrete implementations of real number specifications [5], and to finite implementations of unbounded natural number specifications, which are necessarily inexact. Also, it has been suggested as a laboratory for analysing and understanding emergent behaviour of complex systems [3].

Retrenchment has its critics, but we have argued elsewhere [4] that these criticisms are invalid in the context of real world engineering developments. Here we claim that (a rigorously posed form of) retrenchment is appropriate for casting analog computation in a refinement-like framework. It would be used to analyse the size, nature, and propagation of errors.

Inputs and outputs. The usual classical refinement correctness rules allow inputs to and outputs from the operations, but require these to be the same at the abstract and concrete levels. In previous work [12], we have generalised these rules to allow refinement of i/o, too. This necessitated the introduction of a 'finalisation' step, that can be interpreted as the definition of the 'observation' made on the system. There is an 'initialisation' step, that we have extended to interpret inputs analogously. The finalisation of the most abstract level is usually the identity (we see the 'naked' abstract i/o); more concrete implementations have more sophisticated finalisations (eg, we see a bit stream, but view it, finalise it, as an integer) [10].

The correctness rule (again, ignoring non-determinism) is

$$AFin(A) = CFin(R(A)) \qquad (4)$$

This work has also been extended to the retrenchment arena.

A form of i/o refinement is necessary to move between physical i/o variables and computational i/o variables. For example, in the case of our NMR adder [24]: the physical level is the NMR; the computational level is the XOR gate; the initialisation is interpreting a frequency and a phase delay as a bit; the finalisation is observing an integrated signal as a bit.

For this form of initialisation/finalisation to work in the analysis, it has to be possible *in principle* to provide all the inputs 'up front', and to observe

---

[3] Classical digital hardware is extremely engineered to ensure an exact boolean implementation; this exactness cannot necessarily be assumed in the more general case.

(a record of) all the outputs at the end. This cannot be done for the individual layers of the heterotic computation, where the output from one layer becomes the input to the other (it is closer to a Wegner interaction machine architecture [29]) but can for the overall computation, so we need to be careful about how we set up the analysis, and precisely what we define as i/o. This step is crucial in our heterotic framework, since, as stated earlier, the encoding and decoding processes (formalised as initialisation and finalisation) are non-trivial in general.

Linking outputs to inputs. We have an additional step in the NMR example [24], where the physical inputs and outputs are of different types, but the output from one step becomes the input to the next. We perform a 'signal transduction' step here (integrals over Fourier transforms transduced to phases, that preserves the initialisation/finalisation interpretations). This does not have an analogue in the refinement scenario, because that does not include any link between the outputs of one operation and the inputs of the next. This is important in the context of heterotic computing, as there is potentially significant computation applied to outputs to produce the next inputs. This computation is performed by the 'other' part of the computer.

Heterotic refinement. The base and controller levels can be implemented ('refined') separately. For analog computing, we expect the base level to be implemented in an analogue medium (NMR, quantum cluster state – hence 'retrenched') and the controller level to be digital, but that is not a necessary restriction.

Example 1: NMR, where the base level is the NMR gates; the controller level is mere 'signal transduction' – this shows that there is no sharp separation between the i/o refinement and the computation (in this case it can be done in either).

Example 2: the quantum cluster state and the parity controller. The state is set up initially, and the only 'operation' performed in the base layer is measurement; which measurement to perform is determined in the classical controller level based on previous measurement results. The measurement itself changes ('collapses') the state, which is part of the computation.

Such concrete models could be used as the basis for developing a suitable form of refinement calculus. Possibly the closest pre-existing work relating to this is the use of weakest precondition semantics to study Grover's algorithm developed by d'Hondt and Panagaden [14] — in particular, the way that a hybrid quantum/probabilistic setting is modelled by the density matrix formalism. This gives a specific case of the type of underlying logical rules that need to be preserved by the refinement calculus, by analogy with the way that traditional program refinement preserves the Hoare logic. However, in each concrete setting, the behaviour/logic preserved by the refinement process will be different, and the formal calculus produced in each case will be heavily dependent on the underlying categorical models.

Moreover, for non-discretised systems, this relevant refinement calculus would need to be extended to a retrenchment approach to allow a well-defined and principled form of 'inexact refinement' This would include analysis of propagation of errors (due to noise, and to 'drift'), and techniques for correction and control of these errors.

4 NMR Computing within the Framework

NMR technology provides a well-developed experimental system in which a variety of computational implementations can be studied and combined. There is a rich NMR variable space available to be explored, and the interaction between classical and quantum layers in the NMR system can be adapted to many combinations of heterotic compositions. Our existing work [24,7] on hybrid classical systems provides a good starting point. To illustrate, we outline how the physical NMR system is adapted to perform classical gates, and describe the possibilities for extending this.

4.1 Classical NMR Variables

There are three experimental categories of NMR variables that can be used in a computational context: r.f. pulses and pulse sequences; spin system parameters; choice of material/state of condensed matter.

Error compensation in r.f. pulses and pulse sequences. The unwanted effects of imperfect, say π, or $\pi/2$ pulses in NMR experiments have always been at the centre of attention in NMR pulse sequence design. Numerous error compensation techniques exist, most prominently phase cycling is used in almost all circumstances. As the name implies, phase cycling (of transmitter and receiver phases) consists of repeating and co-adding spectra obtained with suitably cycled sets of phases, sometimes up to 64 spectra. This experimental NMR approach to error compensation is acceptable in many analytical NMR applications as poor signal-to-noise ratio makes signal averaging necessary anyway. Here we cannot afford the accumulation of multiple spectra. Instead we need to carefully assess the imperfections in 'single shot' NMR spectra, and determine, for example, the merits of 'composite pulses' [21] in reducing experimental errors.

Characterising spin system parameters. Computational NMR implementations usually require the presence of more than one spin species with different resonance frequencies. The presence of a range of frequencies affects the performance of different pulse sequences in different ways. Some of these effects are experimentally unavoidable, but have an effect in accumulated errors in the results of concatenated pulse sequences. Some pulse sequences that appeal in theory may have to be discarded for practical reasons. The computational-relevant consequences of these effects need to be determined.

Choice of material/state of condensed matter. NMR experiments using isotropic liquids are (usually) straightforward to implement and samples are easy to prepare. But isotropic liquids do not permit the control of resonance frequencies as continuous variables (unless one employs field gradients, which may often not be desirable). Solid state (anisotropic) systems, using either single crystal or the more readily accessible polycrystalline power, can provide this extra variable to be exploited for heterotic computation.

4.2   Analysis of Classical and Quantum Layers in NMR Computation

Quantum and classical NMR computational systems can be used as heterotic layers effectively.

The modular NMR tools (pulse sequences, spin systems, state of condensed matter) can be built into specific computational implementations, where they can be analysed to understand and classify the relative merits of classical *vs* quantum NMR implementations. A concrete example is the Deutsch-Jozsa algorithm. Several NMR quantum implementations have been described [9,11,22], and one can also think of equivalent classical NMR implementations [25]. Experimental NMR imperfections need to be taken into account, as do specific errors inherent in the classical and/or quantum systems themselves.

The various advantages and disadvantages of classical *vs* quantum NMR implementations with regard to parallelism, superposition, and interference are important. Both classical and quantum implementations of NMR experiments require preparation of the spin system(s) and appropriate read-out procedures. The relative cost of each of these is important in determining how one can exploit the best of both layers. This can be examined by evaluating the performance and preparation needed for a computational task using either a single n-spin system or n single spin species.

5   Stressing the Framework with Continuous Variable Computing

The framework described in §3 is a 'stepwise' approach. More complicated heterotic systems will be 'continuing' systems, where the base system's state continues to evolve whilst the control layer computes.

An appropriate experimental basis for studying these more complex heterotic systems is provided by continuous variable computing, both quantum and classical. Existing models of analog computation could be exploited to extend or modify the framework to accommodate continuous variables, or clearly define the framework's applicability as limited to non-evolving base computation.

Despite providing many of the earliest practical computers, analog computation (both classical and quantum) has had much less thorough theoretical development compared to digital computation. Before it can be cast in a heterotic framework, some gaps in the theory would need to be filled. In particular, an approach to analysing the propagation of errors is needed, as is the determinations of universal gate sets in a substrate-respecting manner (cf [28]). Although underdeveloped for computation, continuous variables (both quantum and classical) are often used in communications channels. It follows that using continuous variables as a control layer in the heterotic computer can take advantage of the more highly developed communications technologies, thus a continuous variable control layer should be easier to achieve than a continuous variable substrate. This architecture is known as a hybrid scheme in quantum computation (eg [26]) and is considered one of the most promising scalable routes to useful quantum computation. The inverse, using a discrete quantum control layer on a continuous quantum base layer, as specified in detail in [28] for a micro-maser experimental system, is suitable for studying continuous variables in the heterotic framework.

6   Discussion and Conclusions

We have described a novel computational framework, heterotic computation, that can be used to combine computational systems from different implementation paradigms in a principled and controlled manner, to produce a computational system qualitatively different from either in isolation. We have outlined a semantic and refinement framework that could be used to support such an approach, and indicated how classical NMR computation can be advanced to exploit the framework.

This is only the first step in hybrid computation. We have discussed an area that would need enhancement to the framework, where the base layer continues its computation whilst the controlling layer is working. This will be the case for a range of dynamical systems; one of the things the controlling layer will need to decide is when to probe/perturb the base layer, to exploit its dynamics. Additionally, further forms of parallelism also need to be added to the framework.

We believe the heterotic approach is needed to ensure that the many forms of unconventional computation can be exploited fully. Each individual paradigm no longer need be distorted to achieve Turing-completeness. Instead, different components can be combined to form a more powerful system, with each component doing what it does naturally, and best.

References

1. Abramsky, S., Coecke, B.: A categorical semantics of quantum protocols. In: Proc. IEEE Symp. Logic in Comp. Sci., pp. 415–425 (2004)
2. Anders, J., Browne, D.: Computational power of correlations. Phys. Rev. Lett. 102, 050502 (2009)
3. Banach, R., Jeske, C., Fraser, S., Cross, R., Poppleton, M., Stepney, S., King, S.: Approaching the formal design and development of complex systems: The retrenchment position. In: WSCS, IEEE ICECCS 2004 (2004)
4. Banach, R., Jeske, C., Poppleton, M., Stepney, S.: Retrenching the purse. Fundamenta Informaticae 77, 29–69 (2007)
5. Banach, R., Poppleton, M.: Retrenchment: an engineering variation on refinement. In: Bert, D. (ed.) B 1998. LNCS, vol. 1393, pp. 129–147. Springer, Heidelberg (1998)
6. Banach, R., Poppleton, M., Jeske, C., Stepney, S.: Engineering and theoretical underpinnings of retrenchment. Sci. Comp. Prog. 67(2-3), 301–329 (2007)
7. Bechmann, M., Sebald, A., Stepney, S.: From binary to continuous gates – and back again. In: Tempesti, G., Tyrrell, A.M., Miller, J.F. (eds.) ICES 2010. LNCS, vol. 6274, pp. 335–347. Springer, Heidelberg (2010)

8. Blakey, E.: Unconventional complexity measures for unconventional computers. Natural Computing (2010), doi:10.1007/s11047-010-9226-9
9. Chuang, I.L., Vandersypen, L.M.K., Zhou, X., Leung, D.W., Lloyd, S.: Experimental realization of a quantum algorithm. Nature 393(6681), 143–146 (1998)
10. Clark, J.A., Stepney, S., Chivers, H.: Breaking the model: finalisation and a taxonomy of security attacks. ENTCS 137(2), 225–242 (2005)
11. Collins, D., et al.: NMR quantum computation with indirectly coupled gates. Phys. Rev. A 62(2), 022304 (2000)
12. Cooper, D., Stepney, S., Woodcock, J.: Derivation of Z refinement proof rules: forwards and backwards rules incorporating input/output refinement. Tech. Rep. YCS-2002-347, Department of Computer Science, University of York (December 2002)
13. Cory, D.G., et al.: NMR based quantum information processing: Achievements and prospects. Fortschritte der Physik 48(9–11), 875–907 (2000)
14. d'Hondt, E., Panangaden, P.: Quantum weakest preconditions. Math. Struct. Comp. Sci. 16(3), 429–451 (2006)
15. DiVincenzo, D.P.: The physical implementation of quantum computation. Fortschritte der Physik 48(9-11), 771–783 (2000), arXiv:quant-ph/0002077v3
16. Dusold, S., Sebald, A.: Dipolar recoupling under magic-angle spinning conditions. Annual Reports on NMR Spectroscopy 41, 185–264 (2000)
17. Hines, P.: The categorical theory of self-similarity. Theory and Applications of Categories 6, 33–46 (1999)
18. Hines, P.: A categorical framework for finite state machines. Mathematical Structures in Computer Science 13, 451–480 (2003)
19. Jones, J.A.: NMR quantum computation. Progress in Nuclear Magnetic Resonance Spectroscopy 38(4), 325–360 (2001)
20. Lambek, J., Scott, P.: An introduction to higher-order categorical logic. Cambridge University Press, Cambridge (1986)
21. Levitt, M.H.: Composite pulses. In: Grant, D.M., Harris, R.K. (eds.) Encyclopedia of Nuclear Magnetic Resonance, vol. 2, pp. 1396–1441. Wiley, Chichester (1996)
22. Linden, N., Barjat, H., Freeman, R.: An implementation of the Deutsch-Jozsa algorithm on a three-qubit NMR quantum computer. Chemical Physics Letters 296(1-2), 61–67 (1998)
23. Mac Lane, S.: Categories for the working mathematician. Springer, Heidelberg (1971)
24. Roselló-Merino, M., Bechmann, M., Sebald, A., Stepney, S.: Classical computing in nuclear magnetic resonance. IJUC 6(3-4), 163–195 (2010)
25. Sebald, A., Bechmann, M., Calude, C.S., Abbott, A.A.: NMR-based classical implementation of the de-quantisation of Deutsch's problem (work in progress)
26. Spiller, T.P., Nemoto, K., Braunstein, S.L., Munro, W.J., van Loock, P., Milburn, G.J.: Quantum computation by communication. New J. Phys. 8(2), 30 (2006)
27. Stepney, S.: The neglected pillar of material computation. Physica D: Nonlinear Phenomena 237(9), 1157–1164 (2008)
28. Wagner, R.C., Everitt, M.S., Jones, M.L., Kendon, V.M.: Universal continuous variable quantum computation in the micromaser. In: Calude, C.S., Hagiya, M., Morita, K., Rozenberg, G., Timmis, J. (eds.) Unconventional Computation. LNCS, vol. 6079, pp. 152–163. Springer, Heidelberg (2010)
29. Wegner, P.: Why interaction is more powerful than algorithms. CACM 40, 80–91 (1997)

A Formal Approach to Unconditional Security Proofs for Quantum Key Distribution

Takahiro Kubota[1], Yoshihiko Kakutani[1], Go Kato[2], and Yasuhito Kawano[2]

[1] Department of Computer Science,
Graduate School of Information Science and Technology,
The University of Tokyo
{takahiro.k11_30,kakutani}@is.s.u-tokyo.ac.jp
[2] NTT Communication Science Laboratories, NTT Corporation
{kato,kawano}@theory.brl.ntt.co.jp

Abstract. We present an approach to automate Shor-Preskill style unconditional security proof of QKDs. In Shor-Preskill's proof, the target QKD, BB84, is transformed into another QKD based on an entanglement distillation protocol (EDP), which is more feasible for direct analysis. We formalized heir method as program transformation in a quantum programming language, QPL. The transform is defined as rewriting rules which are sound with respect to the security in the semantics of QPL. We proved that rewriting always terminates for any program and that the normal form is unique under appropriate conditions. By applying the rewriting rules to the program representing BB84, we can obtain the corresponding EDP-based protocol automatically. We finally proved the security of the obtained EDP-based protocol formally in the quantum Hoare logic, which is a system for formal verification of quantum programs. We show also that this method can be applied to B92 by a simple modification.

Keywords: QKD, BB84, B92, unconditional security, automatic verification, formal methods.

1 Introduction

A quantum key distribution protocol (QKD) allows two distant parties to share a secret key. The law of physics guarantees an important advantage of QKD over the classical schemes, that is, an adversary cannot gain non-negligible amount of information even if her computational power is unlimited. Several QKDs have been presented [1–6]. However, a security proof of a QKD has a difficulty that does not appear in the classical case. As for BB84 QKD [1], the first proof of the security was given by Mayers [7] but it is quite complex. Researchers have been aiming to discover simple unconditional security proofs for QKDs. Lo and Chau presented a QKD based on an entanglement distillation protocol (EDP) [8]. The proof of the security of this protocol is simple, although use of entanglements, which are difficult to implement, is undesirable as a practical protocol. Shor and

C.S. Calude et al. (Eds.): UC 2011, LNCS 6714, pp. 125–137, 2011.
© Springer-Verlag Berlin Heidelberg 2011

Preskill presented a simple proof of BB84 [9], which are divided into two steps, *the transformation step* and *the analysis step*. In the transformation step, the security of BB84 is reduced to that of an EDP-based protocol whose security can be proved in the same way as the security of Lo-Chau's protocol. It employs Calderbank-Shor-Steane (CSS) quantum error correcting codes (QECC) [10]. Their method is versatile and applied to other security proofs [11, 12].

In the field of formal methods, verification techniques on mathematical logic have been developed. Such techniques were applied to various kinds of classical protocols and have produced visible results. For instance, there are formal methods for security validation [13] of Kerberos [14], a practical authentication protocol with at least 11 security properties to verify. Formal methods have two key features. The first is formal systems to precisely describe target protocols' behavior and properties. This enables us to find problems easily and describe many different protocols in one unified framework. Formal methods enables us to reuse the proof techniques. Therefore, it is important to formalize verification of protocols, even if each of them already has a security proof. The second is the use of automated software tools, which either prove protocols satisfy desired properties or find flaws in protocols. Once we have ascertained the soundness of a tool, it is enough for verification to check if formalizations of protocols are valid. There are many classical protocols and some of them do not have a security proof. Automation frees us from writing a tedious proof for each protocol. These features of formal methods will benefit us also in quantum protocols because many protocols will be provided in the future and one who wants to develop a protocol is not necessarily an expert of security proofs. From the experience of the verification of classical protocols, difficulty of verification increases according to the complexity of protocols. An automatic approach in a formal framework is more feasible than a heuristic approach for the case where a large complex protocol is proposed in the future.

Our final goal is to provide a framework for full automation of unconditional security proofs for QKDs. As a first step, in this paper, we formalize Shor-Preskill style security proofs for QKDs [9]. We defined a simple programming language based on Selinger's QPL [15] to describe QKDs formally. We then introduced rewriting rules on programs which are sound with respect to unconditional security. We proved that rewriting always terminates for any program and that the normal form is unique under appropriate conditions. By applying the rewriting rules to the program representing BB84, we can obtain the corresponding EDP-based protocol automatically. We finally proved the security of the obtained EDP-based protocol formally in the quantum Hoare logic [16]. We show also that this method can be applied to B92 QKD [3] with some additional rules.

2 Preliminaries

Actually, the word BB84 does not specify one unique protocol, because there are choices to error correction and privacy amplification steps. In this paper, since our target is Shor and Preskill's proof, we employ the same protocol that they consider [9]. It employs two classical linear codes C_1, C_2 that satisfy $\{\mathbf{0}\} \subsetneq C_2 \subsetneq C_1 \subsetneq \{0,1\}^n$, where n is the length of codewords in C_1 and C_2.

BB84

1. Alice generates two random $(4 + \delta)n$-bit strings $d_{A,1}, ..., d_{A,(4+\delta)n}$ and $b_{A,1}, ..., b_{A,(4+\delta)n}$.
   Bob generates a random $(4 + \delta)n$-bit string $b_{B,1}, ..., b_{B,(4+\delta)n}$.
2. Alice creates a $(4 + \delta)n$-qubit string $q_{B,1}, ..., q_{B,(4+\delta)n}$ according to the randomness: for each $q_{B,i}(1 \le i \le (4+\delta)n)$, Alice generates $|0\rangle$ if $d_{A,i} = 0, b_{A,i} = 0$, $|1\rangle$ if $d_{A,i} = 1, b_{A,i} = 0$, $|+\rangle$ if $d_{A,i} = 0, b_{A,i} = 1$, $|-\rangle$ if $d_{A,i} = 1, b_{A,i} = 1$.
3. Alice sends $q_{B,i}$'s to Bob via the quantum channel.
4. Bob receives q_{B_i}'s measuring i-th qubit $q_{B,i}$ by $\{|0\rangle, |1\rangle\}$ basis if $b_{B,i} = 0$ or by $\{|+\rangle, |-\rangle\}$ basis if $b_{B,i} = 1$. The result is the $(4 + \delta)n$-bit string d_B.
5. Alice announces $b_{A,i}$'s via the classical channel.
6. For each i-th bit $b_{B,i}$, Bob discards $d_{B,i}$ if $b_{A,i} \ne b_{B,i}$. With high probability, there are at least $2n$ bits left. (If not, Bob tells Alice that they abort the protocol.) Bob tells Alice which bits are left via the classical channel. Alice decides $2n$ bits from the bits left, randomly chooses n bits from them as check bits and then tells Bob those information.
7. Alice and Bob announce the values of their check bits to each other. If the error rate is higher than the threshold, they abort the protocol.
8. Alice chooses a codeword $u_A \in C_1$ randomly, and announces $u_A + x_A$, where x_A is the remaining non-check bits.
9. (Error correction) Bob calculates u'_B from announced $u_A + x_A$ and his own non-check bits x_B. Because x_B may include some errors, Bob obtains u_B by correction of the errors. If this error correction works well, u_B is almost equal to u_A.
10. (Privacy amplification) Alice and Bob determine secret keys k_A, k_B from the cosets of $u_A + C_2, u_B + C_2$.

BB84 is transformed into the following EDP-based protocol, which is a modification of the protocol in [9]. This protocol employs CSS-QECC [10]. In Section 4, we formally verify the security proof of the protocol in the quantum Hoare logic.

EDP-based protocol

1. Alice generates $(4 + \delta)n$ EPR pairs $q = (\frac{|00\rangle + |11\rangle}{\sqrt{2}})^{\otimes(4+\delta)n}$ and $(4 + \delta)n$-bit string $b_{A,1}, ..., b_{A,(4+\delta)n}$. Bob generates a random $(4 + \delta)n$-bit string $b_{B,1}, ..., b_{B,(4+\delta)n}$.
2. According to $b_{A,i}$'s, Alice executes Hadamard transformations on halves of q sent to Bob in the next step.
3. Alice sends the halves of q to Bob via the quantum channel.
4. Bob receives his halves, and executes Hadamard transformations according to $b_{B,i}$'s.
5. Alice announces $b_{A,i}$'s via the classical channel.

6. For each i-th bit $b_{B,i}$, Bob discards his i-th qubit if $b_{A,i} \neq b_{B,i}$. With high probability, there are at least $2n$ pairs left. (If not, Bob tells Alice that they abort the protocol.) Bob tells Alice which bits are left via the classical channel. Alice decides $2n$ bits from the bits left, randomly chooses n bits from them as check bits and then tells Bob those information.

7. Alice and Bob measure their halves of the check bits by the $\{|0\rangle, |1\rangle\}$ basis, and share the results. If the error rate is higher than the threshold, they abort the protocol.

8. Alice calculates the parameters of the related CSS code and send them to Bob. Bob then calculates the syndrome and corrects errors using the parameters. Alice and Bob next decode their qubit strings as the CSS code.

9. Alice and Bob measure their qubits in $\{|0\rangle, |1\rangle\}$ basis to obtain shared secret keys k_A, k_B.

3   Automation of the Transformation Step

In this Section, we present an approach to automate the transformation step in Shor and Preskill's proof. First, we define our quantum programming language to formalize QKDs. We then introduce the rewriting rules for the programs that are sound with respect to the security. We finally prove the properties of the rewriting system which are useful for automatic verification.

3.1   Quantum Programming Language

Our language is a sublanguage of Selinger's QPL [15]. We added procedure call representing subprograms as well as adversaries. The grammar is enough expressive to formalize QKDs.

Definition 1. *The definitions of program $P \in \mathcal{P}$, command C and boolean expression k are as follows.*

$$P ::= \quad \texttt{skip;} \,|\, \texttt{bit } b; \,|\, \texttt{discard } b; \,|\, \texttt{qbit } q; \,|\, \texttt{discard } q; \,|\, \texttt{if } k \texttt{ then } C \,|\, P\,P$$
$$C ::= \quad b := k; \,|\, q := k; \,|\, q, ..., q \mathbin{*}= U; \,|\, b := \texttt{measure } q; \,|\, f(b, ..., b, q, ..., q|b, ..., b);$$
$$k ::= \quad 0 \,|\, 1 \,|\, b \,|\, k + k$$

where b, q, f and U are a program variable for a classical bit, a program variable for a quantum bit, a procedure name, and a unitary matrix, respectively.

Intuitive meanings of the syntax is as follows. The syntax `bit` and `qbit` declare a classical bit and a quantum bit initiated with the value of 0. The syntax `discard` closes the scope of a variable. C represents commands composed of value assignment $b := k$; and $q := k$;, unitary transformation $q, ..., q \mathbin{*}= U$;, projective measurement in $\{|0\rangle, |1\rangle\}$ basis $b := \texttt{measure } q$;, and procedure call $f(b, ..., b, q, ..., q|b, ..., b)$;. $P\,P$ is sequential composition of programs. To specify measurement in other bases than $\{|0\rangle, |1\rangle\}$, the target qubit should be rotated appropriately. For example, to measure a qubit `qb` in $\{|+\rangle, |-\rangle\}$ basis and then store the result in a classical bit b, the program should be as follows with Hadamard matrix H: `qb *= H; b := measure qb; qb *= H;`. The procedure call can be used to describe unspecified subprograms. For example, we have infinite ways to generate a random bit, but the details of implementation is not essential. Execution of a procedure is call-by-reference because quantum states cannot be copied. A procedure takes two kinds of input variables separated with the bar '|'. Only the former variables than the bar '|' will be modified in the execution of the procedure. This grammar makes it clear which variables are modified in a procedure. In any implementation of a procedure, local variables can occur and no free variable other than input occurs.

An adversary Eve is also represented by a procedure. For example, Eve's attack to the qubits sent from Alice to Bob is `Eve_Attack(ke[],qe[],qb[]|);`. Inputs `ke[]`,`qe[]`,`qb[]` are Eve's classical and quantum memories and qubits from Alice to Bob, respectively. The body of the procedure is not specified but instantiated with any programs as long as the conditions of variables are satisfied, i.e. no variable other than input and local variables occurs in the body, and only the variables set former than the bar '|' are modified.

We also use array notation for classical or quantum bitstrings. The notation `p[i]` indicates the i-th element of an array named `p`. All elements in an array are of the same type. To make programs more readable, we introduce syntax sugar defined as follows.

$k\ ==\ k' \equiv k + k' + 1,$

unless k then $C \equiv$ if $k + 1$ then C,

if k then C_1 else $C_0 \equiv$ if k then C_1 unless k then C_0

bit $b_0, ..., b_{n-1};\ \equiv$ bit $b_0; , ...,$ bit $b_{n-1};$

qbit $q_0, ..., q_{n-1};\ \equiv$ qbit $q_0; , ...,$ qbit q_{n-1}

discard $b_0, ..., b_{n-1}; \equiv$ discard $b_0; , ...,$ discard $b_{n-1};$

$\mathtt{Rnd}(x_0, ..., x_{n-1}|); \equiv \mathtt{Rnd}(x_0|); , ..., \mathtt{Rnd}(x_{n-1}|);$

$p[m..n] \equiv p[m], ..., p[n]$

$p[] \equiv p[1..l]$ where l is the maximal index.

for $i := m$ to n do P end

$\equiv P[m/i]\ P[m + 1/i], ..., P[n/i].$

where $P[j/i]$ is the program obtained by replacing i with j.

As for the semantics of our language, a program P is interpreted to a super-operator $[\![P]\!]$ on density matrices that are environments for program variables [15]. Each superoperator $[\![P]\!]$ maps a given density matrix A to the density matrix $[\![P]\!](A)$ after the execution of the program P. For example, the semantics of unitary transformation $q_1, ..., q_n * = U$ is as follows: $[\![q_1, ..., q_n * = U]\!](A) = (U \otimes I)A(U^\dagger \otimes I)$

3.2 Formalization of BB84

QKDs can be formalized in our language. In this Subsection, we show the program representing BB84. In the program, the names Alice, Bob, and Eve are just comments indicating whose action each line is.

The procedures `Syndrome` and `Correct` represent syndrome calculation and error correction in C_1 and the procedure `Key` represents privacy amplification by calculating the coset of C_2. The procedure `Unshuffle(a[1..n]|b[1..n])` sorts the elements in array `a[]` according to `b[]`: the value of `a[i]` comes former if `b[i]` is 1.

```
          : bit ka[1..'m]; bit kb[1..'m]; qbit qb[1..'(4+d)n];
          : bit ba[1..'(4+d)n]; bit bb[1..'(4+d)n]; bit c[1..'2n];
          : bit da[1..'(4+d)n]; bit db[1..'(4+d)n]; bit e[1..'log(n)];
          : bit ua[1..'n]; bit ub[1..'n]; bit x[1..'n];
          : bit sx[1..'n]; bit ke[1..'N]; qbit qe[1..'N];
Alice : Rnd(da[1..'(4+d)n]|);
Alice : Rnd(ba[1..'(4+d)n]|);
Alice : for i := 1 to '(4+d)n do qb[i] := da[i]; end
Alice : for i := 1 to '(4+d)n do if ba[i] then qb[i] *= H; end
Eve   : Eve_Attack(ke[],qe[],qb[]|);
Bob   : Rnd(bb[1..'(4+d)n]|);
Bob   : for i := 1 to '(4+d)n do
Bob   :   if bb[i] then qb[i] *= H;
Bob   :   db[i] := measure qb[i];
Bob   : end
Bob   : for i := 1 to '(4+d)n do
Bob   :   if ba[i]==bb[i] then g[i]:= 0 else g[i]:= 1
Bob   : end
Alice : Unshuffle(da[1..'(4+d)n]|g[]);
Bob   : Unshuffle(db[1..'(4+d)n]|g[]);

Alice : Half_Rnd(c[1..'2n]|);
Alice : Unshuffle(da[1..'2n]|c[]);
Bob   : Unshuffle(db[1..'2n]|c[]);
Alice : e[] := 0;
Alice : for i := 'n+1 to '2n do
Alice :   unless da[i]==db[i]
Alice :     then Add_One(e[]|);
Alice : end
Alice : Abort(|e[]);
Alice : Rnd_in_C1(ua[1..'n]|);
Alice : for i := 1 to 'n do x[i] := da[i] + ua[i]; end
Bob   : for i := 1 to 'n do ub[i] := db[i] + x[i]; end
Bob   : Syndrome(sx[]|ub[1..'n]);
Bob   : Correct(ub[1..'n]|sx[]);
Alice : Key(ka[1..'m]|ua[1..'n]);
Bob   : Key(kb[1..'m]|ub[1..'n]);
Eve   : Eve_Guess(ke[],qe[]|ba[],bb[],c[],e[],x[]);
      : discard qb[]; discard ba[]; discard bb[];
      : discard c[]; discard da[]; discard db[];
      : discard e[]; discard ua[]; discard ub[];
      : discard x[]; discard sx[];
```

3.3   Rewriting Rules

We introduce our rewriting rules that are sound with respect to unconditional security: if a protocol after rewriting is secure, the protocol before is secure. Therefore, if we obtain an EDP-based protocol after rewriting BB84, we have only to prove the security of the former to obtain that of the latter. Each rewriting rule is of the form

$$\frac{P}{Q}\text{\#side condition,}$$

where P and Q are programs (we may call P the upper and Q the lower program). It denotes that some part of a program matches P, it can be replaced with Q if the side condition is satisfied. Rewriting rules are divided into three sorts: *general rules*, *protocol rules* and *exchange rules*. General rules are basic rules and protocol rules are specific rules for security proofs. For example, the following are general rules for if statement.

$$\frac{\texttt{b} := 0; \texttt{if b then } P}{\texttt{b} := 0;} \qquad \frac{\texttt{b} := 1; \texttt{if b then } P}{\texttt{b} := 1; \ P}$$

They mean that when a conditional branch occurs by the program variable b, the program P is executed only if the value of b is 1.

In the above rule, a program variable b is in fact a meta-variable ranging over program variables. So, we can apply the rule to the case of another program variable. We use this notation implicitly in the rest of the paper.

Exchange rules are a little different from general and protocol rules. They are non-directed, that is, the upper and the lower can be rewritten to each other. We use them to transform a program so that we can apply other rewriting rules. An example of an exchange rule is as follows with programs P and Q.

$$\frac{P\ Q}{Q\ P}\text{\#No variable in } P \text{ is modified in } Q, \text{ and vice varsa.}$$

It means that we can change the order of two programs if the side condition is satisfied, that is, executions of P and Q do not influence each other.

Properties of procedures are also described as rewriting rules. For example, we use $\texttt{Shuffle(q[1..}n\texttt{]|c[])}$ procedure which shuffles the qubit array $\texttt{q[1..}n\texttt{]}$

according to the classical bit string c[] and Unshuffle(q[1..n]|c[]) proce-
dure which is the inverse of Shuffle(q[1..n]|c[]). The following exchange
rules specify the properties of them.

$$\frac{\texttt{Shuffle(q[1..}n\texttt{]|c[]);}}{\texttt{Unshuffle(q[1..}n\texttt{]|c[]);}}$$
$$\texttt{skip;}$$

$$\frac{\texttt{Unshuffle(q[1..}n\texttt{]|c[]);}}{\texttt{Shuffle(q[1..}n\texttt{]|c[]);}}$$
$$\texttt{skip;}$$

We do not show all the rules because of the limitation of pages. One can find
the whole of the general rules and exchange rules in [17]. The general and exchange rules are trivial rather than the protocol rules. Details of the protocol
rules needed by the security proof of BB84 are discussed later.

Procedures in this paper can be implemented. For example, random assign-
ment procedure Rnd(b|) is implemented as follows.

$$\texttt{Rnd(b|)} \equiv \texttt{qbit q; q *= H; b := measure q; discard q;}$$

Implementations of procedures are not critical for the security proof if they
satisfy the conditions derived from the rewriting rules.

Formally, we define the relation $\equiv$ on programs as the smallest equivalence
relation such that $X \equiv Y$ holds if X can be rewritten to Y by some exchange
rule. The rewriting is defined on the equivalence classes $\mathcal{P}/\equiv$ of programs as
follows.

Definition 2. *Let X, Y be elements of $\mathcal{P}/\equiv$. We write $X \to Y$ if there exist
$X', Y' \in \mathcal{P}/\equiv$ such that $X \equiv X', Y \equiv Y'$ and X' is rewritten to Y' by general
or protocol rules.*

The rewriting on $\mathcal{P}/\equiv$ are sound in the following sense.

Theorem 3. *For some implementation of procedures, $[\![-]\!]$ is well-defined on
$\mathcal{P}/\equiv$, that is, $P \equiv Q$ implies $[\![P]\!] = [\![Q]\!]$. Let X, Y be elements of $\mathcal{P}/\equiv$. The
rewriting rules preserve the canonical semantics of programs, that is, implemen-
tations of procedures exist that satisfy $[\![X]\!] = [\![Y]\!]$ if $X \to Y$ holds.*

3.4   Properties of the Rewriting System

Since more than one rules are applicable to a program in general, we expect
that BB84 is always rewritten to the EDP-based protocol by the rewriting. To
guarantee this property, we require that the protocol rules have a priority over
the general rules and are totally ordered among themselves.

We proved the two important properties: the strong normalization property
(SN) and Church-Rosser property (CR). SN is the property that rewriting always
terminates up to exchange for any start-point program and rewriting strategy.
A program is said to be normal form when it is no more rewritable. CR is the
property that rewriting always confluences (but does not necessarily guarantee
termination of rewriting). When both SN and CR hold, rewriting from a program
always terminates and its normal form is unique.

We present here the proof sketch of SN and CR. As for SN, we define a size function f which maps an equivalence class of programs to a natural number so that $X \to Y$ implies $f(X) > f(Y)$. Namely, a program is rewritten to a program whose size is smaller than before. Since the set of natural numbers is well-founded, there is no infinite sequence of rewriting. As for CR, because we have the priority on the rules, it is sufficient to check the confluence of the general rules.

To apply our framework to a protocol other than BB84, we have to modify the protocol rules. We need to prove again SN and CR for modified rules in general, but the priority condition saves us proving CR.

3.5   Protocol Rules

BB84 is converted to an EDP-based protocol automatically using rewriting rules. In this Section, we present protocol rules used in the transformation. All steps of the transformation is rewriting between equivalent programs. If we formalize the equivalence just as it was argued by Shor and Preskill [9], the number of the protocol rules is about 8. The number is reduced to 4 if we reformulate the rules and change the order of application. The 4 rules, which specify the procedures, are as follows.

1.1
```
Rnd(da[i]);
qb[i] := da[i];
─────────────────
qbit qa[i];
EPR(qa[i],qb[i]);
da[i] := measure qa[i];
discard qa[i];
```

1.2
```
for i := 1 to n do
  db[i] := measure qb[i];
  ub[i] := db[i] + x[i];
end
Syndrome(sx[]|ub[1..n]);
Correct(ub[1..n]|sx[]);
──────────────────────────────────
bit z[1..n]; bit sz[1..k_z]; Rnd(z[1..n]);
CSS_Syndrome(sx[],sz[],qb[1..n]|x[],z[]);
CSS_Correct(qb[1..n]|sx[],sz[]);
for i := 1 to n do
  db[i] := measure qb[i];
  ub[i] := db[i] + x[i];
end
discard z[1..n]; discard sz[1..n];
```

1.3
```
bit db[1..n]; bit ub[1..n];
for i := 1 to n do
  db[i] := measure qb[i];
  ub[i] := db[i] + x[i];
end
Key(kb[1..m],ub[1..n]);
discard db[];discard ub[];discard qb[];
──────────────────────────────────────
CSS_Decode(qb[1..n]|x[],z[]);
for i := 1 to m do
  kb[i] := measure qb[i];
end
discard qb[];
```

1.4
```
Rnd(z[1..n]);
Rnd_in_C1(ua[1..n]);
for i := 1 to n do
  da[i] := measure qa[i];
  x[i] := da[i] + ua[i];
end
─────────────────────────────── #
CSS_Projection(x[],z[],qa[1..n]);
for i := 1 to n do
  da[i] := measure qa[i];
  ua[i] := da[i] + x[i];
end
```
qa[i] is random for all i ($1 \le i \le n$).

First, by rule **1.1**, generation of random numbers by procedure Rnd is converted to that by measuring of EPR pairs. Rule **1.4** is applied next. In the upper, ua[1..n] is chosen at random in C_1 and z[1..n] is random. In the lower, by projecting qa[], Alice obtains x[], z[] as parameters of the CSS code. Rule **1.2** is applied next. In the upper, Bob first measures his qubit qb[i] in $\{|0\rangle, |1\rangle\}$ basis and phase errors are eliminated by this measurement. Bob then corrects bit errors by the code C_1. In the lower, Bob can equivalently correct both phase and bit errors by CSS code that employs C_1, C_2. Rule **1.3** is then applied. In the upper, procedure Key calculates the coset ub[]$+C_2$ for ub[] $\in C_1$. In the lower, qb[] is decoded as a CSS codeword and the corresponding coset of C_2 in C_1 is obtained [10].

A program that does not correspond to an actual protocol may occur in intermediate steps of the transformation. However, such a fact is not problematic if only the first and the last programs denote proper QKDs.

To apply our method to other protocols, the set of the rewriting rules are modified in general but we only need to add two rules for the transformation step of B92 QKD to the corresponding EDP-based protocol.

1.6

```
Rnd(db[1..(8 + δ)n]);
for i := 1 to (8 + δ)n do
  unless db[i] then qb[i] *= H;
  g[i] := measure qb[i]; g[i] := g[i] + 1;
end
discard qb[1..(8 + δ)n];
```

1.5

```
Rnd(da[i]);
qb[i] := 0;
if da[i] then qb[i] *= H;
  qbit qa[i];
qa[i] := 0; qb[i] := 0;
qa[i],qb[i] *= X; da[i] := measure qa[i];
discard qa[i];
```

```
Rnd(bb[1..(8 + δ)n]);
for i := 1 to (8 + δ)n do
  qbit qc[i]; qc[i] := 0;
  qb[i],qc[i] *= Y; g[i] := measure qc[i];
  discard qc[i]; db[i] := measure qb[i];
end
discard qb[1..(8 + δ)n];
```

4 Formal Verification of EDP-Based Protocol

In this Section, we formalize the security proof of the EDP-based protocol in quantum Hoare logic. The logic is based on Kakutani's quantum Hoare logic [16]. In this framework, properties of the programs are asserted as Hoare triples.

4.1 Quantum Hoare Logic

A Hoare triple $\{\Phi\}P\{\Psi\}$ intuitively means that if the precondition Φ holds, the postcondition Ψ holds after the execution of the program P. The logic is sound in the following sense: if $\{\Phi\}P\{\Psi\}$ is deduced by the deduction rules [16] then $A, v \models \Phi$ implies $[\![P]\!](A), v \models \Phi$ holds for any density matrix A and assignment v, where $\models$ is the satisfaction relation. Because the quantum Hoare logic supports the syntax of QPL, we can directly verify the program obtained in the previous section. We first give the syntax of the formulae.

Definition 4. *Formulae of the quantum Hoare logic is defined as follows.*

$$c ::= r \mid \alpha \mid f(c, ..., c)$$
$$t ::= c \mid \text{pr}(s) \mid f(t, ..., t)$$
$$\Phi ::= t \leq t \mid \text{int}(t) \mid t\Phi \mid \Phi \oplus \Phi \mid {}^{x, ..., x}M\Phi \mid \neg\Phi \mid \Phi \wedge \Phi \mid \forall\alpha.\Phi$$

where r, f, α and s range over $\mathbb{R}$, functions on $\mathbb{R}$ predicate variables and subspaces over program variables. We write $\Phi \vee \Psi$ for $\neg(\neg\Phi \wedge \neg\Psi)$ and $\Phi \supset \Psi$ for $\neg\Phi \vee \Psi$, and $\exists\alpha.\Phi$ for $\neg\forall\alpha.\neg\Phi$.

The semantics of the formulae is given by the satisfaction relation $A, v \models \Phi$, where A is a density matrix and v is a function from predicate variables to $\mathbb{R}$ [16]. This intuitively means that Φ holds under the environment given by A, v. The term $\text{pr}(s)$ is the weight of the subspace of A denoted by a condition s. For example, the semantics about $\oplus$ is as follows.

$$A, v \models \Phi_1 \oplus \Phi_2 \text{ iff } \exists A_1, A_2. \ A = A_1 + A_2 \text{ and } A_1, v \models \Phi_1 \text{ and } A_2, v \models \Phi_2$$

4.2   Formal Verification in the Quantum Hoare Logic

We formally verify the EDP-based protocol introduced in Subsection 2. In our following argument, we divide the protocol into three parts P, Q and R: P is the steps from (1) to (7), Q is the step (8) and R is the step (9). The outline of the proof follows the proof discussed in Shor-Preskill's paper [7]. For formal verification in the quantum Hoare logic, implementations of the procedures other than Eve's are fixed or some axioms that hold for the procedures are assumed.

Then, we go to the security analysis step. First, we validate the following Hoare triples. For any natural number n, there exist positive real numbers α, β and γ ($\gamma \geq \beta$) satisfying the following. Let Π be the subspace where the error correction in the step Q will be successful and let $\overline{\Pi}$ be the complementary space of Π. $\mathbf{e} \geq cn$ denotes the subspace where the number of errors in the check bits $\mathbf{e}$ is less than the threshold of protocol abortion cn (i.e. the shared qubits pass the check).

$$\{ \ \} \ P \ \{\text{pr}(\overline{\Pi} \wedge \mathbf{e} \leq cn) \leq 2^{-\gamma n}\} \tag{1}$$

This denotes that the probability is negligible that shared qubits pass the check despite the error correction will not be successful. The validity of this Hoare triple is proven from the facts: Bob's halves are randomly performed Hadamards, check bits are chosen at random, and Eve neither knows which bits are performed Hadamards nor which bits are used as check bits.

$$\{ \ \} \ P \ \{\text{pr}(\Pi \wedge \mathbf{e} \leq cn) = \text{pr}(\Pi)\} \tag{2}$$

This denotes that the error correction is never executed unless the shared qubits pass the check.

$$\{\text{pr}(\Pi) = p\} \ Q \ \{\text{F}(\mathsf{qa}, \mathsf{qb}) \geq p\} \tag{3}$$

$F(qa, qb)$ is the fidelity between qa and qb. This is proven from the definition of CSS error correcting codes [10].

$$\{1 - F(qa, qb|e \leq cn) \leq 2^{-\beta n}\} \; R \; \{I(ka; ke|e \leq cn) \leq 2^{-\alpha n}\}$$

$I(ka; ke \mid e \leq cn$) is the mutual information between ka and ke under the condition $e \leq cn$. This denotes that if the fidelity between qa and qb is high, the mutual information between the secret key ka and the adversary's guess ke is low. This is a well-known theorem argued by Lo and Chau [8].

$$\{ \} \; P \; Q \; R \; \{\mathrm{pr}(ka = kb) \geq 2^{-(\gamma-\beta)n} \rightarrow \mathrm{pr}(e \leq cn) \geq 2^{-(\gamma-\beta)n}\}$$

The contraposition of the postcondition denotes that if the probability that the shared qubits pass the check is negligible, the probability that Alice and Bob have the same key is also negligible. Finally, the property to verify is formalized as follows.

$$\mathrm{pr}(ka = kb) \geq 2^{-(\gamma-\beta)n} \rightarrow I(ka; ke|e \leq cn) \leq 2^{-\alpha n} \tag{4}$$

So the following Hoare triple should be deduced. Using the deduction rules, we formally deduce the following by the use of the Hoare triples (2), (3), (4), (5) and (6).

$$\{ \} \; P \; Q \; R \; \{\mathrm{pr}(ka = kb) \geq 2^{-(\gamma-\beta)n} \rightarrow I(ka; ke|e \leq cn) \leq 2^{-\alpha n}\}$$

5 Discussion

5.1 Related Work

Recently, by Nagarajan et al., a model checking method has been used to analyze the security of BB84 [18]. In that paper, although the security is analyzed automatically in a probabilistic model checker, PRISM [19], it is assumed that an adversary can only operate the eavesdropping attack: with some probability, intercept a qubit sent from one participant to the other, measure it by the canonical basis, and finally resend it to the destination. Unfortunately, a realistic adversary can execute unitary transformations to construct entanglements. It is still valuable to automate proofs of the unconditional security of QKDs.

The quantum Hoare logic [16] used in Section 4 is a general framework for verification of properties on quantum programs. Our work suggests a way to apply formal methods for quantum programs to quantum protocols.

5.2 Conclusion

From the experience of the classical cryptographic protocols, we know that protocol verification tend to include a tedious and error-prone task. Formal methods and automation of proofs are helpful for precise verification. Our work is the first application of a formal method to unconditional security proofs of QKDs. In the paper, we presented a framework to formalize Shor-Preskill style security proofs of QKDs.

First, we defined a simple programming language and formalized BB84 as a program. Second, we introduced rewriting rules on programs. Our rewriting preserves the semantics of programs. Using the rewriting rules, BB84 is converted automatically to an EDP-based QKD that is more feasible for direct security analysis. We have proved that our rewriting system has the SN and CR properties, namely, the rewriting always terminates for any program and the normal form is unique. Last, we formalized the proof of the unconditional security of the EDP-based protocol in the quantum Hoare logic.

The result of this work suggests that term-rewriting will be useful to verify security of quantum protocols. Our framework automatically converts not only BB84 but also B92 to the corresponding EDP-based QKDs. After Shor and Preskill's proof [9], researchers have discovered equivalences between protocols [11, 20]. In fact, the transformation of B92 in this paper is a formalization of the protocol transformation discussed in [11]. Our framework is naturally expected to be applied to other protocol transformations and security proofs. Moreover, if we reformulate the rewriting in the reverse direction, a new BB84-like protocol may be obtained from an EDP-based protocol. Such a study is a possible application of formal techniques to quantum protocols.

References

1. Bennett, C.H., Brassard, G.: Quantum cryptography: Public-key distribution and coin tossing. In: IEEE International Conference on Computers, Systems and Signal Processing, pp. 175–179 (1984)
2. Ekert, A.K.: Quantum cryptography based on bell's theorem. Phys. Rev. Lett. 67(6), 661–663 (1991)
3. Bennet, C.H.: Quantum cryptography using any two nonorthogonal states. Phys. Rev. Lett. 68(21), 3121–3124 (1992)
4. Inoue, K., Waks, E., Yamamoto, Y.: Differential phase shift quantum key distribution. Phys. Rev. Lett. 89(3), 037902 (2002)
5. Stucki, D., Brunner, N., Gisin, N., Scarani, V., Zbinden, H.: Fast and simple one-way quantum key distribution. Appl. Phys. Lett. 87(19), 194108–194108–3 (2005)
6. Nagy, M., Akl, S.G.: Entanglement verification with an application to quantum key distribution protocols. Parallel Processing Letters, 227–237 (2010)
7. Mayers, D.: Unconditional security in quantum cryptography. J. ACM 48, 351–406 (2001)
8. Lo, H.-K., Chau, H.F.: Unconditional security of quantum key distribution over arbitrarily long distances. Phys. Rev. Lett. 283(5410), 2050–2056 (1999)
9. Shor, P.W., Preskill, J.: Simple proof of security of the bb84 quantum key distribution protocol. Phys. Rev. Lett. 85(2), 441–444 (2000)
10. Calderbank, A.R., Shor, P.W.: Good quantum error-correcting codes exist. Phys. Rev. A 54(2), 1098–1105 (1996)
11. Tamaki, K., Koashi, M., Imoto, N.: Unconditionally secure key distribution based on two nonorthogonal states. Phys. Rev. Lett. 90(16), 167904 (2003)
12. Hwang, W.-Y., Wang, X.-B., Matsumoto, K., Kim, J., Lee, H.-W.: Shor-preskill-type security proof for quantum key distribution without public announcement of bases. Phys. Rev. A 67(1), 012302 (2003)

13. Blanchet, B., Jaggard, A.D., Scedrov, A., Tsay, J.-K.: Computationally sound mechanized proofs for basic and public-key kerberos. In: Proceedings of the 2008 ACM Symposium on Information, Computer and Communications Security, ASIACCS 2008, pp. 87–99. ACM, New York (2008)
14. Neuman, C., Yu, T., Hatman, S., Raeburn, K.: The kerberos network authentication service (v5) (July 2005), http://www.ietf.org/rfc/rfc4120
15. Selinger, P.: Towards a quantum programming language. Mathematical Structures in Computer Science 14, 527–586 (2004)
16. Kakutani, Y.: A logic for formal verification of quantum programs. In: Datta, A. (ed.) ASIAN 2009. LNCS, vol. 5913, pp. 79–93. Springer, Heidelberg (2009)
17. Kubota, T.: Formalization and Automation of Unconditional Security Proof of QKD. Master thesis, the University of Tokyo (February 2011)
18. Nagarajan, R., Papanikolaou, N., Bowen, G., Gay, S.: An Automated Analysis of the Security of Quantum Key Distribution. ArXiv Computer Science e-prints (February 2005)
19. http://www.prismmodelchecker.org/
20. Lo, H.-K.: Proof of unconditional security of six-state quantum key distribution scheme, http://arxiv.org/cits/quant-ph/0102138

Geometric Computations by Broadcasting Automata on the Integer Grid

Russell Martin, Thomas Nickson, and Igor Potapov

Department of Computer Science,
University of Liverpool, Ashton Building,
Ashton St, Liverpool L69 3BX, U.K
{Russell.Martin,T.Nickson,Potapov}@liverpool.ac.uk

Abstract. In this paper we introduce and apply a novel approach for self-organiz- ation, partitioning and pattern formation on the non-oriented grid environment. The method is based on the generation of nodal patterns in the environment via sequences of discrete waves. The power of the primitives is illustrated by giving solutions to two geometric problems using the broadcast automata model arranged in an integer grid (a square lattice) formation. In particular we show linear time algorithms for: the problem of finding the centre of a digital disk starting from any point on the border of the disc and the problem of electing a set of automata that form the inscribed square of such a digital disk.

1 Introduction

In many cases it is deemed that large numbers of simple robots can achieve certain tasks with greater efficiency than a single complex robot. This leads to the question of how to co-ordinate these robots in their task whilst retaining their simplicity of design, function and the robustness that is inherent to distributed systems. Problems currently being worked on are common subdivisions of more complicated and pragmatic tasks which are found to tackle a large set of problems simply through the coordinated use of such sub problems. These are often, but not restricted to, pattern formation, aggregation, chain formation, self assembly, coordinated movement, hole avoidance, foraging etc.

The problem of swarm (or a set of mobile robots) configuration into regular grids is well known for exploration tasks and environmental or habitat monitoring [5]. However, arranging robots into a regular grid structure has a number of other benefits in terms of self organization and efficient communication. In this paper we illustrate the possibility of employing the property of regular structure for complex geometric constructions via non-oriented broadcasting in the static cluster of robots (broadcasting automata).

We first define a model of swarms of robots arranged on the grid. Each robot is represented by a broadcasting automaton with finite memory that is unable to observe its neighbourhood, but can communicate through the non-oriented broadcasting of messages with its neighbours. As a model broadcasting automata has widely been used for designing communication protocols [1,4] and provides a

C.S. Calude et al. (Eds.): UC 2011, LNCS 6714, pp. 138–151, 2011.

realistic abstraction for network interaction.In the broadcasting automata model direct communication between automata is only possible via the broadcast of messages to all neighbouring automata within a certain communication range. In the case of the square lattice topology (as well as triangular and hexagonal lattices), non-oriented broadcasting can be used to efficiently solve a variety of geometric problems by utilizing the effects found in both real physical systems (i.e. waves and interference patterns) and computational systems (i.e. information processing by finite state automata).

The waves generated by activating processes in a digital environment can be used for designing a variety of wave algorithms. In fact, even very simple finite functions for the transformation and analysis of passing information provides more complex dynamics than classical wave effects. We generalize the notion of the standing wave which is a powerful tool for partitioning a cluster of robots on a non-oriented grid. In contrast to classical waves where interference patterns are generated by nodal lines (i.e. lines formed by points with constant values), an automata network can have more complex patterns which are generated by periodic sequences of states in time.

In this paper we introduce and apply a novel approach for self-organization, partitioning and pattern formation on the non-oriented grid environment based on the generation of nodal patterns in the environment via sequences of discrete waves. The power of the primitives are illustrated by giving solutions to two geometric problems: the problem of finding the centre of a digital disk starting from any point on the border of the disc and the problem of electing a set of automata that form the inscribed square of the same digital disk.

2 Broadcasting Automata Model on $\mathbb{Z}^2$

In general, a model of broadcasting automata is defined as a **network of finite automata** which is represented by a pair (G, Λ) where: G is a graph and Λ is a deterministic I/O automaton which is at each vertex, v, of the graph G.

The communication between automata is organized by message passing, where **messages** are from the alphabet Σ and are generated as the output symbols of the automata. The set of vertices connected to the automaton $a \in \Lambda$ at some vertex $v_a \in V$ is given by $\Gamma(a)$ and is the **set of neighbours** for that automaton, a. Messages generated by an automaton, $a \in \Lambda$, are passed to the automaton's **transmission neighbourhood**, which is a subset of its set of neighbours, $\Gamma_T(a) \subseteq \Gamma(a)$. Messages are generated and passed instantaneously at discrete time steps, resulting in synchronous steps. We will assume that if several messages are transmitted to a particular automaton a, it will receive only a set of unique messages, For any multiset of transmitting messages received by a in a single round, the information about the quantity of each type will be lost.

Let us consider two variants of broadcasting automata: **synchronous** and **asynchronous** (or reactive) models. In the case of the synchronous model, every automaton from the moment of activation, which occurs upon the receipt of a signal, follows discrete time steps and reacts on a set of received events (including an empty set, which corresponds to a no event situation). In the case of the asynchronous (reactive) model, every automaton follows discrete time steps but can only react to a non-empty set of events. Although such automata have no shared notion of time, a form of implied synchronicity may be derived from the constant amount of time taken for each round of communication. Both models are computationally equivalent, but the algorithms designed in each model may require different amount of resources such as the size of the messaging alphabet, number of broadcasts and the overall execution time.

Proposition 1. *Any algorithm for broadcasting automata in the reactive model with a message alphabet of size $|\Sigma|$ can be simulated by the synchronous model with a singular alphabet and $|\Sigma|$ slowdown. Any algorithm for broadcasting automata in the synchronous model with a message alphabet of size $|\Sigma|$ can be simulated by the reactive model with an alphabet $|\Sigma'|$ where $|\Sigma'| = |\Sigma| + 1$.*

Broadcasting automata on $\mathbb{Z}^2$. In this paper we consider a model of broadcasting automata on the non-oriented square lattice (integer grid). We note, however, a similar model can be defined on any other lattice or graph structure. On the square lattice it is possible to vary the transmission neighbourhood, Γ_T, by varying the transmission range. In particular, with a transmission radius equal to 1 or 1.5 (which covers those nodes at an exact distance of 1 and $\sqrt{2}$), the so-called **Von Neumann neighbourhood** and **Moore neighbourhoods** are generated, respectively.

In Figure 1 the source of the transmission is shown as the circle at the centre of the surrounding automata, those that are black are within the transmission range and thus in the transmission neighbourhood. The large outer circles represent the transmission range which can be changed to alter the automata that are included in the transmission neighbourhood. If the transmission radius is equal to 1, as in Figure 1 diagram i) then only four of the eight automata can be reached. If the radius is made slightly larger and is equal to 1.5, it can encompass all eight automata in its neighbourhood. As we will show later, iterative broadcasting within Von Neumann and Moore neighbourhoods can distribute messages in the form of a diamond wave and a square wave[1].

3   Computational Primitives for Broadcasting Automata

In this section we discuss a number of computational primitives that can be applied to non-oriented broadcasting automata on the grid in order to generate geometrical constructions. In the **synchronous** model waves are passed from automaton to automaton according to the following rules:

1) An automaton a receives a message from an activating source at a time t;

---

[1] It is also possible to show, in a more or less straightforward way, that broadcasting automata on $\mathbb{Z}^n$ (for any $n > 0$), with a single initial source of transmission, two radii of broadcasting (1 and 1.5) and a large alphabet of messages, can simulate a Turing Machine.

2) At a time $t + 1$, a sends a message to all automata within its transmission neighbourhood $\Gamma_T(a)$;

3) At a time $t + 2$, a ignores all incoming messages for this round.

In the **asynchronous model** simulation we require at least 3 symbols $\{u_0, u_1, u_2\}$:

1) An automaton a receives a message u_i from an activating source;

2) The automaton a broadcasts a message $u_{(i+1) \mod 3}$ to all automata within its transmission neighbourhood $\Gamma_T(a)$;

3) Ignore next incoming message $u_{(i+2) \mod 3}$.

Step 3 in each model prevents a node from receiving back the wave that it just passed to its neighbours by ignoring all transmissions received the round after transmission and ensuring that the front of the wave is always carried away from the source of transmission. Waves are passed using two different transmission neighbourhoods referred to as square and diamond waves, which are equivalent to Moore and Von Neumann neighbourhoods respectively, see Figure 1. Whilst within the model an automaton is unable to directly access

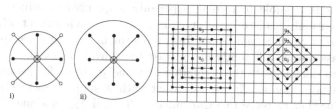

Fig. 1. Diagram i) represents the propagation pattern for a diamond wave (Von Neumann neighbourhood) and diagram ii) shows the propagation pattern for a square wave (Moore neighbourhood). Also wave propagation with the asynchronous model is shown on the square grid: Moore neighbourhood (left) and Von Neumann neighbourhood (right).

the state of its neighbours it is still possible to design many primitives in a similar style as designed for the cellular automata model [6], including synchronization procedures, finding the edge elements on a line (Lemma 1), shortest branch of a tree (Lemma 2), etc. These basic tools will be utilized to illustrate more complex algorithmic methods in Section 4.

Lemma 1. *Given a line of broadcast automata of size n and a single source A at a point on the line it is possible to elect the automaton which is on an edge of the line in $n + 3$ time steps, where k is the number of automata from the initiating automaton to its closest side, and with an alphabet, Σ, of size $|\Sigma| = 3$, in the asynchronous model.*

Lemma 2. *It is possible to elect the end point of the shortest branch of a tree in time $3R$, where R is the length of the shortest branch and the alphabet required is $|\Sigma| = 2$.*

3.1   Nodal Patterns of Discrete Interference

In physics, a standing wave is a wave that remains in a constant position. This phenomenon can occur in a stationary medium as a result of interference between two waves travelling in opposite directions. A standing wave in a transmission line is a wave in which the distribution of current value is formed by the super-position of two waves propagating in opposite directions. The effect is a series of nodes (zero displacement) and anti-nodes (maximum displacement) at fixed points along the transmission line. With standing waves in a two dimensional environment the nodes become nodal lines, lines on the surface at which there is no movement, that separate regions vibrating with opposite phase [7].

In this paper we investigate standing waves in a discrete environment, where the original physical system is generalized in several ways. First, the transmitted waves will be discrete, i.e. the values of the wave which are passing a point p will correspond to a sequence of symbols that are observed in p. Assuming that a point in discrete environment has even simple computational power, like finite memory, the idea of nodal points and nodal lines can be extended. Apart from recognizing points with classical standing waves (i.e. having a constant value over time), it is possible to recognize periodic sequences of values over time. This will eventually lead to a richer computational environment where nodal lines of different periodic values can form patterns in $\mathbb{Z}^n$ and where points may react and pass a wave differently in contrast to a physical model.

In order to demonstrate the technique of nodal patterns for geometric com-putations in non-oriented $\mathbb{Z}^2$ we will start with a simpler abstraction in one dimension line. Let us place two transmitters, T_1 and T_2, on a one dimensional line, each broadcasting words u^*, v^* respectively, where $u, v \in \Sigma$, $|u| = |v| = s$ is the length of the word. The **broadcasting** of symbols begins at transmission points with symbols broadcast away from the source at each time step, cycling through all symbols in the word. We now have two infinite words $(u)^*$ and $(v^R)^*$, the reverse of v, that are shifted towards each other every next time step. In this case any point on the line, which contains two symbols from u and v, forms a pair $p_{i,j} = (u_i, v_j)$. So once a point l contains two symbols, we can define a sequence of pairs $p_{(i+t) \mod s,(j+t) \mod s}$ contained in l over discrete time $t = 1, \ldots, \infty$. It is easy to see that such sequence will be periodic with a period less or equal to s, where $|u| = |v| = s$, and it represents a history of symbol pairs from u and v which are meeting at the point l over time.

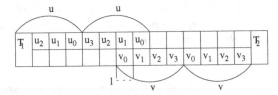

Fig. 2. Shows two transmitters T_1 and T_2 broadcasting words u and v respectively

The **nodal pattern** of a point l is a finite subsequence of s pairs $p_{(i+t) \mod s}$, $(j+t) \mod s$ over some $t = t'..t' + s$. Let us also assume that nodal patterns are equivalent up to a cyclic shift, so it will not be dependent to the initial time t'. Nodal patterns are labelled using the difference between the indices defined as $P_{|i-j|}$, where i and j are the indices from any two of the pairings of symbols $p_{i,j}$ for some point on the plane at some time. Nodal patterns have s possible labels, $P_0, P_1, \ldots, P_{s-1}$, one for each of the possible pairings. Such indices are now defined as the **node index** of the automaton, which signify the distinct pairing observed by the automata.

Nodal patterns are exemplified in Figure 2 where $s = 4$, $u = u_0, u_1, u_2, u_3$ and $v = v_0, v_1, v_2, v_3$. At point l the pairs $((u_1, v_0), (u_2, v_1), (u_3, v_2), (u_0, v_3))$ form a nodal point corresponding to $p_{(1+t) \mod s, (0+t) \mod s}$ over time t. The nodal pattern is now labelled $P_{|(1+t)-(0+t)|} = P_{|1+t-0-t|} = P_{|1-0|} = P_1$.

Patterns p_1 and p_2 are called **distinct** if p_1 cannot be constructed from p_2 by applying any cyclic shift to it. Where two patterns are formed by ordered pairs and a word $u = v$ is without repetitive symbols we have s distinct nodal patterns. In the case of unordered pairs such that $(u_i, v_j) = (v_j, u_i)$ the number of distinct nodal patterns is smaller and defines a specific sequence described in the Propositions 2, 3.

Proposition 2. *Given a non-periodic word u, $|u| = s$. The number of possible distinct nodal patterns generated by broadcasting words u and v, where $v = u$, is $s/2 + 1$ if s is even and $(s + 1)/2$ is s is odd.*

Example 1. Given a sequence $u = v = 1, 0, -1, 0$ and so $|u| = |v| = s = 4$ it can be shown that there are $(s/2) + 1 = (4/2) + 1 = 3$ distinct nodal points. Enumerating all possible patterns (modulo removed for clarity) and where t is over time.

$$P_0 = ((1, 1), (0, 0), (-1, -1), (0, 0)); \quad P_1 = ((1, 0), (0, -1), (-1, 0), (0, 1))$$
$$P_2 = ((1, -1), (0, 0), (-1, 1), (0, 0)); \quad P_3 = ((1, 0), (0, 1), (-1, 0), (0, -1)).$$

Patterns P_1 and P_3 are the same assuming that the order of the symbols is not relevant.

Proposition 3. *In the sequence of s nodal patterns $P_0, P_1, \ldots, P_{s-1}$ derived from words $u = v$ where $|u| = |v| = s$, two nodal patterns P_k and $P_{s-k \mod s}$ are equivalent up to a cyclic shift.*

3.2   Nodal Patterns with Two Sequential Transmitters

Standing waves and nodal patterns may be formed when u^* ($|u| = s$) is transmitted from one source T_0 and is retransmitted from another source T_1 after reaching it. Any point of a grid having both signals from T_0 and T_1 will have identified its nodal pattern as the "difference" between them.

Upon moving the waves to two dimensions, $\mathbb{Z}^2$, it is important to define the distance between the points in terms of (square or diamond) wave propagation for proper calculation of the nodal patterns, as it is not the same as in Cartesian geometry.

Definition 1. *Square waves adhere to the distance function d_s for distances from x to y where $d_s : \mathbb{Z}^n \times \mathbb{Z}^n \to \mathbb{Z}$ and is defined by the correspondence $d_s((x_1, \ldots, x_n),$*
$(y_1, \ldots, y_n)) = \max_{1 \leq i \leq n}\{|x_i - y_i|\}, (x_1, \ldots, x_n), (y_1, \ldots, y_n) \in \mathbb{Z}^n.$

Definition 2. *Diamond waves adhere to the distance function d_d for distances from x to y where $d_d : \mathbb{Z}^n \times \mathbb{Z}^n \to \mathbb{Z}$ and is defined by the correspondence $d_d((x_1, x_2, \ldots, x_n),$*
$(y_1, y_2, \ldots, y_n)) = \sum_{i=1}^{n} |x_i - y_i|, (x_1, x_2, \ldots, x_n), (y_1, y_2, \ldots, y_n) \in \mathbb{Z}^n.$

Informally speaking, the distance functions defined above provide the time taken for the diamond/square wave front to reach an automata at coordinate $(y_1, y_2, \ldots, y_n)$ from a transmitter at $(x_1, x_2, \ldots, x_n)$.

Let us consider two ways of computing nodal patterns for this process in terms of synchronous and asynchronous cases with transmitters T_0 and T_1. In both cases we aim to get the same distribution of nodal patterns on $\mathbb{Z}^2$ without the need to continuously propagate values from the sources. The local distribution of these patterns will allow us later to locate points in a non-oriented environment.

Definition 3. *An index i of a nodal pattern P_i in a point $\rho \in \mathbb{Z}^2$ is defined as the absolute difference (modulo s) between the shortest time when ρ is reached from T_0 and the sum of times of reaching T_1 from T_0 and ρ from T_1:*

$$i = |d(T_0, \rho) - (d(T_0, T_1) + d(T_1, \rho))| \mod s$$

where d is understood as d_s or d_d for square or diamond wave respectively.

Synchronous model with a single message. In case of transmitting a single message, waves here are activation waves, which the automata use to start internal clocks. Transmissions begin from T_0 where the activation wave propagates through the use of a square (or diamond) wave arriving at T_1 which is activated when reached by the first wave after a constant delay s. Any point that has received the first signal will start its internal clock (which counts modulo s), and then after receipt of the second signal the clock is stopped and the value of clocks corresponds to the index of the nodal pattern for this point which is $|(d(T_0, \rho) \mod s) - ((d(T_0, T_1) \mod s) + (d(T_1, \rho) \mod s))|$.

Asynchronous model with multiple messages. In the case of the asynchronous model, the same distribution of nodal patterns can be simulated by sending a wave from T_0 where, on the wave front, every point that receives a symbol u_i immediately transmits the symbol $u_{(i+1) \mod s}$. The pseudo-synchronization of wave propagation is achieved by assuming that every transmission takes the same constant time. Then transmitter T_1 operates in the same way once reached by u_i, transmitting the next symbol corresponding to u_{i+1} but using a different alphabet $\{v_1, \ldots, v_s\}$ to avoid problems whereby transmitting in the same alphabet could have a blocking effect on the wave. Each node should now contain a pair of symbols $(u_{i'}, v_{i''})$ which is enough to define the pattern $P_{|i'-i''|}$, where $i' = d(T_0, \rho) \mod s$ and $i'' = (d(T_0, T_1) + d(T_1, \rho)) \mod s$.

Theorem 1. *Let T_0 and T_1 be any two points in $\mathbb{Z}^2$ with coordinates (j_0, k_0) and (j_1, k_1), respectively. Assume that nodal patterns P_i were formed by square waves, i.e. $i = |d_s(T_0, \rho) - (d_s(T_0, T_1) + d_s(T_1, \rho))| \mod s$. For any point $\rho \in \mathbb{Z}^2$ the membership to one of the following sets $\{D_1, D_3, D_5, D_7\}$, $\{D_2, D_4, D_6, D_8\}$ or $\{D_9\}$ is uniquely identified by a number of distinct nodal patterns P_i in the Moore neighbourhood of ρ: $\{D_9\}$ — has two nodal patterns, $\{D_1, D_3, D_5, D_7\}$ — has one nodal pattern, $\{D_2, D_4, D_6, D_8\}$ — has three nodal patterns, see Figure 3:*

$$D_1 = \{(x,y)|y \geq -x + (k_1 + j_1), y \geq x + (k_0 - j_0)\}$$
$$D_2 = \{(x,y)|y \geq -x + (k_1 + j_1), y \leq x + (k_0 - j_0), y \geq x + (k_1 - j_1)\}$$
$$D_3 = \{(x,y)|y \leq x + (k_2 - j_2), y \geq -x + (k_2 + j_2)\}$$
$$D_4 = \{(x,y)|y \leq -x + (k_1 + j_1), y \leq x + (k_0 - j_0), y \geq x + (k_0 + j_0)\}$$
$$D_5 = \{(x,y)|y \leq -x + (k_0 + j_0), y \leq x + (k_1 - j_1)\}$$
$$D_6 = \{(x,y)|y \leq x + (k_0 - j_0), y \leq -x + (k_0 + j_0), y \geq x + (k_1 - j_1)\}$$
$$D_7 = \{(x,y)|y \geq x + (k_0 - j_0), y \leq -x + (k_0 + j_0)\}$$
$$D_8 = \{(x,y)|y \geq x + (k_0 - j_0, y \leq -x + (k_1 + j_1), y \geq -x + (k_0 + j_0)\}$$
$$D_9 = \{(x,y)|y \leq x + (k_0 - j_0, y \geq -x + (k_0 + j_0), y \leq -x + (k_1 + j_1), y \geq x + (k_1 - j_1)\}.$$

Proof (Sketch.). In a system with two transmitters, T_0 and T_1, nodal patterns are formed by broadcasting the square wave from T_0 which, once reached by the wave, will then be broadcast by T_1. The main observation is that broadcasting a square wave generates quadrants defined by the lines $x = y$ and $x = -y$, assuming that the transmitter is the origin, whereby within each quadrant the front of the wave expands such that each element of the orthogonal axis, to the one wholly contained by the quadrant (ie x,-x,y,-y), within the quadrant will contain the same member of the alphabet as all the others. The direction of the waves dictate the structure of the neighbourhoods for the automata. Then by computing an index i of a nodal pattern P_i at each point ρ: $i = |d_s(T_0, \rho)$ mod $s - (d_s(T_0, T_1)$ mod $s + d_s(T_1, \rho)$ mod $s)|$ it is easy to observe that any point ρ in $D_{\{1,3,5,7\}}$ is surrounded by points with identical nodal pattern. Each area $D_{\{1,3,5,7\}}$ share common direction for both the waves and all waves share a common speed. So the difference between the first and second wave is constant. Hence, any region with arrows that move in the same direction will contain a single nodal pattern.

Those in areas D_9 are waves which are both heading towards each other. This causes the intersection of the waves and their alphabets to only occur in differences of two. If one set of symbols match and generate a nodal pattern (i.e. (u_i, v_j)) then it will not be the next set of symbols that match (i.e. (u_i+1, v_j+1)) but every other pair (i.e. $(u_i + 2, v_j + 2)$). This generates the pattern of every other nodal pattern being present.

Those in areas $D_{\{2,4,6,7\}}$ have waves that move in orthogonal directions. This leads to diagonal patterns of the same node. If one were to track the intersections of the lines $x = x'$ and $y = y'$ as y' and x' increase, $0 \leq x', y' \leq n$, the result would be a diagonal line of the form $x = y$. The very same happens but with a multitude of the lines represented by the fronts of the wave.

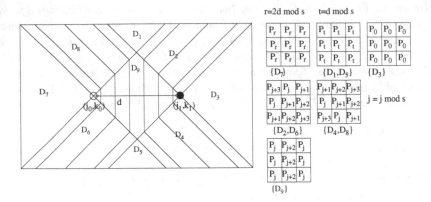

Fig. 3. The distribution of three distinct nodal patterns in case of two "transmitter", where $s = 4$

Partial Vision of the Neighbourhood. In the current model each element is only aware of its own node index, and information about its neighbours is not directly observable. However, it is possible to organize a partial vision which provides access to a set of distinct values of nodes from the Moore and Von Neumann neighbourhoods via non-oriented broadcast. In order to get this set of distinct values an automaton should send a special message (nodal request) to all neighbourhoods which, in turn, will send their nodal indices on the next step. Alternatively it is possible to initiate an iterative process of such requests through the whole network of broadcasting automata on $\mathbb{Z}^2$ where each next automata will react on this information by transmitting its own nodal value. This type of transmission we call as a **node wave**.

It is possible to test membership to one of the partitions (defined in Theorem 1) generated through waves broadcast by two transmitters. Having partitioned the plane with two transmitters, T_0 and T_1, T_1 waits a constant delay s before broadcasting a nodal wave which allows all automata information about its neighbourhood. After these steps the automaton has enough information to decide to which of the partitions it belongs.

Apart from the above partitioning it is possible to elect vertical, horizontal (set L_0) and diagonal lines (set L_1) starting from a single source by transmitting square and diamond waves after a constant delay.

Proposition 4. *Given a vertex, $T_0 = (j, k)$ on $\mathbb{Z}^2$, it is possible to elect a set of points $L_0 = \{(x, y) | y = k, x \in \mathbb{Z}\} \cup \{(x, y) | x = j, y \in \mathbb{Z}\}$ or the set $L_1 = \{(x, y) | y = x + k - j \in \mathbb{Z}\} \cup \{(x, y) | x = -y + k + j \in \mathbb{Z}\}$ in linear time.*

4   Geometric Problems on the Digital Disc

A set ζ of points in $\mathbb{Z}^2$ is a **digital disk** if there exists a Euclidean circle, with a centre at an integral point, that encloses the pixels of ζ but excludes its complement. Let us consider a model of broadcasting automata on a digital disk which has a diameter D. We define a procedure for finding the centre of the digital disk in linear time using the notion of waves as described in previous sections. It is possible to find the centre of the digital disk as a single point or as a set of two points, depending on whether the radius of the digital disk is odd or even, respectively. In this section we abbreviate the partitions previously mentioned to $a = \{D_9\}$, $b = \{D_1, D_3, D_5, D_7\}$ and $c = \{D_2, D_4, D_6, D_8\}$. The algorithm for finding the centre can begin from any arbitrary point, T_1, on the edge of the digital disc and is implementable in both asynchronous and synchronous models. Depending on the location of the initial point, one of three algorithms is applicable. Finding the correct algorithm to apply is reduced to checking the initial point's neighbourhood to one of three possible sets in the following way.

Definition 4. *Eight points* $\{0, 1, ..., 7\}$ *on the circumference of a digital circle,* ζ *corresponds to the following eight angles 0,45,90,135,180,225,270,315.*

Lemma 3. *Given an automaton on the edge of* ζ *it is possible to check the automaton's membership to one of three sets:* $\{0, 2, 4, 6\}$, $\{1, 3, 5, 7\}$ *and all other points on the edge of the circle in a time* $O(D)$ *for both models.*

Proof. From automaton T_1, on the edge of ζ, Proposition 4 is applied via the transmission of a square then diamond wave, resulting in horizontal, vertical lines of elected automata and electing at most two paths. All elected automata on the edge of the disc, ζ, become new points T_N which is a set of points encompassing up to three automata, $\{T_2, T_3, T_4\} \in T_N$. As soon as the automaton or automata on the edge of the disc, ζ, have been elected by T_1, the automata, now denoted as T_N, begin transmission of a square wave. As the transmission of these waves from all automata in T_N, may occur simultaneously on the disc, points at which waves meet each other proceed no further on the disc, due to the automata's inability to receive and broadcast at the same time, cancelling each other. Points of wave cancellation are shown as dotted lines in Figure 4. The partitions formed by the transmissions from new points in T_N can now be detected by the initial point T_1 through the transmission of a neighbourhood detection wave which gives nodal patterns to automata through the transmission of its own square wave and causing neighbouring automata to transmit their states which allows the detection of T_1's neighbouring nodal patterns. By Theorem 1, possible neighbourhood partitions for the initial point T_1 are now be categorised as $\{a\}, \{a, c\}$ and $\{a, b, c\}$ which are the points $\{0, 2, 4, 6\}$ and $\{1, 3, 5, 7\}$ and all other points respectively. The procedure requires only three waves of transmissions, each wave require the time that is no more then the diameter of the circle as well as some constant time between transmissions and the constant time for the neighbourhood recognition.

An Algorithm for Locating the centre of a Digital Disk

1) An automaton on the edge of the disc ζ, T_1, checks its location by the creation

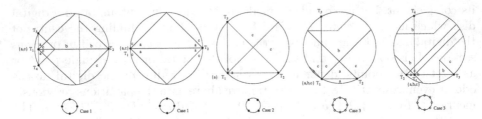

Fig. 4. The above Figure shows the three possible cases stemming from the five possible variants that require differing solutions based on their location. The two differing sets of the 8 points and those points which lay in none of these. The two diagrams for Case 1 correspond to the situations whereby, for case 1, T_1 generates a 3-branched tree with two equidistant branches or a single chord respectively. Whereas for case 3 the situations are those such that T_1 generates a 2- or 3-branched tree respectively.

of the unique local neighbourhood sets: $\{a\}$, $\{a,c\}$ or $\{a,b,c\}$, see Lemma 3.

2) In case of neighbourhood set $\{a,c\}$ apply the algorithm for case 1. In case of neighbourhood set $\{a\}$ apply algorithm case 2.

3) In the case of neighbourhood set $\{a,b,c\}$, T_1 is the root of a tree with two or three branches. The third branch may appear if the automata, T_1, finds itself on a 'ledge', such that there are automata on three sides of its Von Neumann neighbourhood, formed from the digitization of the circle. The end point of the shortest branch of such tree, placing the automata in a position whereby there are only two automata in its Von Neumann neighbourhood, is found by Lemma 2 which is relabelled T_1 and then apply case 3.

Cases 1 and 2 are basic because the location of the point T_1 is known exactly. The least trivial case is Case 3 where further partitioning is required for locating the centre.

Algorithm for Case 1 (set $\{0,2,4,6\}$)

1) T_1 sends message m_0 to T_2 through the chord which will be sent back from T_2 after some constant delay $k = |\Sigma|$.

2) T_1 sends message m_1 which has a delay of 3 after a constant delay $k = |\Sigma|$.

3) The automata on the chord elected through receipt of both message m_0 and m_1 at the same time will be the centre of the digital disc ζ.

Algorithm for Case 2 (set $\{1,3,5,7\}$)

1) A new point T_4 is generated along the diagonal through the use of diamond neighbourhood detection wave as described in Proposition 4.

2) T_1 sends message m_0 to T_2 through the chord which will be sent back from T_2 after some constant delay $k = |\Sigma|$.

3) T_1 sends message m_1 which has a delay of 3 after a constant delay $k = |\Sigma|$.

4) The automata on the chord elected, through receipt of both message m_0 and m_1 at the same time, will be the centre of the digital disc ζ.

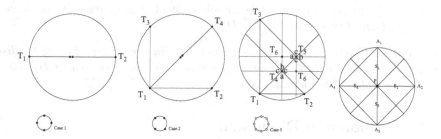

Fig. 5. Constructions required to find the centre of the circle for the three differing cases. Centres are indicated by black dots. Figure on the right corresponds to production of the inscribed square from centre point.

Algorithm for Case 3

1) Point T_2 is identified as the shortest chord of the tree constructed from T_1 via Lemma 2.

2) Point T_3 is identified as the longest chord of the tree constructed from T_1 by broadcasting a signal c only to those automata that have received two a's but no b's after Lemma 2, the longest chord.

3) Transmissions from T_3 followed by T_1 elect the point T_5, the only point on the digital disc that has a neighbourhood containing the nodal patterns $\{a, b, c\}$.

4) Transmissions from T_2 followed by T_1 elect the point T_4, the only point on the digital disc that has a neighbourhood containing the nodal patterns $\{a, b, c\}$.

5) Application of proposition 4 from T_4 and T_5 will generate horizontal and vertical lines. The points at which these two line cross are elected as T_6. Only one of the two points marked T_6 will be in partition a given by the initial construction of T_1 and T_3, this is the centre point of the circle.

Theorem 2. *It is possible to find the centre of a digital disc ζ with diameter D starting from a point on the circumference of ζ, T_0, in both models, in $O(D)$ time* [2].

Theorem 3 directly follows from the construction of the next algorithm and also holds when starting from a point is on the circumference of ζ.

Algorithm for electing elements of the inscribed square:

1) From point P transmit two waves (square and diamond) to elect the four points A_1, A_2, A_3 and A_4 (see Figure 5).

2) Transmit a square wave from points A_1, A_2, A_3 and A_4. The interference pattern from these waves will form four squares, shown as S_1, S_2, S_3, S_4 which will be of the distinct type D_9 according to Theorem 1.

3) A node wave must now be sent from P which informs the automata of their neighbours and allows the automaton to place itself in the set D_9 if it is within

---

[2] As there is a finite number of passing of waves which all time-bounded by at most $O(D)$, the algorithm cannot exceed linear growth by D.

the inscribed square. Note that the complement of the square will also have a distinct pattern of type $\{D_2, D_4, D_6, D_8\}$.

Theorem 3. *Given an initial transmitting node in the centre of a digital disc. It is possible to elect automata forming an inscribed square in time $O(D)$, where D is the diameter of the digital disk.*

5   Conclusion and Discussion

We have shown that non-oriented broadcasting of messages on the square grid can form stable interference patterns. Such patterns can be used for efficient partitioning, self-location problems and geometric computations on the static cluster (of robots or automata) via transmission of discrete square and diamond waves, where shapes are defined by the radius of broadcasting and the topology of the grid structures. These shapes can be much more complex then square and diamond waves when we can choose larger radii, higher dimensions and other structures of grid topology. For example, broadcasting with a radius three on the square lattice resulting in the octagon shape of the wave which also, in its turn, provides more complex partitioning of the lattice. Based on our ongoing experimental and theoretical work with lager radii we have observed that they can form quite complex shapes and patterns for square, triangular and hexagon grids as well for high dimensional structures.

Moreover, the sequence of the string that is transmitted by a wave can also be increased in complexity (i.e. different periodic or non-periodic strings, numerical sequences, etc.) along with the corresponding aggregation function. For example, in the case of natural wave intersection the aggregation function is simply the addition of amplitudes which we only extended in this paper through access to a finite history.

The proposed algorithms based on digital waves and nodal patterns can also be extended by using iterative application of informational waves on already generated nodal pattern. So in each subsequent round waves can be transmitted with different delays, speed and shapes based on already formed pattern. Currently we are exploring many other generalization and constraints of the proposed approach which will be presented in future publications by the authors.

The current paper extends the existing area of wave algorithms by introducing new methods, framework and models for further theoretical analysis and practical implementation of complex self-orientation and self-organization mechanisms.

References

1. Alderighi, S., Mazzei, T.: Broadcast Automata: a Parallel Scalable Architecture for Prototypal Embedded Processors for Space Applications. HICSS (5), 208–217 (1997)
2. Czyzowicz, J., Gąsieniec, L., Pelc, A.: Gathering Few Fat Robots in the Plane. In: Shvartsman, M.M.A.A. (ed.) OPODIS 2006. LNCS, vol. 4305, pp. 350–364. Springer, Heidelberg (2006)

3. Peleg, D., Efrima, A.: Distributed algorithms for partitioning a swarm of autonomous mobile robots. Theoretical Computer Science 410(14), 1355–1368 (2009)
4. Hendriks, M.: Model Checking the Time to Reach Agreement. In: Pettersson, P., Yi, W. (eds.) FORMATS 2005. LNCS, vol. 3829, pp. 98–111. Springer, Heidelberg (2005)
5. Lee, G., Chong, N.: A geometric approach to deploying robot swarms. Annals of Mathematics and Artificial Intelligence, 257–280 (2008)
6. Kari, J.: Theory of cellular automata: a survey. Theor. Comput. Sci. 334 (2005)
7. Standing Waves (January 2010), http://en.wikipedia.org/wiki/Standing_wave

Computations with Uncertain Time Constraints: Effects on Parallelism and Universality

Naya Nagy and Selim G. Akl*

School of Computing, Queen's University,
Kingston, Ontario, Canada K7L 3N6
{nagy,akl}@cs.queensu.ca
http://www.cs.queensu.ca/home/akl

Abstract. We introduce the class of problems with uncertain time constraints. The first type of time constraints refers to uncertain time requirements on the input, that is, *when* and *for how long* are the input data available. A second type of time constraints refers to uncertain *deadlines* for tasks. Our main objective is to exhibit computational problems in which it is very difficult to find out (read 'compute') *what* to do and *when* to do it. Furthermore, problems with uncertain time constraints, as described here, prove once more that it is impossible to define a 'universal computer', that is, a computer able to compute all computable functions. Finally, one of the contributions of this paper is to promote the study of a topic, conspicuously absent to date from theoretical computer science, namely, the role of physical time and physical space in computation. The focus of our work is to analyze the effect of external natural phenomena on the various components of a computational process, namely, the *input phase*, the *calculation phase* (including the *algorithm* and the computing *agents* themselves), and the *output phase*.

Keywords: real-time computation, unconventional computation, Turing Machine, universal computer, parallel computing, physical time, physical space, external phenomena.

1   Introduction

Time plays an important role in real-life computations. In some applications, the data to be processed have to be collected at specific time intervals around the clock. In other cases, a computational task must meet a given deadline.

This paper presents two problems that exhibit time constraints. The first problem has a time constraint on the input of data, whereas the second problem has a constraint on the output. Each problem is solved on both a sequential machine and a parallel machine. Computing the time constraints in each case is computationally demanding. As such, the parallel machine can use its additional power to compute the time constraints on time and to monitor the environment

---

* This research was supported by the Natural Sciences and Engineering Research Council of Canada.

C.S. Calude et al. (Eds.): UC 2011, LNCS 6714, pp. 152–163, 2011.

exhaustively. The sequential machine performs worse than the parallel machine and therefore fails where the parallel machine succeeds.

The problems presented here define a new paradigm. They are easy to solve when it is known what to do and when to do it. Yet, computing what to do and when to do it is what presents the main challenge.

One of the consequences of our work is to demonstrate, yet again, that the concept of a universal computer cannot be realized. No computer, fixed a priori, can successfully perform the computations described in this paper, even if given an unbounded amount of memory and time. This result on the non-universality in computation, aside from exposing a fundamental principle, opens a new area of investigation into the role of time and space in computing. We note here that in conventional computing, both time and space are: (i) internal to the computer (time is *running time*, space is *memory space*, neither is related to the outside world, and both are valid only when performing a computation), (ii) static (time depends on the algorithm while nothing depends on time, and placement of data in memory does not affect the results), (iii) unbounded (at least in theory), (iv) entirely under the control of the user (a better algorithm improves time and/or space requirements, and often there is a time-space tradeoff), and (v) solely used in performance analysis.

In unconventional computing, by contrast, time and space take a new significance. As the examples in this paper (among many others [2], [3], [4], [5]) show, physical time and physical space in the environment external to the computer play a crucial role in the definition and requirements of the computational problems, and ultimately in the success or failure of the computation. The present paper proposes to begin an exploration of the role of physical time and physical space in computing and the properties of computational problems that depend on them. This will introduce new paradigms to the theory of computation and promote the study of problems whose importance can only grow with the widespread and diverse uses of computers. Preliminary results on this topic were presented in [6]. There it was shown that "computational ubiquity" is required in several important computations.

We emphasize here that there exist many computational paradigms that have the appearance of being in some way or another related to *time*. These include, for example, on-line computing, π-calculus, trace theory, streaming algorithms, stochastic scheduling, neural networks, complex systems, time-relativistic computation, hyperbolic cellular automata, adaptive algorithms, learning algorithms, genetic algorithms, evolutionary algorithms, emergent algorithms, reconfigurable circuits, and of course, real-time algorithms. The crucial observation here is that 'time' in *all* of these paradigms is *computer time*, that is, time internal to the machine currently performing the computation. In *none* of these paradigms does the computer have any awareness of true physical time in its environment, or of the physical space surrounding it, and therefore *none* of the external physical phenomena taking place during the computation has (or is supposed to have) any direct influence on the variables, the algorithm, the machine itself, or its output.

2 Dynamic Time Constraints

A task scheduler is responsible for scheduling some arbitrary set of tasks in an efficient way. Here, a *task* is meant to be some computation or program. The scheduler has to decide on the order in which the tasks are to be executed, what resources are to be allocated to a specific task, and when a task is to start or to finish. Depending on the characteristics of the task set, the scheduler may work very differently, from one scheduling problem to another.

In particular, *static tasks* are tasks defined at the outset, before any computation or task execution starts. The scheduler has full knowledge of these tasks well in advance. By contrast, *dynamic tasks* arrive to the scheduler during the computation in an unpredictable way.

The time requirements of a task are defined as its deadline, or as the time constraints on the input of data, or in general as any time constraint concerning the task's connection to the outside world. *Static time requirements* on a task are requirements that are well defined before the task starts executing. If the time constraints of a task are defined during the task's execution, or are computed by the task itself, then they are called *dynamic time requirements*.

To date, in all problems with time constraints appearing in the literature [9], [10], the time constraint is defined outside of the computation. When a task arrives, it has all time constraints fully defined already. It is *known* when and how to acquire input and what time constraints apply on the output, or on completion of the task.

The dynamic aspect of a scheduling problem lies solely in whether the tasks are fully known to the scheduler before the computation even starts, or whether tasks are created and destroyed during the computation. Thus, the scheduler is faced with four (task, time requirements) scenarios.

The **first scenario** is defined by static tasks and static time requirements. Thus, the number of tasks to be scheduled is known prior to the computation. Also, the characteristics of the tasks are given at the outset. Time constraints are defined once and for all, and do not depend on, or vary over time, nor are they influenced by the execution of a task or by the scheduler itself. Descriptions of how to schedule static tasks are given in [9].

The **second scenario** is defined by dynamic tasks and static time requirements. Dynamic systems exhibiting dynamic scheduling problems are treated in [10]. In this case, the tasks and their characteristics are not fully known at the beginning of the computation. Tasks are generated and destroyed during the computation; they arrive to the scheduler in an unpredictable way and have to be scheduled on the fly. Still, the task itself is fully defined at its generation. A new task, when taken into the scheduler, comes with its own clear requirements of resources, clear requirements on the input data and specifically with a well defined deadline (hard or soft).

The **third scenario** is defined by static tasks and dynamic requirements. The algorithms addressed in this paper fall into this category. The problem here is even more interesting: The task to be performed has uncertain characteristics at the outset. The time requirements of the task are not defined at the generation of the task. Time constraints on the acquisition of data and/or deadlines will be defined during the execution of the task. Of special interest is the case when considerable computational effort is required to establish the time for an input operation, or the deadline of an output. In this situation, the power of the computational device becomes crucial in order to learn which computational step, necessary for the task to complete, is to be performed, and when to perform it.

Let us consider a task with no clearly defined deadline. Its deadline will be computed while executing the task itself. In this case, the deadline will be known at some point during the computation and may fall into one of the situations: (i) The deadline is far into the future and the task can be executed without taking its deadline into consideration; (ii) The deadline is close, but the task can still be executed; (iii) The deadline is so close that it is impossible for the task to be completed; (iv) The deadline has already passed. In some situations, the computation reaches an unexpected state: Completion was unsuccessful because the computation did not know the deadline. Alternatively, the main computation was easy to perform, but the time constraints were too difficult to compute.

The **fourth scenario** is defined by dynamic tasks and dynamic time requirements. This category has not been studied yet, as it has many variable characteristics and is therefore difficult to formalize.

3   Background

In this section, we define the models of computation to be used in our analyses, and review a recent result on non-universality in computation for which further evidence is provided in this paper.

Sequential Model. The sequential computational model used in this paper is the standard Random Access Machine (RAM). A RAM consists of a processor able to perform computations. Also, the processor has access to a memory and can read or write any arbitrary memory location. The memory is connected to the outside world for receiving input and for returning output.

Parallel Model. The parallel algorithms described in the next sections are meant to run on a Parallel Random Access Machine (PRAM). The PRAM comprises a set of n processors that access a common memory. Each processor is identical to the processor of the RAM. Several processors can read from the same memory location, an operation referred to as a *concurrent read*. Also, several processors may write into the same memory location, thereby executing a *concurrent write*. The convention adopted in this paper for the concurrent write is that the values, simultaneously written in a memory location, are added together and their sum is stored in that memory location.

Time Unit. All analyses in this paper are based on using a time unit to measure the running time of an algorithm. We define a time unit as the time taken by the RAM to execute a step in a computation. An algorithmic step, in turn, is defined as consisting of three phases: reading a fixed-size input, performing a basic calculation (that is, an elementary arithmetic or logical operation, such as addition or logical AND) on one or two data elements, and finally producing a fixed-size output.

Furthermore, in order to avoid any possible confusion, we make two explicit assumptions. First, the RAM is assumed to operate at the speed of light in vacuum; in other words, there does not exist a sequential model of computation faster than the RAM. Second, each processor on the PRAM is a RAM operating at the speed of light in vacuum; in other words, there does not exist a model of computation faster than the PRAM. These two assumptions allow us to avoid comments of the type: "Why not use a faster sequential computer that outperforms the parallel computer?".

A Time-Aware Model of Computation. We introduce a new model of computation that takes into consideration the time at which an operation is executed. It will allow us to state computational problems where external physical phenomena impose time constraints on the input and on the output. The new time-aware model of computation is essentially a RAM defined by the quadruple $U = (C, O, M, P)$ where C is a sequence representing a clock that keeps track of elapsed physical time, O is a set of elementary operations, M is a memory of unbounded size, and P is a processor capable of executing the elementary operations. Details are omitted here due to space limitations.

The new model of computation is intended to fill both a theoretical and a practical void. The need for such a model will become increasingly important for computations that are increasingly unconventional [4]. In this paper, we show how the model allows us to study problems with uncertain time requirements. Here, time requirements depend on events or computations that occur *during* the execution of a task. Henceforth we use the new model in one of its two incarnations, namely, the time-aware RAM and the time-aware PRAM.

Non-Universality. One of the purposes of this paper is to provide further examples in support of the previously established theoretical result whereby a universal model of computation does not exist [2]–[6]. So, what is a 'universal computer'? The simplest and most appropriate definition is that a 'universal computer' is one capable of performing in finite time, through simulation, any computation that is possible on any other computer. The 'universal computer' is such that once it has been defined, it is fixed once and for all; one cannot go back at a later time and change its specifications. Hundreds of quotes from well respected sources elaborate on the above definition, without caveat, exception, or qualification (a selection is provided in [1]); the following quote is typical: "It is possible to build a universal computer: a machine that can be programmed to perform any computation that any other physical object can perform." [12]

Several models of computation have been claimed to be 'universal', such as the Turing Machine, the Random Access Machine, the Cellular Automaton, and so on. In fact, none is. Contrary to the above quotes, no universal model is possible, even if it is provided with an unlimited supply of time and memory, and the ability to interact with the outside world. Indeed, for any model of computation X that purports to be universal, there exists a computational problem Y that X

fails to solve, yet Y is capable of being solved on another model of computation X'. For example, if X (being a putative 'universal computer' and thus fixed once and for all) is capable of z operations per time unit, and Y requires for its solution that z' operations be executed per time unit, where $z' > z$, then X fails to solve Y. If X', on the other hand, is capable, by definition, of performing z' operations per time unit, then X' succeeds to solve Y. However X', in turn, fails when presented with a computation necessitating z'' operations per time unit, where $z'' > z'$. The process continues indefinitely.

An Example. The following problem illustrates non-universality in computation. For a positive even integer n, let n distinct integers be stored in an array A with n locations $A[0]$, $A[1]$, ..., $A[n-1]$, one integer per location. Thus $A[j]$, for all $0 \leq j \leq n-1$, represents the integer currently stored in the jth location of A. Using pairwise swaps, it is required to sort the n integers in place into increasing order, such that after swap i of the sort, for all $i \geq 1$, no three consecutive integers satisfy $A[j] > A[j+1] > A[j+2]$, for all $0 \leq j \leq n-3$, and when the sort terminates we have $A[0] < A[1] < \cdots < A[n-1]$.

An $n/2$-processor PRAM solves the problem handily. A PRAM with fewer processors fails, and so does the RAM. In the particularly nasty case where the input is of the form $A[0] > A[1] > \cdots > A[n-1]$, the RAM and any PRAM with fewer than $n/2$ processors fail after the first swap. It is interesting to note here that a Turing Machine with $n/2$ heads succeeds in solving the problem, yet its simulation by a standard (single-head) Turing Machine fails to satisfy the requirements of the problem. This goes against the common belief that any computation by a variant of the Turing Machine can be effectively simulated by the standard model. It is also useful to point out that no machine that claims to be 'universal' can solve the sorting problem described above (or any problem of this kind on which a mathematical constraint is imposed that needs to be satisfied throughout the computation), even if such a machine were endowed with the formidable power of time travel [7]. The non-universality result applies to all models of computation, sequential as well as parallel, theoretical as well as practical, conventional as well as unconventional.

4   Dynamic Time Requirements on Data Acquisition

If faced with a task to be executed, we are accustomed to know *what* input data we will need in order to perform the task and *when* to acquire the data necessary for the task's computations, in other words *when* to listen to the environment. This is actually an abstraction, a simplification of many real-life tasks. Even very simple paradigms can illustrate the idea that the nature of the input data (*what*) or the time requirements (*when*) of some data *cannot* be known unless some event happens *during* the task's computation.

We will call this paradigm, the *paradigm of the unexpected event*, where the cause of the event needs to be investigated. During a computation or the execution of a task, an unexpected event happens. By 'unexpected', it is meant not planned in advance and not predicted by the computation itself. The need immediately arises to investigate the cause of the event. This would mean to inspect the environment or, depending on the situation, inspect the computer itself. Yet, we are interested in the state of the environment *before* the event happened. This means that we need values of parameters describing the environment measured at a moment *prior* to the one in which we are now performing the investigation.

The following example expresses the difficulty to perform a simple computation if time constraints are to be defined during the computation. In particular, the time constraints here refer to the acquisition of data.

A computer equipped with a sensor has to monitor the environment in which it is located. The environment is defined by two different parameters, namely, temperature T and pressure p. These parameters vary unpredictably with time t; thus, T and p are expressed as functions of t, that is, $T = T(t)$ and $p = p(t)$.

The computation starts at $t_0 = 0$ and continues for a length of time given by $length = 10$ time units, that is, the computer has to operate for this length of time. Computation must terminate at time $t_f = t_0 + length$. When designing the algorithm, we may refer to any time unit of the computer's operational time, namely, $t_0 = 0$, $t_1 = 1$, $t_2 = 2$, ..., $t_{10} = t_f = 10$.

Requirements. The computer has to measure, exactly once, one parameter of the environment and output this value at the end of its computation, that is, at time t_f. At the beginning of the computation, that is, at time t_0, the computer does not yet know what needs to be measured and when. The time at which the parameter is to be measured, denoted t_{input}, is anywhere during the computation, specifically, $t_0 \leq t_{input} \leq t_f$.

As a dynamic characteristic, t_{input} is not known at the beginning of the computation. Information to compute t_{input} will be available during the computation. In addition, it is not known at the beginning of the computation which of the two parameters is to be measured. This information is represented by a binary value *which*, and will also be obtained later on during the computation. If *which* $= 0$ then the sensor has to measure the temperature and if *which* $= 1$ then the sensor has to measure the pressure.

These two variables t_{input} and *which* will not be available easily, nor directly. There is no information in the environment, that the computer could read, until the middle of the computation, that is, at time $t_{middle} = t_0 + \frac{length}{2} = 0 + \frac{10}{2} = 5 = t_5$.

Now, at t_{middle}, suddenly several variables will make the computation of t_{input} and *which* possible. Three variables denoted x_0, x_1, and x_2 are now available in the environment for the computer to read in. In this example, the number of variables has been set to three, in order for it to be smaller than the six time units left for computation, from t_{middle} to t_f, inclusive. This will give the sequential computer some small chance to successfully complete the task. If there are more variables, the sequential computer will never be successful.

The time t_{input} and the binary *which* will be computed from the sum of x_0, x_1, and x_2 according to the following formulas: $t_{input} = (x_0 + x_1 + x_2) \bmod 11$, and *which* $= (x_0 + x_1 + x_2) \bmod 2$.

Clearly, the problem here is to compute t_{input} and *which* on time. Although the monitoring is trivial otherwise, the difficulty is in knowing what to do and when. We show in the following subsections that the type of the computer employed to solve the problem will make the difference between success and failure.

Idle Sequential Solution. The first algorithmic solution runs on a sequential computer, that is, on a time-aware RAM. The latter does not start monitoring its environment unless it fully knows what to do. In this case, the computer is working "on request", or according to the principle: *Don't do anything unless you absolutely have to* ...

As specified earlier, the RAM is able to read a fixed-size input, perform a fixed-size calculation, and produce a fixed-size output, all in one time unit. It starts its computation at $t_0 = 0$, but remains idle until it receives a measurement request at $t_{\text{middle}} = t_5$. At t_{middle}, the computer computes the parameters of the request. It takes the computer 5 time units to compute the values for t_{input} and *which*. The following situations can happen:

1. $t_{\text{input}} > t_{\text{middle}} + 5 - 1$. The computer can satisfy the measurement request. At t_{input}, the computer measures the requested parameter, temperature or pressure, and outputs the value at t_f.
2. $t_{\text{middle}} \leq t_{\text{input}} < t_{\text{middle}} + 5$. The computer has not finished calculating t_{input} and fails to read the input at the required time.
3. $t_{\text{input}} < t_{\text{middle}}$. The computer certainly fails to read the input at the required time. It cannot go back in time to measure the environment parameter at t_{input}.

This computer has little chance of success, as it will try to get input only after all information concerning the input, namely, t_{input} and *which*, is computed. This leaves only $t_f - t_{\text{middle}} - 5 + 1$ time units at the end, in which the environment is actually listened to. If t_{input} falls in this range, the computer is successful, otherwise it fails. The overall success rate of this computer is:

$$success = (t_f - t_{\text{middle}} - 5 + 1)/(t_f - t_0) = 1/10 = 10\%.$$

Active (Smart) Sequential Solution. In this case, the computer will try to do better by anticipating the request. It will monitor the environment, even though it does not know exactly what the requirements will be. The principle of this computer is: *Do as much as you can in advance* ...

The computer is again a time-aware RAM as in the previous case. Instead of being idle while waiting for the request, it busily measures the environment. This is done from the moment the computation starts at t_0, until the request is received at t_{middle}. Because the parameter of interest is unknown, the best the computer can do is to choose just one parameter arbitrarily, for example the temperature. The value of the temperature is recorded for all time units: t_0, t_1, t_2, t_3 and t_4. Then the request is received during $t_{\text{middle}} = t_5$, t_6, and t_7. The parameters t_{input} and *which* are computed during t_8 and t_9. During t_{10} it is still possible to read the correct environment parameter, if it so happens that $t_{\text{input}} = t_{10}$. The following situations can happen:

1. $t_{\text{input}} > t_{\text{middle}} + 5 - 1$. The computer is able to measure the required environment parameter and output the value of interest after it received the request.
2. $t_{\text{middle}} \leq t_{\text{input}} < t_{\text{middle}} + 3$. The computer is busy reading the variables x_0, x_1, and x_2 and has not recorded any history of the environment. The task fails.
3. $t_{\text{input}} < t_{\text{middle}}$ and the value of the temperature is required. The computer has recorded the history of the temperature and can output the desired recorded value.
4. $t_{\text{input}} < t_{\text{middle}}$ and the value of the pressure is required. The computer has not recorded the history of the pressure and is unable to meet the request.

This computer has a better chance of being successful. If the temperature is indeed required, the computer's chances are high. The exact chance of success is $(t_f - t_0 - 3)/(t_f - t_0) = 7/10$. If the value of the pressure is required, the success rate is the same as for the idle solution $(t_f - t_{\text{middle}} - 5 + 1)/(t_f - t_0) = 1/10$. The overall success rate is the average of the two rates computed above:

$$success = \frac{1}{2} \times \left(\frac{t_f - t_0 - 3}{t_f - t_0} + \frac{t_f - t_{\text{middle}} - 5 + 1}{t_f - t_0}\right) = \frac{1}{2}\left(\frac{7}{10} + \frac{1}{10}\right) = 40\%.$$

Parallel Solution. The last and most successful algorithmic solution is offered by a parallel computer. The latter monitors the environment exhaustively in order to answer the request, no matter what is asked or when it is to be performed. The principle of this computer is: *Do absolutely everything and be prepared for the worst* ...

The parallel computer is a time-aware PRAM, that is, a collection of time-aware RAMs operating synchronously. It can perform several steps (each executed by a distinct processor, and each consisting of a measurement, a computation, and an output) simultaneously, in one time unit. For our example, it needs to have three processors, denoted P_0, P_1, and P_2.

The PRAM can measure both the temperature and pressure of the environment at the same time. It is able to record the full history of the environment. When t_{input} is computed during $t_{\text{middle}} = t_5$, the computer will still be able to fully monitor the environment and also compute t_{input} and *which*. The following situations can happen:

1. $t_{input} > t_{\text{middle}} + 4 - 1$. The computer will measure the requested parameter at t_{input} and write it out at the end of the computation.
2. $t_{input} < t_{\text{middle}} + 4$. The computer inspects its recorded history of the environment and outputs the desired parameter.

This monitor always answers the request successfully. Therefore, the success rate is 100 %. It should be noted here that, in general for n monitored parameters, the number of processors on the parallel computer needs to be at least $n + 1$.

5   Dynamic Time Requirements on Data Output

This section explores the situation where the deadline of a task is not defined at the outset, but will be computed *during* the execution of the task. Thus it is of the utmost importance that the computer executing the task be able to compute the deadline before it has passed.

Requirements. A computer equipped with a sensor is required to measure some parameter of the environment. For definiteness, consider the parameter to be the temperature T. The computer is active for a certain length of time $\delta + length$. This time is divided into two intervals δ and $length$, that will be defined later. It starts its activity at time t_0 and consequently finishes at time $t_f = t_0 + \delta + length$. The task of the computer is to measure the temperature T at the beginning, that is, at time t_0. Then, after some delay, the measured temperature value is to be output at time t_{output}. The delay is not allowed to be null, it has to be larger than δ. Thus, $t_0 + \delta < t_{output} \leq t_f$. With the output of the measured temperature, the task finishes.

Time t_{output} is not given at the outset, but has to be calculated by the computer according to the following rules. The n input variables x_0, x_1, ..., x_{n-1} are available in the environment throughout the computation. The output time is defined as: $t_{output} = t_0 + \delta + [(x_0 + x_1 + \cdots + x_{n-1}) \bmod length] + 1$.

The problem will be solved on a sequential computer and on a parallel computer. This will show again that the size of the machine matters. A sequential machine, or a parallel machine without adequate resources, fails to perform the task, whereas a sufficiently powerful parallel machine successfully completes the task. In our examples, we consider the following values: $t_0 = 0$, $\delta = 2$, $length = 8$, and $n = 10$.

Sequential Solution. The sequential computer is a time-aware RAM. It measures the temperature at t_0 and then proceeds to read the input variables x_0, x_1, ..., x_9. By the time the computer has managed to calculate the deadline t_{output}, that moment has already passed. Therefore, the sequential computer will not be able to output the result, for no other reason than the fact that it did not know the deadline in time.

Parallel Solution. The parallel computer is a time-aware PRAM. It has $n+1 = 11$ processors, P_0, P_1, P_2, ..., P_{10}, in order to be able to perform all measurements at the very beginning, that is, at t_0. The 10 inputs x_0, x_1, ..., x_9 are read simultaneously and concurrently written in a given location in the common memory. As a result, that memory location now contains the sum $x_0+x_1+\cdots+x_9$ (as specified in Section 3). Then during the delay δ, the deadline t_{output} is computed. After this, the computer only has to wait for the right time to output the value of the temperature.

The parallel computer can meet the deadline because it is able to compute t_{output} on time. A parallel computer with enough processors can compute the deadline in δ time units. On the other hand, if the PRAM has fewer processors, for example $\lfloor (n + 1)/2 \rfloor = 5$, then this computer is not guaranteed to succeed. If t_{output} is very close to $t_0 + \delta$ then the computer will miss the deadline.

6   Conclusion

The examples in this paper describe problems whose time constraints are difficult to compute. They show that a parallel computer of an appropriate size is essential for the successful completion of some computational task. In particular, a standard Turing Machine is not able to perform such tasks successfully. Furthermore, the Turing Machine cannot simulate a more powerful machine capable of carrying these tasks to completion. This confirms, once again, the previously established result [11], [13]–[17] that the Turing Machine is in fact *not* universal.

Furthermore, any parallel computer can face a problem in which computing the time constraints of a task is above its computational capacity, while a more powerful parallel computer can be defined to solve the given problem. This means that a parallel computer of fixed size cannot be specified that is able to successfully complete any problem of the type described in this paper. For any parallel computer, a problem can be defined such that this computer is not able to compute its time constraints on time. Therefore, even a parallel computer is not universal. This also confirms the more general result, first given in [2], that *no computer capable of a finite number of operations per time unit can be universal.* This result holds even if the computer has access to the outside world for input and output, is endowed with an unbounded memory and is allowed all the time it needs to simulate a successful computation by another computer.

We close with some directions for future work. An obvious feature of computing systems is that if a system needs to perform another N operations to compute a value, and can perform at most n operations per time unit, where $n \leq N$, then it can complete the computation only if its deadline is at least N/n time units away. For basic computations this is already enough to explain why parallel computers do better than sequential ones, and why more processors are better than fewer: When the N operations are independent, and more processors are available, the value of n is greater. As we show in this paper, however, the question is more subtle than this, since we consider 'self-referential' situations in which the value of N includes the number of steps needed to compute N itself. This suggests a wide range of subsidiary questions that might be tackled. For example, suppose that a computation consists of two parts, namely, computing a non-negative integer N, then performing N elementary operations. The overall complexity is the sum of the complexities of the two parts. In some cases, like the ones described in the present paper, performing the first part efficiently leads to a successful computation. In general: When and how does the overall complexity of a computation depend on the complexity of 'self-referentially' computing the number of operations it needs to perform?

As demonstrated previously in [2], it is not always the case that observations are independent of one another. For example, measuring temperature may cause pressure to change. What will be the effect of such behavior on our parallel implementation?

Moreover, it may be useful to consider other components of a task that allow for a dynamic definition. Thus, components that vary along the computation, according to results of the computation, often occur in realistic applications. One example of such components mentioned in this paper is the inner computation of a task, that is, the calculation phase, which may be subject to constraints, perhaps affected by the computation itself.

Finally, a formalization of schedulers of dynamic tasks with dynamic time requirements would lead to settings that are more general than the one described here.

References

1. Akl, S.G.: Universality in computation: Some quotes of interest. Technical Report No. 2006-511, School of Computing, Queen's University, Kingston, Ontario (2006)
2. Akl, S.G.: Three counterexamples to dispel the myth of the universal computer. Parallel Processing Letters 16, 381–403 (2006)
3. Akl, S.G.: Even accelerating machines are not universal. International Journal of Unconventional Computing 3, 105–121 (2007)
4. Akl, S.G.: Evolving Computational Systems. In: Rajasekaran, S., Reif, J.H. (eds.) Handbook of Parallel Computing: Models, Algorithms, and Applications, pp. 1–22. Taylor and Francis, CRC Press, Boca Raton (2008)
5. Akl, S.G.: Unconventional Computational Problems with Consequences to Universality. International Journal of Unconventional Computing 4, 89–98 (2008)
6. Akl, S.G.: Ubiquity and Simultaneity: The Science and Philosophy of Space and Time in Unconventional Computation. Invited talk, Conference on the Science and Philosophy of Unconventional Computing. The University of Cambridge, Cambridge (2009)
7. Akl, S.G.: Time travel: A New Hypercomputational Paradigm. International Journal of Unconventional Computing 6, 329–351 (2011)
8. Bruda, S.D., Akl, S.G.: The Characterization of Data-Accumulating Algorithms. Theory of Computing Systems 33, 85–96 (2000)
9. Buttazzo, G.C.: Hard Real-Time Computing Systems: Predictable Scheduling Algorithms and Applications. Kluwer Academic Publishers, Boston (1997)
10. Buttazzo, G.C.: Soft Real-Time Computing Systems: Predictable Scheduling Algorithms and Applications. Springer, New York (2005)
11. Calude, C.S., Păun, G.: Bio-Steps Beyond Turing. BioSystems 77, 175–194 (2004)
12. Deutsch, D.: The Fabric of Reality. Penguin Books, London (1997)
13. Etesi, G., Németi, I.: Non-Turing Computations via Malament-Hogarth Space-Times. International Journal of Theoretical Physics 41, 341–370 (2002)
14. Nagy, M., Akl, S.G.: Quantum Measurements and Universal Computation. International Journal of Unconventional Computing 2, 73–88 (2006)
15. Siegelmann, H.T.: Neural Networks and Analog Computation: Beyond the Turing limit. Birkhäuser, Boston (1999)
16. Stannet, M.: X-machines and the Halting Problem: Building a Super-Turing Machine. Formal Aspects of Computing 2, 331–341 (1990)
17. Wegner, P., Goldin, D.: Computation Beyond Turing Machines. Communications of the ACM 46, 100–102 (1997)

BFS Solution for Disjoint Paths in P Systems

Radu Nicolescu and Huiling Wu

Department of Computer Science, University of Auckland,
Private Bag 92019, Auckland, New Zealand
r.nicolescu@auckland.ac.nz, hwu065@aucklanduni.ac.nz

Abstract. This paper continues the research on determining a maximum cardinality set of edge- and node-disjoint paths between a source cell and a target cell in P systems. With reference to the previously proposed solution [3], based on depth-first search (DFS), we propose a faster solution, based on breadth-first search (BFS), which leverages the parallel and distributed characteristics of P systems. The runtime complexity shows that, our BFS-based solution performs better than the DFS-based solution, in terms of P steps.

Keywords: P systems, edge-disjoint paths, node-disjoint paths, depth-first search, breadth-first search, network flow.

1 Introduction

P systems is a computational model inspired by the structure and interactions of cell membranes, introduced by Păun [17]. The essential specification of a P system includes a membrane structure, objects and rules. All cells evolve synchronously by applying rules in a non-deterministic and (potentially maximally) parallel manner. Thus, P systems is a strong candidate as a model for distributed and parallel computing.

There are many important applications that need to find alternative paths between two nodes. Alternative paths are fundamental in biological remodelling, e.g., of nervous or vascular systems. Multipath routing provides effective bandwidth in networks. Disjoint paths are sought in streaming multi-core applications to avoid sharing communication links between processors [18]. The maximum matching problem in a bipartite graph can be transformed to the disjoint paths problem. In case of non-complete graphs, Byzantine Agreement requires at least $2k+1$ node-disjoint paths, between each pair of nodes to ensure that a distributed consensus can occur, with up to k failures [9].

In a native P system solution for the disjoint path problem, the input graph is given by the P system structure itself and the system is fully distributed, i.e. there is no central node and only local channels (between structural neighbours) are allowed. Dinneen et al. proposed the first P solution [3], as a distributed version of the Ford-Fulkerson algorithm [5], based on depth-first search (DFS). This solution searches cells sequentially, which is not always efficient. To exploit the parallel potential of P systems, we propose a faster P solution—a distributed

C.S. Calude et al. (Eds.): UC 2011, LNCS 6714, pp. 164–176, 2011.

version of the Edmonds-Karp algorithm [4], which concurrently searches as many paths as possible in breadth-first search (BFS). A full version of this paper appears as a CDMTCS report [14].

2 Preliminary

Essentially, a static P system is specified by the membrane structure, objects and rules. The membrane structure can be modeled as: a rooted tree (cell-like P systems [17]), a directed acyclic graph (hyperdag P systems [11], [12], [13]), or an arbitrary digraph (neural P systems [10], [15]). The objects are symbols from a given alphabet, strings, or more complex structures (such as tuples). P systems combine rewriting rules to change objects in the region and communication rules to move objects. Here, we define a simple P system, with *priorities*, *promoters* and *duplex* channels as a system, $\Pi = (O, \sigma_1, \sigma_2, \ldots, \sigma_n, \delta)$, where:

1. O is a finite non-empty alphabet of *objects*;
2. $\sigma_1, \ldots, \sigma_n$ are cells, of the form $\sigma_i = (Q_i, s_{i,0}, w_{i,0}, R_i)$, $1 \le i \le n$, where:
   - Q_i is a finite set of *states*;
   - $s_{i,0} \in Q_i$ is the *initial state*;
   - $w_{i,0} \in O^*$ is the *initial multiset* of objects;
   - R_i is a finite *ordered* set of rewriting/communication *rules* of the form: $s\ x \to_\alpha s'\ x'\ (y)_\beta|_z$, where: $s, s' \in Q_i$, $x, x', y, z \in O^*$, $\alpha \in \{min, max\}$, $\beta \in \{\uparrow, \downarrow, \updownarrow\}$.

3. δ is a set of *digraph* arcs on $\{1, 2, \ldots, n\}$, without symmetric arcs, representing *duplex* channels between cells.

The membrane structure is a digraph with duplex channels, so parents can send messages to children *and* children to parents, but the disjoint paths strictly follow the parent-child direction. Rules are prioritized and are applied in *weak priority* order [16]. The general form of a rule, which transforms state s to state s', is $s\ x \to_\alpha s'\ x'\ (y)_{\beta_\gamma}|_z$. This rule consumes multiset x, and then (after all applicable rules have consumed their left-hand objects) produces multiset x', in the same cell (*"here"*). Also, it produces multiset y and sends it, by *replication* (*"repl"* mode), to all parents (*"up"*), to all children (*"down"*) or to all parents and children (*"up and down"*), according to the target indicator $\beta \in \{\uparrow, \downarrow, \updownarrow\}$. $\alpha \in \{min, max\}$ describes the rewriting mode. In the *minimal* mode, an applicable rule is applied once. In the *maximal* mode, an applicable rule is used as many times as possible and all rules with the same states s and s' can be applied in the maximally parallel manner. Finally, the optional z indicates a multiset of promoters, which enable rules but are not consumed.

3 Disjoint Paths

Given a *digraph*, $G = (V, E)$, a *source* node, $s \in V$, and a *target* node, $t \in V$, the edge- and node-disjoint paths problem looks for a maximum cardinality set of edge- and node-disjoint s-to-t paths. A set of paths are *edge-disjoint* or *node-disjoint* if they have no common arc or no common intermediate node, respectively. Node-disjoint paths are also edge-disjoint paths, but the converse is not true. Cormen et al. [1] give a detailed presentation of these topics. The edge-disjoint paths problem can be transformed to a maximum flow problem by assigning unit capacity to each edge [5]. Given a set of already *established* edge- or node-disjoint paths P, we recall the definition of the *residual* digraph $G_r = (V_r, E_r)$:

- $V_r = V$ and
- $E_r = (E \setminus E_P) \cup E'_P$, where E_P is the set of arcs (u, v) that appear in the P paths and $E'_P = \{(v, u) \mid (u, v) \in E_P\}$.

Briefly, the residual digraph is constructed by reversing the already established path arcs. An *augmenting* path is an s-to-t path in the residual digraph, G_r. Augmenting paths are used to extend the existing set of established disjoint paths. If an augmenting arc reverses an existing path arc (also known as a *push-back* operation), then these two arcs "cancel" each other, due to zero total flow, and are discarded. The remaining path fragments are relinked to construct an extended set of disjoint paths. This round is repeated, starting with the new and larger set of established paths, until no more augmenting paths are found [5].

Figure 1 illustrates a residual digraph and an augmenting path: (a) shows a digraph, where two edge-disjoint paths, 0.1.4.7 and 0.2.5.7, are present; (b) shows the residual digraph, formed by reversing path arcs; (c) shows an augmenting path, 0.3.5.2.6.7, which uses a reverse arc, $(5, 2)$; (d) discards the cancelling arcs, $(2, 5)$ and $(5, 2)$; (e) relinks the remaining path fragments, 0.1.4.7, 0.2, 5.7, 0.3.5 and 2.6.7, resulting in three edge-disjoint paths, 0.1.4.7, 0.2.6.7 and 0.3.5.7.

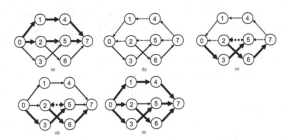

Fig. 1. Finding an augmenting path in a residual digraph

Typically, the augmenting paths search uses a DFS algorithm [5] or a BFS algorithm [4]. A *search path* in the residual graph (also known as a *tentative augmenting path*) starts from the source and tries to reach the target. A *successful search path* becomes a new *augmenting path* and is used to increase the number of disjoint paths. Conceptually, this solves the edge-disjoint paths problem. However, the node-disjoint paths require additional refinements—usually by *node splitting* [8]. Each *intermediate* node, v, is split into an *entry* node, v_1, and an *exit* node, v_2, linked by an arc (v_1, v_2). Arcs that were directed into v are redirected into v_1 and arcs that were directed out of v are redirected out of v_2. Figure 2 illustrates this node splitting, where all intermediate nodes are split—this is a bipartite digraph.

4 Disjoint Paths in P Systems

Classical algorithms use the digraph as data and keep global information. In contrast, our solutions are *fully distributed*. There is no central cell to convey global information among cells, i.e. cells only communicate with neighbors via local channels (between structural neighbours). Unlike traditional programs, which keep full path information globally, our P systems solutions record paths predecessors and successors locally in each cell, similar to distributed routing tables in networks. To construct such routing indicators, we assume that each cell σ_i is "blessed" with a unique *cell ID* object, ι_i, functioning as a *promoter*.

Although many P systems accept cell division, we feel that this feature should not be used here and intentionally discard it. Rather than actually splitting P cells, we simulate this by ad-hoc rules. This approach could be used in other distributed networks, where nodes cannot be split [3]. Essentially, node splitting prevents more than one unit flow through an intermediate node [8].

In our case, node splitting can be simulated by: (i) constraining in and out flow capacities to one and (ii) having two *visited* markers for each cell, one for a *virtual entry* node and another for a *virtual exit* node, extending the visiting idea of classical search algorithms. Figure 2 illustrates a scenario when one cell, y, is visited *twice*, first on its entry and then on its exit node [3]. Assume that path, $\pi = s.x.y.z.t$, is established. Consider a search path, τ, starting from cell, s, and reaching cell, y, in fact, y's entry node. This is allowed and y's entry node is marked as visited. However, to constrain its in-flow to one, y can only push-back τ on its in-flow arc, (x, y). Cell x's exit node becomes visited, x's out-flow becomes zero and τ continues on x's outgoing arc, (x, z). When τ reaches cell z, z's entry node becomes visited and z pushes τ back on its in-flow arc, (y, z). Cell y's exit node becomes visited, y's out-flow becomes zero and τ continues on y's outgoing arc, (y, t). Finally, the search path, τ, reaches the target, t, and becomes $\tau = s.y.x.z.t$. After removing cancelling arcs and relinking the remaining arcs, we obtain two node-disjoint paths, $s.x.z.t$ and $s.y.t$.

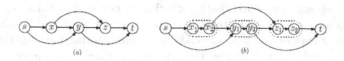

Fig. 2. Simulating node splitting [3]

5   Distributed BFS-Based Strategy

As mentioned in Section 3, augmenting paths can be searched using DFS or BFS. Conceptually, DFS explores as far as possible along a single branch, before backtracking, while BFS explores as many branches as possible concurrently—P systems can exploit this parallelism. Our BFS-based algorithms work in successive rounds.

Each round starts with a set of already *established disjoint paths*, which is empty at the start of the first round. The *source* cell, σ_s, broadcasts a "wave", to find new augmenting paths. Current "frontier" cells send out *connect signals*. The cells which receive and *accept* these connect signals become the new frontier, by appending themselves at the end of current search paths. The advancing wave periodically sends *progress indicators* back to the source: (a) *connect acknowledgments* (at least one search path is still extending) and (b) *path confirmations* (at least one search path was successful, i.e. a new augmenting path was found). While travelling towards the source, each path confirmation reshapes the existing paths and the newly found augmenting path, creating a larger set of paths.

If no progress indicator arrives in the expected time, σ_s assumes that the search round ends. If at least one search path was successful (at least one augmenting path was found), σ_s broadcasts a *reset signal*, which prepares the next round, by resetting all cells (except the target). Otherwise, σ_s broadcasts a *finalize signal* to all cells and the search terminates.

In each round, an *intermediate* cell, σ_i, can be visited only once. Several search paths may try to visit the same intermediate cell *simultaneously*, but only one of them succeeds. Figure 3 (a) shows such a scenario: cells 1, 2 and 3 try to connect cell 4, in the same step; but only cell 1 succeeds, via arc $(1, 4)$. As detailed in our report [14], technically, we solve this *choice* problem using a catalyst-like object [16] (h).

The *target* cell, σ_t, faces an additional decision problem. When several search paths arrive, simultaneously or sequentially, σ_t must quickly decide which augmenting path can be established and which one must be ignored (in the current round). We solve this problem using a *branch-cut* strategy, based on branch IDs [14]. Given a search path, τ, its *branch ID* is the cell ID of its *first intermediate* cell after the source, taken by τ. Frontier cells forward branch IDs on top of their connect signals. The following result is straightforward:

Proposition 1. *In any search round, search paths which share the same branch ID are incompatible; only one of them can succeed and become an augmenting path.*

Figure 3 (b) shows four search paths arriving at cell 6: $\pi = 0.1.6$, $\tau_1 = 0.1.3.6$, $\tau_2 = 0.1.5.6$ and $\tau_3 = 0.2.4.6$; their branch IDs are 1, 1, 1 and 2, respectively. Assume that π arrives first and succeeds as an augmenting path; in this case, σ_t records π's branch ID, 1. Consider the fate of the other search paths, τ_1, τ_2 and τ_3, which attempt to reach the target 6, later in the same round. τ_1 and τ_2 fail, because they share π's recorded branch ID, 1. However, τ_3 succeeds as a new augmenting path, because it has a different branch ID, 2.

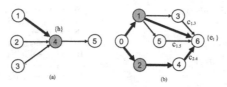

Fig. 3. BFS challenges: (a) A choice made between several search paths connecting the same cell, (b) Search paths sharing the same branch ID are incompatible

A branch ID, i, is recorded via object e_i and "piggybacks" on connection signals, $c_{i.j}$, forwarded by frontier cell σ_j [14]. Technically, it is important that recording objects e_i are used as *promoters*, which enable rules, without being consumed [7].

5.1   Rules for Edge-Disjoint Paths

The P system rules for edge- and node-disjoint paths are slightly different, due to the simulated node-splitting approach, but the basic principle is the same.

Algorithm 1 (P system algorithm for edge-disjoint paths)

Input: All cells start with the *same set of rules* and *without any topological awareness* (they do not even know their local neighbours). All cells start in the *same initial state*, S_0. Initially, each cell, σ_i, contains a *cell ID* object, ι_i, which is *immutable* and used as a *promoter*. Additionally, the source cell, σ_s, and the target cell, σ_t, are decorated with objects, a and z, respectively.

Output: All cells end in the *same final state*. On completion, all cells are *empty*, with the following exceptions: (1) The source cell, σ_s, and the target cell, σ_t, are still decorated with objects, a and z, respectively; (2) The cells on *edge-disjoint paths* contain path link objects, for *predecessors*, p_j, and *successors*, s_k.

We next present the rules and briefly explain them. Our report offers more complete descriptions of state transitions, symbols and rules [14].

0. Rules in state S_1:
 1 $S_0\ a \rightarrow_{\min} S_1\ ah(o)_{\downarrow}$
 2 $S_0\ o \rightarrow_{\min} S_1\ h(o)_{\downarrow}$
 3 $S_0\ o \rightarrow_{\max} S_1$

1. Rules in state S_1:
 1 $S_1\ o \rightarrow_{\max} S_1$
 2 $S_1\ d \rightarrow_{\max} S_1$
 3 $S_1\ b \rightarrow_{\max} S_1$
 4 $S_1\ e_j \rightarrow_{\max} S_1$
 5 $S_1\ g \rightarrow_{\min} S_4\ (g)_{\downarrow}$
 6 $S_1\ v \rightarrow_{\max} S_1$
 7 $S_1\ w \rightarrow_{\max} S_1$
 8 $S_1\ x \rightarrow_{\max} S_1$
 9 $S_1\ y \rightarrow_{\max} S_1$
 10 $S_1\ f_{j.k} \rightarrow_{\max} S_1$
 11 $S_1\ t_j \rightarrow_{\max} S_1$
 12 $S_1\ r_j \rightarrow_{\max} S_1$
 13 $S_1\ a \rightarrow_{\min} S_2\ a$

14 $S_1\ c_j p_j \rightarrow_{\min} S_1\ up_j$
15 $S_1\ c_{j.k} p_k \rightarrow_{\min} S_1\ p_k$
16 $S_1\ zhc_{j.k} \rightarrow_{\min} S_5\ zhp_k e_j (f_{i.k})_{\updownarrow}|_{\iota_i}$
17 $S_1\ zhc_j \rightarrow_{\min} S_5\ zhup_j (f_{i.j})_{\updownarrow}|_{\iota_i}$
18 $S_1\ hl_{j.k} s_k \rightarrow_{\min} S_2\ ht_k e_j s_k\ (r_k)_{\uparrow}$
19 $S_1\ hc_j \rightarrow_{\min} S_2\ hut_j\ (r_j)_{\uparrow}$
20 $S_1\ hc_{j.k} \rightarrow_{\min} S_2\ ht_k e_j\ (r_k)_{\updownarrow}$

2. Rules in state S_2:
 1 $S_2\ b \rightarrow_{\min} S_1 (b)_{\downarrow}$
 2 $S_2\ g \rightarrow_{\min} S_4 (g)_{\downarrow}$
 3 $S_2\ ah \rightarrow_{\min} S_3\ ahw(c_i)_{\downarrow}|_{\iota_i}$
 4 $S_2\ he_j \rightarrow_{\min} S_3\ he_j (l_{j.i})_{\uparrow} (c_{j.i})_{\downarrow}|_{\iota_i}$
 5 $S_2\ hu \rightarrow_{\min} S_3\ hu (l_{i.i})_{\uparrow}\ (c_{i.i})_{\downarrow}|_{\iota_i}$
 6 $S_2\ f_{j.k} \rightarrow_{\max} S_2$
 7 $S_2\ c_{j.k} \rightarrow_{\max} S_2$
 8 $S_2\ l_{j.k} \rightarrow_{\max} S_2$

3. Rules for state S_3:
 1 $S_3\ b \rightarrow_{\min} S_1 (b)_{\downarrow}$
 2 $S_3\ g \rightarrow_{\min} S_4 (g)_{\downarrow}$
 3 $S_3\ axyy f_{j.i} \rightarrow_{\min} S_3\ ads_j x|_{\iota_i}$
 4 $S_3\ axyy r_i \rightarrow_{\min} S_3\ ax|_{\iota_i}$

5 $S_3 \, axyyyf_{j.i} \rightarrow_{\min} S_3 \, ads_jx|_{\iota_i}$

6 $S_3 \, axyyyr_i \rightarrow_{\min} S_3 \, ax|_{\iota_i}$

7 $S_3 \, adxyyy \rightarrow_{\min} S_1 \, a(b)_\downarrow$

8 $S_3 \, axyyy \rightarrow_{\min} S_4 \, a(g)_\downarrow$

9 $S_3 \, awvv \rightarrow_{\min} S_4 \, a(g)_\downarrow$

10 $S_3 \, awvf_{j.i} \rightarrow_{\min} S_3 \, ads_jx|_{\iota_i}$

11 $S_3 \, awvr_i \rightarrow_{\min} S_3 \, ax|_{\iota_i}$

12 $S_3 \, x \rightarrow_{\min} S_3 \, y$

13 $S_3 \, t_jf_{k.i} \rightarrow_{\min} S_3 \, p_js_k(f_{i.j})_\updownarrow|_{\iota_i}$

14 $S_3 \, af_{j.i} \rightarrow_{\min} S_3 \, as_j|_{\iota_i}$

15 $S_3 \, p_js_j \rightarrow_{\min} S_3$

16 $S_3 \, r_it_j \rightarrow_{\min} S_3 \, t_j(r_j)_\updownarrow|_{\iota_i}$

17 $S_3 \, w \rightarrow_{\min} S_3 \, wv$

18 $S_3 \, r_j \rightarrow_{\max} S_3$

19 $S_3 \, c_{j.k} \rightarrow_{\max} S_3$

20 $S_3 \, f_{j.k} \rightarrow_{\max} S_3$

21 $S_3 \, l_{j.k} \rightarrow_{\max} S_3$

4. Rules for state S_4:

1 $S_4 \, g \rightarrow_{\max} S_4$

2 $S_4 \, e_j \rightarrow_{\max} S_4$

3 $S_4 \, f_{j.k} \rightarrow_{\max} S_4$

4 $S_4 \, c_{j.k} \rightarrow_{\max} S_4$

5 $S_4 \, l_{j.k} \rightarrow_{\max} S_4$

6 $S_4 \, t_j \rightarrow_{\max} S_4$

7 $S_4 \, r_j \rightarrow_{\max} S_4$

8 $S_4 \, w \rightarrow_{\max} S_4$

9 $S_4 \, v \rightarrow_{\max} S_4$

10 $S_4 \, u \rightarrow_{\max} S_4$

11 $S_4 \, h \rightarrow_{\max} S_4$

12 $S_4 \, o \rightarrow_{\max} S_4$

5. Rules for state S_5:

1 $S_5 \, c_jp_j \rightarrow_{\min} S_5 \, p_j$

2 $S_5 \, c_{j.k} \rightarrow_{\min} S_5|e_j$

3 $S_5 \, c_{j.k}p_k \rightarrow_{\min} S_5 \, p_k$

4 $S_5 \, hc_{j.k} \rightarrow_{\min} S_5 \, hp_ke_j(f_{i.k})_\updownarrow|_{\iota_i}$

5 $S_5 \, hc_j \rightarrow_{\min} S_5 \, hp_j(f_{i.j})_\updownarrow|_{\iota_i}$

6 $S_5 \, g \rightarrow_{\max} S_4$

7 $S_5 \, b \rightarrow_{\max} S_5$

8 $S_5 \, f_{j.k} \rightarrow_{\max} S_5$

9 $S_5 \, l_{j.k} \rightarrow_{\max} S_5$

10 $S_5 \, t_j \rightarrow_{\max} S_5$

11 $S_5 \, r_j \rightarrow_{\max} S_5$

12 $S_5 \, u \rightarrow_{\max} S_5$

The following paragraphs outline how these rules are used by each major cell group: the source cell, frontier cells, other intermediate cells and the target cell.

Scripts for the source cell. The source cell, σ_s, starts by broadcasting an object, o, to all cells (rule 0.1); each receiving cell creates a local catalyst-like object, h (rule 0.2). Next, σ_s starts the search wave via connection requests, c_s (rules 1.13, 2.3). Then, σ_s waits one step for first progress indicators, r_s, $f_{j.s}$ (rules 3.10, 3.11, 3.17), and two steps for relayed progress indicators (rules 3.3, 3.4, 3.12). If no expected relayed progress indicator arrives, then σ_s waits one more step (rules 3.5, 3.6), before signalling the end of the round. If no new disjoint path was found, σ_s broadcasts a finalize signal, g (rules 3.8, 3.9). Otherwise, σ_s broadcasts a reset signal, b (rules 3.7).

Scripts for a frontier cell. An intermediate cell, σ_i, which was not yet visited (in this round), becomes a frontier cell by accepting either (1) a connect signal from σ_s, c_s (rules 1.14, 1.19), or (2) a connect signal from another frontier cell, σ_k, $c_{j.k}$ or $l_{j.k}$, where j is the branch ID (rule 1.15, 1.20, for the forward mode; rule 1.18, for the push-back mode). In case (1), σ_i is the first intermediate cell on the current search path and sends out connect signals: $c_{i.i}$, to its children; $l_{i.i}$, to its parents (rule 2.5). In case (2), σ_i sends out connect signals: $c_{j.i}$, to its children, and $l_{j.i}$, to its parents (rule 2.4). In both cases (1) and (2), σ_i records a temporary path predecessor, t_s, respectively t_k. When σ_i sends its connect signals, the frontier advances.

Scripts for other intermediate cell. A previous frontier cell, σ_i, relays progress indicators: connect acknowledgments, r_i (rule 3.16) and path confirmations, $f_{k.i}$ (rule 3.13). On receiving path confirmations, σ_i transforms a temporary path predecessor, t_j, into an established predecessor, p_j, and records the path successors, s_k. Cell σ_i may already contain (from a previous round) another predecessor-successor pair, $p_{j'}$, $s_{k'}$. If $j = k'$, then p_j and $s_{k'}$ are deleted, as one end of the cancelling arc pair, (j, i) and (i, j); similarly, if $k = j'$, then s_k and $p_{j'}$ are deleted (rule 3.15).

Scripts for the target cell. The target cell, σ_t, accepts either (1) a connect signal from σ_s, c_s (rules 1.17, 5.1, 5.5), or (2) a connect signal from a frontier cell σ_k, $c_{j.k}$, where the branch ID, j, is recorded as e_j (rules 1.16, 5.2, 5.3, 5.4). In both cases (1) and (2), σ_t records a path predecessor, p_s, respectively p_k.

Table 4 shows Algorithm 1 tracing fragments for stages (a), (c) and (e) of Figure 1. Objects, p_j and s_k, indicate predecessors and successors, respectively, on established disjoint paths; objects t_j indicate (temporary) predecessors on current search and augmenting paths. In stage 1(a), the two established paths, 0.1.4.7 and 0.2.5.7, are recorded by the following cell contents: $\sigma_0 : \{s_1, s_2\}$, $\sigma_1 : \{p_0, s_4\}$, $\sigma_2 : \{p_0, s_5\}$, $\sigma_4 : \{p_1, s_7\}$, $\sigma_5 : \{p_2, s_7\}$, $\sigma_7 : \{p_4, p_5\}$. In stage 1(c), the successful search path 0.3.5.2.6.7 is recorded as: $\sigma_3 : \{t_0\}$, $\sigma_5 : \{t_3\}$, $\sigma_2 : \{t_5\}$, $\sigma_6 : \{t_2\}$, $\sigma_7 : \{p_6\}$ (the target records p_6 directly). In stage 1(e), there are three paths, 0.1.4.7, 0.2.6.7 and 0.3.5.7, which are recorded as: $\sigma_0 : \{s_1, s_2, s_3\}$, $\sigma_1 : \{p_0, s_4\}$, $\sigma_2 : \{p_0, s_6\}$, $\sigma_3 : \{p_0, s_5\}$, $\sigma_4 : \{p_1, s_7\}$, $\sigma_5 : \{p_3, s_7\}$, $\sigma_6 : \{p_2, s_7\}$, $\sigma_7 : \{p_4, p_5, p_6\}$.

Table 4. Algorithm 1 tracing fragments for stages (a), (c) and (e) of Figure 1

Stage\Cell	σ_0	σ_1	σ_2	σ_3
1(a)	$S_3\ \iota_0\ a\ s_1 s_2\ \ldots$	$S_3\ \iota_1\ p_0 s_4\ \ldots$	$S_3\ \iota_2\ p_0 s_5\ \ldots$	$S_3\ \iota_3\ \ldots$
1(c)	$S_3\ \iota_0\ a\ s_1 s_2\ \ldots$	$S_1\ \iota_1\ p_0 s_4\ \ldots$	$S_3\ \iota_2\ p_0 s_5 t_5\ \ldots$	$S_3\ \iota_3\ t_0\ \ldots$
1(e)	$S_4\ \iota_0\ a\ s_1 s_2 s_3$	$S_4\ \iota_1\ p_0 s_4$	$S_4\ \iota_2\ p_0 s_6$	$S_4\ \iota_3\ p_0 s_5$

Stage\Cell	σ_4	σ_5	σ_6	σ_7
1(a)	$S_3\ \iota_4\ p_1 s_7\ \ldots$	$S_3\ \iota_5\ p_2 s_7\ \ldots$	$S_3\ \iota_6\ \ldots$	$S_5\ \iota_7\ z\ p_4 p_5\ \ldots$
1(c)	$S_1\ \iota_4\ p_1 s_7\ \ldots$	$S_3\ \iota_5\ p_2 s_7 t_3\ \ldots$	$S_3\ \iota_6\ t_2\ \ldots$	$S_5\ \iota_7\ z\ p_4 p_5 p_6\ \ldots$
1(e)	$S_4\ \iota_4\ p_1 s_7$	$S_4\ \iota_5\ p_3 s_7$	$S_4\ \iota_6\ p_2 s_7$	$S_4\ \iota_7\ z\ p_4 p_5 p_6$

The preceding arguments indicate a bisimulation relation between our BFS-based algorithm and the classical Edmonds and Karp BFS-based algorithm for edge-disjoint paths [4]. The following theorem encapsulates these arguments:

Theorem 1. *When Algorithm 1 terminates, path predecessor and successor objects indicated in its output section indicate a maximal cardinality set of edge-disjoint paths.*

5.2 Rules for Node-Disjoint Paths

Algorithm 2 (P system algorithm for node-disjoint paths)

Input: As in the edge-disjoint paths algorithm of Algorithm 1.

Output: Similar to the edge-disjoint paths algorithm. However, the predecessor and successor objects indicate *node-disjoint paths*, instead of edge-disjoint paths.

We next present the rules and briefly comment on them. Our report offers more detailed descriptions of these symbols and rules [14].

0. Rules in state S_1:
1 $S_0\ a \rightarrow_{\min} S_1\ ah(o)_\downarrow$
2 $S_0\ o \rightarrow_{\min} S_1\ h(o)_\downarrow$
3 $S_0\ o \rightarrow_{\max} S_1$

1. Rules in state S_1:
1 $S_1\ o \rightarrow_{\max} S_1$
2 $S_1\ d \rightarrow_{\max} S_1$
3 $S_1\ b \rightarrow_{\max} S_1$
4 $S_1\ e_j \rightarrow_{\max} S_1$
5 $S_1\ g \rightarrow_{\min} S_4\ (g)_\downarrow$
6 $S_1\ v \rightarrow_{\max} S_1$
7 $S_1\ w \rightarrow_{\max} S_1$
8 $S_1\ u \rightarrow_{\max} S_1$
9 $S_1\ m \rightarrow_{\max} S_1$
10 $S_1\ n \rightarrow_{\max} S_1$
11 $S_1\ f_{j.k} \rightarrow_{\max} S_1$
12 $S_1\ t_{j.k} \rightarrow_{\max} S_1$
13 $S_1\ t_j \rightarrow_{\max} S_1$
14 $S_1\ r_{j.k} \rightarrow_{\max} S_1$
15 $S_1\ a \rightarrow_{\min} S_2\ a$
16 $S_1\ c_j p_j \rightarrow_{\min} S_1\ p_j$
17 $S_1\ c_{j.k} p_k \rightarrow_{\min} S_1\ p_k$
18 $S_1\ zhc_{j.k} \rightarrow_{\min} S_5\ zhp_k e_j (f_{i.k})_\updownarrow|_{\iota_i}$
19 $S_1\ zhc_j \rightarrow_{\min} S_5\ zhp_j (f_{i.j})_\updownarrow|_{\iota_i}$
20 $S_1\ hl_{j.k} s_k \rightarrow_{\min} S_2\ ht_k e_j s_k n\ (r_{i.k})_\updownarrow|_{\iota_i}$
21 $S_1\ hc_{j.k} q \rightarrow_{\min} S_2\ ht_k e_j m q\ (r_{i.k})_\updownarrow|_{\iota_i}$
22 $S_1\ hc_j \rightarrow_{\min} S_2\ hut_j\ (r_{i.j})_\updownarrow|_{\iota_i}$
23 $S_1\ hc_{j.k} \rightarrow_{\min} S_2\ ht_k e_j\ (r_{i.k})_\updownarrow|_{\iota_i}$

2. Rules in state S_2:
1 $S_2\ b \rightarrow_{\min} S_1(b)_\downarrow$
2 $S_2\ g \rightarrow_{\min} S_4(g)_\downarrow$
3 $S_2\ ah \rightarrow_{\min} S_3\ ahw(c_i)_\downarrow|_{\iota_i}$
4 $S_2\ he_j m \rightarrow_{\min} S_3\ he_j m\ (l_{j.i})_\uparrow|_{\iota_i}$
5 $S_2\ he_j n \rightarrow_{\min} S_3\ he_j n\ (l_{j.i})_\uparrow\ (c_{j.i})_\downarrow|_{\iota_i}$
6 $S_2\ he_j \rightarrow_{\min} S_3\ he_j\ (l_{j.i})_\uparrow\ (c_{j.i})_\downarrow|_{\iota_i}$
7 $S_2\ hu \rightarrow_{\min} S_3\ hu(l_{i.i})_\uparrow\ (c_{i.i})_\downarrow|_{\iota_i}$
8 $S_2\ f_{j.k} \rightarrow_{\max} S_2$
9 $S_2\ c_{j.k} \rightarrow_{\max} S_2$
10 $S_2\ l_{j.k} \rightarrow_{\max} S_2$

3. Rules in state S_3:
1 $S_3\ b \rightarrow_{\min} S_1(b)_\downarrow$
2 $S_3\ g \rightarrow_{\min} S_4(g)_\downarrow$
3 $S_3\ hml_{j.k} s_k \rightarrow_{\min} S_3\ hmnt_k e_j s_k\ (r_{i.k})_\updownarrow|_{\iota_i}$
4 $S_3\ he_j mn \rightarrow_{\min} S_3\ hwe_j\ (l_{j.i})_\uparrow\ (c_{j.i})_\downarrow|_{\iota_i}$
5 $S_3\ axyyf_{j.i} \rightarrow_{\min} S_3\ ads_j x|_{\iota_i}$
6 $S_3\ axyyr_{j.i} \rightarrow_{\min} S_3\ ax|_{\iota_i}$
7 $S_3\ axyyyf_{j.i} \rightarrow_{\min} S_3\ ads_j x|_{\iota_i}$
8 $S_3\ axyyyr_{j.i} \rightarrow_{\min} S_3\ ax|_{\iota_i}$
9 $S_3\ adxyyy \rightarrow_{\min} S_1\ a\ (b)_\downarrow|_{\iota_i}$

10 $S_3\ axyyy \rightarrow_{\min} S_4\ a\ (g)_\downarrow|_{\iota_i}$
11 $S_3\ awvyy \rightarrow_{\min} S_4\ a(g)_\downarrow|_{\iota_i}$
12 $S_3\ awvf_{j.i} \rightarrow_{\min} S_3\ ads_j x|_{\iota_i}$
13 $S_3\ awvr_{j.i} \rightarrow_{\min} S_3\ ax|_{\iota_i}$
14 $S_3\ x \rightarrow_{\min} S_3\ xy$
15 $S_3\ t_{j.k} f_{k.i} \rightarrow_{\min} S_3\ p_j s_k q\ (f_{i.j})_\updownarrow|_{\iota_i}$
16 $S_3\ t_j f_{k.i} \rightarrow_{\min} S_3\ p_j s_k q\ (f_{i.j})_\updownarrow|_{\iota_i}$
17 $S_3\ af_{j.i} \rightarrow_{\min} S_3\ as_j|_{\iota_i}$
18 $S_3\ p_j s_j q \rightarrow_{\min} S_3$
19 $S_3\ r_{k.i} t_{j.k} \rightarrow_{\min} S_3\ t_{j.k}\ (r_{i.j})_\updownarrow|_{\iota_i}$
20 $S_3\ t_j r_{k.i} \rightarrow_{\min} S_3\ t_{j.k}\ (r_{i.j})_\updownarrow|_{\iota_i}$
21 $S_3\ t_{j.l} r_{k.i} \rightarrow_{\min} S_3\ t_{j.l} t_{j.k}\ (r_{i.j})_\updownarrow|_{\iota_i}$
22 $S_3\ w \rightarrow_{\min} S_3\ wv$
23 $S_3\ ar_{j.i} \rightarrow_{\max} S_3\ a|_{\iota_i}$
24 $S_3\ c_{j.k} \rightarrow_{\max} S_3$
25 $S_3\ f_{j.k} \rightarrow_{\max} S_3$
26 $S_3\ l_{j.k} \rightarrow_{\max} S_3$

4. Rules in state S_4:
1 $S_4\ g \rightarrow_{\max} S_4$
2 $S_4\ e_j \rightarrow_{\max} S_4$
3 $S_4\ q \rightarrow_{\max} S_4$
4 $S_4\ f_{j.k} \rightarrow_{\max} S_4$
5 $S_4\ c_{j.k} \rightarrow_{\max} S_4$
6 $S_4\ l_{j.k} \rightarrow_{\max} S_4$
7 $S_4\ t_{j.k} \rightarrow_{\max} S_4$
8 $S_4\ t_j \rightarrow_{\max} S_4$
9 $S_4\ r_{j.k} \rightarrow_{\max} S_4$
10 $S_4\ w \rightarrow_{\max} S_4$
11 $S_4\ v \rightarrow_{\max} S_4$
12 $S_4\ u \rightarrow_{\max} S_4$
13 $S_4\ m \rightarrow_{\max} S_4$
14 $S_4\ n \rightarrow_{\max} S_4$
15 $S_4\ h \rightarrow_{\max} S_4$
16 $S_4\ o \rightarrow_{\max} S_4$

5. Rules in state S_5:
1 $S_5\ c_j p_j \rightarrow_{\min} S_5\ p_j$
2 $S_5\ c_{j.k} \rightarrow_{\min} S_5|_{e_j}$
3 $S_5\ c_{j.k} p_k \rightarrow_{\min} S_5\ p_k$
4 $S_5\ hc_{j.k} \rightarrow_{\min} S_5\ hp_k e_j (f_{i.k})_\updownarrow|_{\iota_i}$
5 $S_5\ hc_j \rightarrow_{\min} S_5\ hp_j (f_{i.j})_\updownarrow|_{\iota_i}$
6 $S_5\ g \rightarrow_{\max} S_4$
7 $S_5\ b \rightarrow_{\max} S_5$
8 $S_5\ f_{j.k} \rightarrow_{\max} S_5$
9 $S_5\ l_{j.k} \rightarrow_{\max} S_5$
10 $S_5\ t_{j.k} \rightarrow_{\max} S_5$
11 $S_5\ t_j \rightarrow_{\max} S_5$
12 $S_5\ r_{j.k} \rightarrow_{\max} S_5$
13 $S_5\ u \rightarrow_{\max} S_5$

When a cell, σ_i, is first reached by a search path, then both its "entry node" and "exit node" become *visited*. If this search path is successful, then σ_i is marked by one object q. In a subsequent round, new search paths can visit σ_i (1) via an incoming arc (forward mode); (2) via an outgoing arc, in the reverse direction (push-back mode) or (3) on both ways. When a search path visits σ_i via an incoming arc, it marks σ_i with one object, m, indicating a visited entry node (rule 1.21); in this case, the search path can only continue with a push-back (rule 2.4). When a search path visits σ_i via an outgoing arc, it marks the cell with one object, n, indicating a visited exit node (rule 1.20); in this case, the search path continues with all other possible arcs (rule 2.5), i.e. all forward searches and also a push-back on its current in-flow arc. Cell, σ_i, can be visited at most once on each of its entry or exit nodes; but, it can be visited both on its entry and exit nodes, in which case it has two temporary predecessors (which simulate the node-splitting technique).

In Figure 5, the search path, 0.4.5.2.1.8.9.3.2.6.7.10, has visited cell 2 twice, once on its "entry" node and again on its "exit" node. Cell 2 has two temporary predecessors, cells 5 and 3, and receives progress indicators from two successors, cells 1 and 6. Progress indicators relayed by cell 6 must be further relayed to cell 3 and progress indicators relayed by cell 1 must be further relayed to cell 5. To make the right choice, each cell records matching predecessor-successor pairs, e.g., cell 2 records pairs (5,1) and (3,6). The following theorem sums up all these arguments and further arguments which appear in our report [14]:

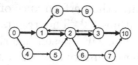

Fig. 5. An example of node-disjoint paths

Theorem 2. *When Algorithm 2 ends, path predecessors and successors objects in its output indicate a maximal cardinality set of node-disjoint paths.*

6 Performance of BFS-Based Solutions

Consider a simple P system with n cells, $m = |\delta|$ arcs, where f_e = the maximum number of edge-disjoint paths, f_n = the maximum number of node-disjoint paths and d = the outdegree of the source cell. Dinneen at al. show that the DFS-based algorithms for edge- and node-disjoint paths run in $O(mn)$ P steps [3]. A closer inspection, not detailed here, shows that this upper bound can be improved.

Theorem 3. *The DFS-based algorithms run in $O(md)$ P steps, in both the edge- and node-disjoint cases.*

In our report, we prove that our algorithms run asymptotically faster ($f_e, f_n \le d$) [14]:

Theorem 4. *Our BFS-based algorithms run in at most $B(m, f) = (3m + 5)f + 4m + 6$ P steps, i.e. $O(mf)$, where $f = f_e$, in the edge-disjoint case, and $f = f_n$, in the node-disjoint case.*

Table 6 compares the asymptotic complexity of our BFS-based algorithms against some well-known maximum flow BFS-based algorithms. Our BFS-based algorithms are faster, because they leverage the potentially unbounded paral-lelism inherent in P systems.

Theorem 4 indicates a worst-case upper bound. However, a typical search path does not use all m arcs. Also, the algorithm frequently finds several augmenting

Table 6. Asymptotic worst-case complexity: classical BFS-based algorithms (steps), P system DFS-based algorithms [3] (P steps) and our P system BFS-based algorithms (P steps)

Edmonds-Karp [4]	$O(m^2 n)$ steps
Dinic [2]	$O(mn^2)$ steps
Goldberg and Tarjan [6]	$O(nm \log n^2 / m)$ steps
P System DFS-based [3]	$O(md)$ P steps
P System BFS-based [here]	$O(mf)$ P steps

paths in the same round, thus the number of rounds is much smaller than f. Therefore, the average runtime is probably much less than the upper bound indicated by Theorem 4.

Figure 7 empirically compares the performance of our BFS-based algorithms against the DFS-based algorithms [3]. The results show that BFS-based algorithms take fewer P steps than DFS-based algorithms. The performance is, as expected, influenced by the complexity of the underlying digraph. We conclude that, the empirical complexity of our BFS-based algorithms is substantially smaller than the asymptotic worst-case complexity indicated by Theorem 4.

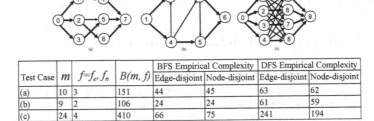

Test Case	m	$f = f_e, f_n$	$B(m, f)$	BFS Empirical Complexity		DFS Empirical Complexity	
				Edge-disjoint	Node-disjoint	Edge-disjoint	Node-disjoint
(a)	10	3	151	44	45	63	62
(b)	9	2	106	24	24	61	59
(c)	24	4	410	66	75	241	194

Fig. 7. Empirical complexity of BFS-based and DFS-based algorithms (P steps)

7 Conclusions

We proposed the first BFS-based P system solutions for the edge- and node-disjoint paths problems. As expected, because of potentially unlimited parallelism inherent in P systems, our algorithms compare favourably with the traditional BFS-based algorithms. Empirical results show that, in terms of P steps, our BFS-based algorithms outperform the previous DFS-based algorithms [3].

Several interesting questions and directions remain open. Can we solve this problem using a restricted P system without states, without sacrificing the current descriptive and performance complexity? What is the average complexity of our BFS-based algorithms? How much can we speedup the existing DFS-based algorithms, by use more efficient distributed DFS algorithms? An interesting avenue is to investigate a limited BFS design, in fact, a mixed BFS-DFS solution, which combines the advantages of both BFS and DFS. Finally, another direction is to investigate disjoint paths solutions on P systems with asynchronous semantics, where additional speedup is expected.

Acknowledgments

The authors wish to thank Tudor Balanescu, Michael J. Dinneen, Yun-Bum Kim, John Morris and three anonymous reviewers, for valuable comments and feedback that helped us improve the paper.

References

1. Cormen, T.H., Stein, C., Rivest, R.L., Leiserson, C.E.: Introduction to Algorithms, 3rd edn. The MIT Press, Cambridge (2009)
2. Dinic, E.A.: Algorithm for solution of a problem of maximum flow in a network with power estimation. Soviet Math. Dokl. 11, 1277–1280 (1970)
3. Dinneen, M.J., Kim, Y.B., Nicolescu, R.: Edge- and node-disjoint paths in P systems. Electronic Proceedings in Theoretical Computer Science 40, 121–141 (2010)
4. Edmonds, J., Karp, R.M.: Theoretical improvements in algorithmic efficiency for network flow problems. J. ACM 19(2), 248–264 (1972)
5. Ford Jr., L.R., Fulkerson, D.R.: Maximal flow through a network. Canadian Journal of Mathematics 8, 399–404 (1956)
6. Goldberg, A.V., Tarjan, R.E.: A new approach to the maximum flow problem. Journal of the ACM 35(4), 921–940 (1988)
7. Ionescu, M., Sburlan, D.: On P systems with promoters/inhibitors. Journal of Universal Computer Science 10(5), 581–599 (2004)
8. Kozen, D.C.: The Design and Analysis of Algorithms. Springer, New York (1991)
9. Lynch, N.A.: Distributed Algorithms. Morgan Kaufmann Publishers Inc., San Francisco (1996)
10. Martín-Vide, C., Păun, G., Pazos, J., Rodríguez-Patón, A.: Tissue P systems. Theor. Comput. Sci. 296(2), 295–326 (2003)
11. Nicolescu, R., Dinneen, M.J., Kim, Y.B.: Structured modelling with hyper-dag P systems: Part A. Report CDMTCS-342, Centre for Discrete Mathematics and Theoretical Computer Science, The University of Auckland, Auckland, New Zealand (December 2008),
    http://www.cs.auckland.ac.nz/CDMTCS/researchreports/342hyperdagA.pdf
12. Nicolescu, R., Dinneen, M.J., Kim, Y.B.: Structured modelling with hyper-dag P systems: Part B. Report CDMTCS-373, Centre for Discrete Mathematics and Theoretical Computer Science, The University of Auckland, Auckland, New Zealand (October 2009),
    http://www.cs.auckland.ac.nz/CDMTCS//researchreports/373hP_B.pdf
13. Nicolescu, R., Dinneen, M.J., Kim, Y.B.: Towards structured modelling with hyperdag P systems. International Journal of Computers, Communications and Control 2, 209–222 (2010)
14. Nicolescu, R., Wu, H.: BFS solution for disjoint paths in P systems. Report CDMTCS-399, Centre for Discrete Mathematics and Theoretical Computer Science, The University of Auckland, Auckland, New Zealand (March 2011),
    http://www.cs.auckland.ac.nz/CDMTCS//researchreports/399radu.pdf

15. Păun, G.: Membrane Computing: An Introduction. Springer-Verlag New York, Inc., Secaucus (2002)
16. Păun, G.: Introduction to membrane computing. In: Ciobanu, G., Pérez-Jiménez, M.J., Păun, G. (eds.) Applications of Membrane Computing. Natural Computing Series, pp. 1–42. Springer, Heidelberg (2006)
17. Păun, G., Centre, T., Science, C.: Computing with membranes. Journal of Computer and System Sciences 61, 108–143 (1998)
18. Seo, D., Thottethodi, M.: Disjoint-path routing: Efficient communication for streaming applications. In: IPDPS, pp. 1–12. IEEE, Los Alamitos (2009)

The Key Elements of Logic Design in Ternary Quantum-Dot Cellular Automata

Primoz Pecar and Iztok Lebar Bajec

University of Ljubljana, Faculty of Computer and Information Science,
Trzaska 25, 1000 Ljubljana, Slovenia
primoz.pecar@fri.uni-lj.si

Abstract. The ternary Quantum-dot Cellular Automata (tQCA) were demonstrated to be a possible candidate for the implementation of a future multi-valued processing platform. Recent papers show that the application of adiabatic pipelining can be used to solve the issues of tQCA logic primitives. The architectures of the resulting tQCAs are similar to their binary counterparts and the physical design rules remain similar to those for the binary QCA domain. The design of complex processing structures is, however, usually based on logic design. The foundation of logic design is a functionally complete set of elementary logic primitives (functions). The currently available tQCA logic primitives, i.e. tQCA majority gate and tQCA inverter gate, do not constitute a functionally complete set. We here present a tQCA implementation of the ternary characteristic functions, which together with the tQCA majority gate and the ternary constants constitute a functionally complete set according to multi-valued Post logic.

Keywords: ternary quantum-dot cellular automaton, multi-valued Post logic, ternary characteristic functions, ternary functionally complete set.

1 Introduction

Quantum-dot cellular automata (QCA) were demonstrated to be a promising processing platform that could bridge the technological limitations of current CMOS technology. The foundations of binary QCA (bQCA) processing were set in the early 1990s with the introduction of the bQCA cell followed by its demonstration in a laboratory environment soon after [11]. Later research led to the development of the first binary processor in QCA [1].

Exploiting binary logic, as the basis of elementary computer structures, is a legacy of the technological limitations that computer designers had to overcome in the past. The desired simplicity of data representation was achievable only with the binary system and its realization with a simple two-state switch. The most effective representative of such a two-state switch is the transistor or CMOS circuit. The pioneers of computer design were well aware of the advantages of multi-valued logic. Both, ternary logic and ternary based processing have been extensively researched over the past five decades [2–4, 6, 16, 18]. The actual

C.S. Calude et al. (Eds.): UC 2011, LNCS 6714, pp. 177–188, 2011.

working platform designs are, however, unable to keep up with the theoretical advancement. The main obstacle is the shortage of building blocks that offer native ternary support. Currently known solutions are built mostly on CMOS technology, whose binary nature prevents effective and economically justifiable ternary computer design.

The earlier mentioned QCAs were demonstrated as a possible solution to the problem. The first advancement of QCA to native ternary processing was made with the redesign of the bQCA cell. The cell's geometry was altered to allow the representation of three logic values and hence named as the ternary QCA (tQCA) cell [7–9]. Adiabatic pipelining was later introduced to solve the issues of the tQCA logic gates [14, 15]. The architectures of the proposed ternary QCAs equal those employed for the implementation of the corresponding binary logic functions, which opens up the possibility to use physical design rules similar to those developed for the binary QCA domain.

This is all encouraging but the design of complex processing elements is still at its first steps. In order to promote efficient composition of complex ternary processing elements one should follow a systematic logic design. The later is based on a set of ternary logic functions, that constitute a ternary functionally complete set. Using it one can implement any arbitrary logic function. It is therefore imperative to identify such a set and implement it in the tQCA platform.

The principal functionally complete set of the binary logic system comprises binary conjunction, disjunction and negation and the corresponding binary QCAs are available. Ternary logic, the simplest multi-valued logic, represents a generalization of binary logic so one cannot simply use the binary functionally complete set [2]. Similarly the designs proposed for the ternary CMOS platform cannot be relied upon. These typically employ implementations, like the TXOR gate, that exploit the platform's physical properties.

Here we present a tQCA implementation of the ternary functionally complete set according to chain-based Post logic [2]. The set comprises the ternary majority gate and ternary characteristic functions. While the majority gate was implemented using proven approaches from bQCA design [15], this was not the case for the characteristic functions. They were developed by observing the behavior of simple tQCA segments and their subsequent composition according to physical design rules, thus illustrating the bottom-up approach [5, 9].

In section 2 we present a brief overview of the tQCA, the adiabatic pipelining concept and the elementary tQCAs. In section 3 we continue with a brief overview of the multi-valued Post logic and the corresponding ternary functionally complete set. In section 4 we describe the tQCA implementation of the ternary characteristic functions.

2   tQCA Overview

In general, a QCA is a planar array of quantum-dot (QCA) cells [11]. The fundamental unit of a ternary QCA is a tQCA cell [7]. It comprises eight quantum dots arranged in a circular pattern and two mobile electrons. The electrons can only

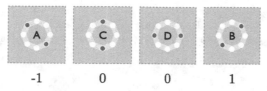

Fig. 1. The four possible arrangements that ensure maximal separation of electrons are mapped to balanced ternary values -1, 0 and 1

reside at quantum dots or tunnel between adjacent quantum dots, but cannot tunnel outside the cell. The Coulomb interaction between the electrons causes them to localize in quantum dots that ensure their maximal separation (energetic minimal state). The four arrangements, which correspond to energetic minimal states (ground states), are marked as A, B, C and D (see Fig. 1). Relying on the principle of ground state computing, the four states can be interpreted as logic values. We here employ the balanced ternary logic, so A is interpreted as logic value -1, B as logic value 1 and C and D as 0. The arrangement D is typically not allowed (desired) for input or output cells [8, 9, 14]. Placing one or more cells in the observed cell's neighborhood, usually causes one of the arrangements to become the favored ground state. The cell to cell interaction is strictly Coulombic and involves only rearrangements of electrons within individual cells, thus it enables computation. With specific planar arrangements of cells it is possible to mimic the behavior of interconnecting wires as well as logic gates [19]. By interconnecting such building blocks more complex devices capable of processing can be constructed.

The reliability of the behavior of a QCA device depends foremost on the reliability of the switching process, i.e. the transition of a cell's state that corresponds to one logic value to a state that corresponds to another and vice versa. It is achieved by means of the adiabatic switching concept, where a cyclic signal, namely adiabatic clock, is used to control the cells' switching dynamic [14, 20]. The signal comprises four phases. The switch phase serves the cells' gradual update of the state with respect to their neighbors. The hold phase is intended for the stabilization of the cells' states when they are to be passed on to the neighbors that are in the switch phase. The release phase and the relax phase support the cells' gradual preparation for a new switch.

Recent research [15] showed that the correct behavior of tQCA logic gates requires a synchronized data transfer, achievable through a pipelined architecture based on the adiabatic clock. The four phased nature of the clock signal allows any tQCA to be decomposed to smaller stages, or subsystems, controlled by phase shifted signals, each defining its own clocking zone. Subsystems that are in the hold phase act as inputs for subsystems that are in the switch phase. A subsystem, after performing its computation locks its state and acts as the input for the following subsystem. As the transaction and processing in the second subsystem is finished it can lock its state while the first prepares for accepting new inputs. With the correct assignment of cells to clocking zones, the direction of data flow can be controlled. Large regions of nearby cells are usually assigned

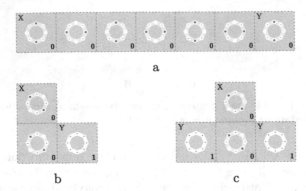

Fig. 2. Using clocking zones for a robust tQCA wire: straight wire (a), corner wire (b) and fan-out (c). Let 0 denote the clocking zone controlled by the base signal and 1 the clocking zone controlled by the base signal shifted by 1 phase.

to the same clocking zone in order to eliminate the challenges that would be caused by attempting to deliver a separate clock signal to every cell.

The latency of a QCA circuit is determined by the number of clocking zones along its critical path. A sequence of four clocking zones causes the delay of one clock cycle. Consequently minimizing the number of clocking zones leads to better designs [13].

The tQCA logic primitive that enables propagation of data from the input cell to the output cell (see Fig. 2) is called tQCA wire. When the input cell's state is A (logic value -1) or B (logic value 1) all cells propagate the same state. However, when the input cell's state is C (logic value 0) the cells propagate the state in an alternating fashion. This effectively means that wires have to be of odd lengths [9]. Having that in mind the tQCA wire can be described as a processing element performing the logic function:

$$y = \mathrm{w}(x) = x, \tag{1}$$

where $x \in \{-1, 0, 1\}$ corresponds to the state of cell X and $y \in \{-1, 0, 1\}$ corresponds to the state of cell Y. The correct behavior of the corner wire and fan-out is ensured by means of a pipeline of two stages, as presented on Figs. 2b and 2c. The first stage ensures the propagation of the input value to the corner, and the second stage ensures its propagation towards the output cell.

3   The Functionally Complete Set

In the 1920s Emil Post introduced a logic system intended for manipulation of multi-valued logic functions, known as chain-based Post logic [17]. A multi-valued logic function is a discrete function whose input and output variables take two or more truth values. Formally, an n variable multi-valued (m-valued) function $f(x_1, \ldots, x_n)$ is a mapping $f : M^n \mapsto M$, with the variables x_i taking truth values from a totally ordered set of m elements, M [2]. In case of $m = 2$

Post logic reduces to Boolean logic, which is currently employed in computer design, however this article focuses on ternary logic system with $m = 3$. Due to the fact that it does not matter how the elements are denoted [18], we here use the balanced set of elements, $M = \{-1, 0, 1\}$.

An arbitrary n variable multi-valued function can be realized as a single primitive or as a composition of primitives. In logic design a primitive is often called a logic gate. The set of all n variable m-valued functions denoted $O_m{}^{(n)}$ consists of $m^{(m^n)}$ functions. In order to make logical design practical it is essential to identify a subset F of functions, i.e. logic gates, whose composition can be used to realize any function from $O_m{}^{(n)}$. Such a set F is denoted a functionally complete set and its elements are denoted elementary logic primitives.

Well known functionally complete sets in Boolean logic are $\{OR, NOT\}$ and $\{AND, NOT\}$, where OR denotes binary disjunction, AND binary conjunction and NOT binary negation. Post generalized binary disjunction to multi-valued logic. The multi-valued disjunction, denoted $\vee$, is introduced as the multi-variable max operator that returns the highest truth value of all n input variables:

$$y = \bigvee_i x_i \equiv \max(x_i), \quad i = 1, 2, ..., n. \tag{2}$$

The symbol $x_i \in M$ denotes the i-th multi-valued input variable. Similarly the generalization of binary conjunction, denoted $\wedge$, is introduced as multi-variable min operator returning the lowest truth value of n input variables:

$$y = \bigwedge_i x_i \equiv \min(x_i), \quad i = 1, 2, ..., n. \tag{3}$$

With the generalization of binary negation achieved as

$$y = \neg x \equiv -x, \tag{4}$$

where $x \in M$ one can obtain sets $\{\vee, \neg\}$, $\{\wedge, \neg\}$ and $\{\vee, \wedge, \neg\}$, but in ternary logic these are not functionally complete. Post solved the problem with a function called cyclic negation, denoted $\underline{x}$. It is given as

$$y = \underline{x} \equiv ((x + 2) \bmod m) - 1, \tag{5}$$

where $x \in M$. Post also proved that the $\wedge$ and $\underline{x}$ constitute a functionally complete set.

After one complete set is identified, others can be found by constructing logic gates by means of which one can implement all the elementary logic primitives belonging to the original set. Every set of functions created in this way is also functionally complete.

The selection of the functionally complete set depends on the properties of the chosen implementation platform. Obviously the elementary logic primitives should be as compact as possible so as to minimize the space they occupy and subsequently their cost and they should exploit the platform's advantages.

In order to find optimal solutions logic design of complex processing structures relies upon various minimization techniques. Well known approaches are the Karnaugh map and the Quine-McCluskey algorithm, which both operate on functions expressed in the normal form [2]. The functionally complete set $\{\wedge, \underline{x}\}$ is not suitable for representing functions in a normal form. For this purpose we would need the functionally complete set consisting of $\wedge$, $\vee$ and characteristic functions. The characteristic function, termed also the unary literal operator of a multi-valued variable, is defined as

$$y = f^z(x) = \begin{cases} 1, & \text{if } x = z \\ -1, & \text{otherwise,} \end{cases} \tag{6}$$

where $x, z \in M$. The disjunctive normal form of the ternary cyclic negation can be given as

$$\underline{x} = (f^{-1}(x) \wedge 0) \vee (f^0(x) \wedge 1) \vee (f^1(x) \wedge -1) \tag{7}$$

which shows that two-variable $\wedge$, two-variable $\vee$ together with literals f^{-1}, f^0, f^1 also constitute a functionally complete set [5].

The tQCA implementation of the ternary $\wedge$ and $\vee$ function is based on the tQCA majority voting gate. The latter is, due to the lack of implementations of other multi-input ternary logic functions, currently the fundamental building block in tQCA design. It is constructed as a crossing of three ternary wires and can be implemented in two possible ways [15] presented in Fig. 3. The tQCA has

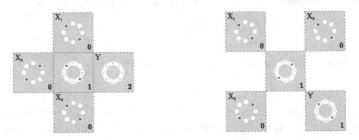

Fig. 3. Two possible pipeline implementations of the ternary majority voting gate

three input cells denoted X_1, X_2 and X_3, a device cell in the center and an output cell Y. It acts as majority voting logic; the output reflects either the logic value that has been present at the majority of the inputs or logic value 0 if the majority cannot be determined (e.g. in the case of the input combination $x_1 = -1$, $x_2 = 1$, $x_3 = 0$). The described behavior can only be achieved through an appropriate assignment of clocking zones. The gate's behavior can be described with the logic function:

$$y = m(x_1, x_2, x_3) = (x_1 \wedge x_2) \vee (x_2 \wedge x_3) \vee (x_1 \wedge x_3), \tag{8}$$

where $x_1, x_2, x_3 \in M$ correspond to the states of input cells X_1, X_2, X_3 and $y \in M$ corresponds to the state of the output cell Y. The ternary two variable $\vee$ and $\wedge$ functions can be expressed as

$$y = x_1 \lor x_2 = m(x_1, x_2, 1)$$
and
$$y = x_1 \land x_2 = m(x_1, x_2, -1),$$

(9)

where $x_1, x_2, y \in M$. That is, the ternary disjunction can be implemented by fixing one input logic value of the ternary majority voting gate to 1, and the ternary conjunction can be implemented by fixing one input logic value to -1.

4 Implementation of the tQCA Characteristic Functions

The implementation of ternary characteristic functions is based on the bottom-up approach, i.e. the concept of combining compact and simple structures. Searching over all possible solutions turns out to be computationally complex [12]. Therefore the search was focused on symmetrical structures with odd number of inputs. Various tQCAs composed of a small number of tQCA cells grouped into one clocking zone were constructed. Their behavior was analyzed using the ICHA simulation approach [10, 15] based on the following parameters: quantum dots had a diameter of 10 nm, the distance between adjacent quantum dots was 20 nm, cell centers were placed on a 110 nm grid. All other relevant parameters were evaluated for a GaAs/AlGaAs material system.

The truth tables for the most promising tQCAs were computed by mapping the output tQCA cell's states to appropriate ternary logic values, as presented in chapter 2. The obtained tables represented the search space for an iterative deepening based design method [5].

Following the described approach three tQCAs were used. The simplest one relies on the fact that two cells arranged diagonally, assigned to the same clocking

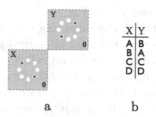

X	Y
A	B
B	A
C	C
D	D

a   b

Fig. 4. The ternary inverter (a) and its behavior (b)

zone, assume alternate states when one is in state A or B and the same state when one is in state C or D, Fig. 4a. The tQCA performs as a unary function $I : \{A, B, C, D\} \mapsto \{A, B, C, D\}$. Comparing the behavior presented in Fig. 4b and equation (4) reveals that the given tQCA evaluates ternary negation where $x \in M$ corresponds to the state of cell X and $y \in M$ corresponds to the state of cell Y.

The most useful tQCA, which made the implementation of ternary characteristic functions feasible is presented in Fig. 5. The tQCA has three input cells

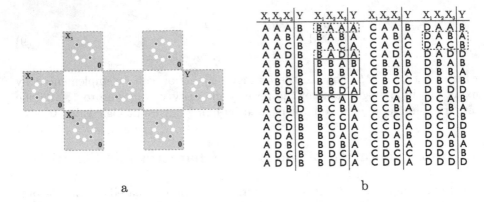

$X_1 X_2 X_3$	Y	$X_1 X_2 X_3$	Y	$X_1 X_2 X_3$	Y	$X_1 X_2 X_3$	Y
A A A	B	B A A	A	C A A	B	D A A	B
A A B	A	B A B	A	C A B	A	D A B	A
A A C	B	B A C	A	C A C	C	D A C	B
A A D	B	B A D	A	C A D	A	D A D	D
A B A	B	B B A	B	C B A	B	D B A	B
A B B	B	B B B	A	C B B	A	D B B	A
A B C	B	B B C	A	C B C	C	D B C	B
A B D	B	B B D	A	C B D	A	D B D	D
A C A	B	B C A	D	C C A	B	D C A	B
A C B	D	B C B	A	C C B	A	D C B	A
A C C	B	B C C	A	C C C	C	D C C	B
A C D	B	B C D	A	C C D	A	D C D	D
A D A	B	B D A	C	C D A	B	D D A	B
A D B	C	B D B	A	C D B	A	D D B	A
A D C	B	B D C	A	C D C	C	D D C	B
A D D	B	B D D	A	C D D	A	D D D	D

a   b

Fig. 5. The tQCA used for the implementation of characteristic functions (a) and its behavior (b). The combinations marked with solid rectangle are the basis for the implementation of f^{-1} and f^1 and the combinations marked with dashed rectangles are the basis for the implementation of f^0.

denoted X_1, X_2 and X_3, three device cells and an output cell Y. The tQCA performs as a three variable function $S_1 : \{A, B, C, D\}^3 \mapsto \{A, B, C, D\}$.

Observing its behavior where cells X_1 and X_2 are fixed to state B and cell X_3 acts as input cell (see black rectangle in Fig. 5b) and comparing with equation (6) reveals that the tQCA can perform as f^{-1} or f^1 literal expressed as

$$y = f^{-1}(x) \equiv S_1(B, B, X)$$
$$\text{and} \tag{10}$$
$$y = f^1(x) \equiv S_1(B, B, I(X)),$$

where $X \in \{A, B, C\}$ corresponds to the logic variable $x \in M$ and $y \in M$ corresponds to the state of cell Y (see Fig. 6).

The construction of the tQCA implementing the f^0 literal proved to be more difficult. The basis of the desired behavior is marked with the dashed rectangles in Fig. 5b. If cell X_2 is fixed to state A and cell X_3 is declared as an inverted input, the state of cell X_1 has to change according to the input state, i.e. in the case of input states A or C the required state is D, otherwise the required state is B. This was achieved with the tQCA presented in Fig. 7a. The tQCA has three input cells denoted X_1, X_2 and X_3 and an output cell Y. Observing its behavior (see Fig. 7b) one can notice that the state of output cell Y is not well defined (marked as N) for all possible input combinations. However, for input combinations where cell X_1 is fixed to state A and cell X_3 is fixed to state D (see dashed rectangles in Fig. 7b) the tQCA performs as a unary function $S_2 : \{A, B, C, D\} \mapsto \{B, D\}$. Therefore, the f^0 literal can be expressed as

$$y = f^0(x) \equiv S_1(S_2(X), A, I(X)), \tag{11}$$

where $X \in \{A, B, C\}$ corresponds to the logic variable $x \in M$ and $y \in M$ corresponds to the state of cell Y (see Fig. 8).

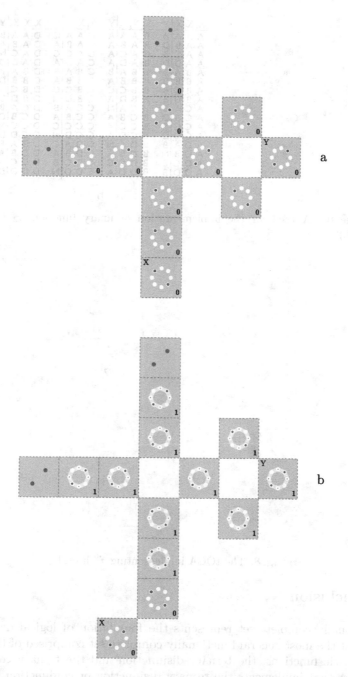

Fig. 6. The tQCAs implementing f^{-1} literal (a) and f^1 literal (b)

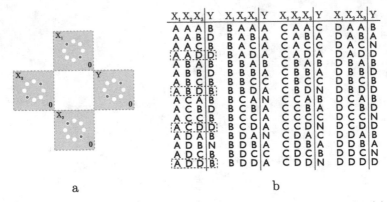

$X_1X_2X_3$	Y	$X_1X_2X_3$	Y	$X_1X_2X_3$	Y	$X_1X_2X_3$	Y
A A A	B	B A A	A	C A A	C	D A A	B
A A B	D	B A B	A	C A B	A	D A B	A
A A C	B	B A C	A	C A C	A	D A C	N
A A D	D	B A D	A	C A D	A	D A D	D
A B A	B	B B A	B	C B A	B	D B A	B
A B B	D	B B B	A	C B B	A	D B B	D
A B C	B	B B C	C	C B C	A	D B C	D
A B D	B	B B D	A	C B D	N	D B D	D
A C A	B	B C A	N	C C A	B	D C A	B
A C B	D	B C B	A	C C B	A	D C B	D
A C C	B	B C C	A	C C C	C	D C C	N
A C D	D	B C D	A	C C D	N	D C D	D
A D A	B	B D A	N	C D A	C	D D A	B
A D B	N	B D B	A	C D B	B	D D B	A
A D C	B	B D C	C	C D C	B	D D C	N
A D D	B	B D D	A	C D D	N	D D D	D

a   b

Fig. 7. The tQCA used for the implementation of unary function S_2 (a) and its behavior (b)

Fig. 8. The tQCA implementing f^0 literal

5   Conclusion

A functionally complete set represents the foundation of logical design. Here we present the most general functionally complete set comprised of the ternary characteristic functions, the ternary disjunction and the ternary conjunction. The tQCA that implements the ternary disjunction or conjunction is actually the ternary majority gate, but the tQCAs that implement the ternary characteristic functions have been developed using the bottom-up concept. The presented architectures are not optimal, in the sense of space requirement and the number of required adiabatic phases, but it is a first step. In the next iterations we will search for more optimized solutions, however as the foundations are set we can also focus on the design of the ternary processor building blocks, as well.

Acknowledgments

The work presented in this paper was done at the Computer Structures and Systems Laboratory, Faculty of Computer and Information Science, University of Ljubljana, Slovenia and is part of a PhD thesis that is being prepared by P. Pecar. It was funded in part by the Slovenian Research Agency (ARRS) through the Pervasive Computing research programme (P2-0395).

References

1. Bernstein, G., Bazan, G., Chen, M., Lent, C., Merz, J., Orlov, A., Porod, W., Snider, G., Tougaw, P.: Practical issues in the realization of quantum-dot cellular automata. Superlattices and Microstructures 20, 447–559 (1996)
2. Dubrova, E., Jamal, Y., Mathew, J.: Non-silicon non-binary computing: Why not? In: 1st Workshop on Non-Silicon Computation, Boston, Massachusetts, pp. 23–29 (2002)
3. Fitting, M., Orlowska, E. (eds.): Beyond two: Theory and applications of multiple-valued logic. Physica-Verlag, Heidelberg (2003)
4. Frieder, G., Luk, C.: Ternary computers: Part 1: Motivation for ternary computers. In: 5th Annual workshop on Microprogramming, Urbana, Illinois, pp. 83–86 (September 1972)
5. Janez, M., Lebar Bajec, I., Pecar, P., Jazbec, A., Zimic, N., Mraz, M.: Automatic design of optimal logic circuits based on ternary quantum-dot cellular automata. WSEAS Trans. Cir. and Sys. 7, 919–928 (2008)
6. Knuth, D.E.: The Art of Computer Programming, 2nd edn., vol. 2. Addison-Wesley, Reading (1981)
7. Lebar Bajec, I., Mraz, M.: Towards multi-state based computing using quantum-dot cellular automata. In: Teucher, C., Adamatzky, A. (eds.) Unconventional Computing 2005: From Cellular Automata to Wetware, pp. 105–116. Luniver Press, Beckington (2005)
8. Lebar Bajec, I., Zimic, N., Mraz, M.: The ternary quantum-dot cell and ternary logic. Nanotechnology 17(8), 1937–1942 (2006)
9. Lebar Bajec, I., Zimic, N., Mraz, M.: Towards the bottom-up concept: extended quantum-dot cellular automata. Microelectronic Engineering 83(4-9), 1826–1829 (2006)
10. Lent, C., Tougaw, P.: Lines of interacting quantum-dot cells: A binary wire. Journal of Applied Physics 74(10), 6227–6233 (1993)
11. Lent, C., Tougaw, P., Porod, W., Bernstein, G.: Quantum cellular automata. Nanotechnology 4, 49–57 (1993)
12. Lusth, J.C., Dixon, B.: A characterization of important algorithms for quantum-dot cellular automata. Inf. Sci. 113, 193–204 (1999)
13. Niemier, M.T., Kogge, P.M.: Problems in designing with QCAs: Layout = timing. International Journal of Circuit Theory and Applications 29, 49–62 (2001)
14. Pecar, P., Mraz, M., Zimic, N., Janez, M., Bajec, I.L.: Solving the ternary QCA logic gate problem by means of adiabatic switching. Japanese Journal of Applied Physics 47(6), 5000–5006 (2008)

15. Pecar, P., Ramsak, A., Zimic, N., Mraz, M., Lebar Bajec, I.: Adiabatic pipelining: A key to ternary computing with quantum dots. Nanotechnology 19(49), 495401 (2008)
16. Porat, D.I.: Three-valued digital systems. Proceedings of IEE 116(6), 947–954 (1969)
17. Post, E.L.: Introduction to a general theory of elementary propositions. American Journal of Mathematics 43(3), 163–185 (1921)
18. Rine, D.C. (ed.): Computer science and multiple-valued logic: Theory and applications, 2nd edn. North-Holland, Amsterdam (1984)
19. Tougaw, P.D., Lent, C.S.: Logical devices implemented using quantum cellular automata. Journal of Applied Physics 75(3), 1818–1825 (1994)
20. Tougaw, P., Lent, C.: Dynamic behaviour of quantum cellular automata. Journal of Applied Physics 80(8), 4722–4736 (1996)

A Universal Flying Amorphous Computer*

Lukáš Petrů and Jiří Wiedermann

Institute of Computer Science, Academy of Sciences of the Czech Republic,
Pod Vodárenskou věží 2, 182 07 Prague 8, Czech Republic
lukas.petru@st.cuni.cz
jiri.wiedermann@cs.cas.cz

Abstract. Amorphous computers are systems that derive their computational capability from the operation of vast numbers of simple, identical, randomly distributed and locally communicating units. The wireless communication ability and the memory capacity of the computational units is severely restricted due to their minimal size. Moreover, the units originally have no identifiers and can only use simple communication protocols that cannot guarantee a reliable message delivery. In this work we concentrate on a so-called flying amorphous computer whose units are in a constant motion. The units are modelled by miniature RAMs communicating via radio. We design a distributed probabilistic communication protocol and an algorithm enabling a simulation of a RAM in finite time. The underlying algorithms make use of a number of original ideas having no counterpart in the classical theory of distributed computing. Our result is the first one showing computational universality of a flying amorphous computer.

1 Introduction

Amorphous computing systems are a relatively recent phenomenon. Apart from the sci-fi literature where various forms of such systems have been envisaged in diverse futuristic scenarios (cf. an interstellar intelligent mobile cloud in [5], or a mentally controlled flying dust serving as an extension of human senses and as an interface to a very distant offspring of today's Internet in [12]), amorphous computing systems as a subject of scientific research emerged by the end of the 1990s (cf. [1],[2],[3],[6],[7],[13],[14],[11]).

The common issue in all visions and research projects was the fact that all of them considered a vast amount of very simple autonomous devices. These devices were randomly placed in the target area — there was no regular topology assumed as, e.g., in the case of cellular automata. Another joint idea was that, using local communication, the respective devices should self–organize in order to perform a coordinated action none of the element alone was able to realize.

Similar devices could, e.g., be spread from an airplane on the surface over a certain region in order to monitor certain parameters of the environment, such as

---

* This research was carried out within the institutional research plan AV0Z10300504 and partially supported by the GA ČR grant No. P202/10/1333.

C.S. Calude et al. (Eds.): UC 2011, LNCS 6714, pp. 189–200, 2011.

temperature, humidity, precipitation, or life manifestations (on an other planet). These measurements can be broadcasted into a base station that will make the data processing. Such gadgets can also be used for object surveillance. Another application is in medical sciences — the devices can be attached to patients and spread over the hospitals in order to monitor the patients' movements and life functions. A nano-size device of such kind can even enter human bodies in order to perform genetic manipulations at the cell level, to strengthen the immune system, to heal up injuries, to cure heart or brain strokes, etc. (cf. [8]). Real bacteria represent a template for such systems already existing in nature.

Obviously, amorphous computing systems can be seen as an extreme case of wireless sensory networks: not only are the nodes of amorphous systems considered under severe size and cost constraints resulting into corresponding constraints on resources such as energy, memory, and computational speed. There are additional constraints usually not considered in the domain of wireless sensory networks: the computational and communication hardware in the processors is stripped down to an absolute minimum that seems to be necessary in order to maintain the required functionality and scalability of the network. For instance, in order to allow potentially unbounded scalability of amorphous computing systems their processors initially do not possess any identity: they are all alike, they are very simple and theoretically can be seen as finite state automata. In order to maximally simplify the wireless communication mechanism in the nodes of amorphous computing systems the non-existence of communication software is assumed. There is no synchronicity assumed and the communication among the nodes is "blind" — a sender cannot learn that its message has been delivered since a receiver node cannot reliably send message acknowledgement to a message sender. The broadcast is unreliable — the simultaneously broadcasting nodes jam each other in such a way that no message can be delivered and moreover, the interference in broadcasting cannot be recognized by the processors.

The prevailing focus of projects mentioned in the introductory part was on engineering or technological aspects of amorphous computing systems, almost completely ignoring theoretical questions related to computational power and efficiency of such systems. Obviously, without knowing their theoretical limits, one cannot have a complete picture of the potential abilities and limitations of such systems. This was the starting point of the project of the present authors devoted to studies of theoretical issues in amorphous computing initiated in 2004. Since that time various models of amorphous systems have been investigated, roughly in the order of their increased generality. In all cases computational universality of the underlying systems has been proved. This points to the versatility of such systems in various computational or robotic applications.

In [15], [17] amorphous computing systems consisting of identical (with no identifiers) simple asynchronous computational processors (typically RAMs with a small number of registers), equipped by a single channel radio communicator with a limited range and a random number generator, were considered.

Such processors were randomly placed in a closed area or volume, and the task was to design a randomized communication protocol leading to a self-organization of a communication network that would enable, with a high probability, to perform universal computations. For the static, non-movable processors this task was solved in [15] and [17]. Later on, in [16] it was shown that similar protocols also work on the level of molecular communication in an aqueous environment (e.g., in a bloodstream); the corresponding processors can take a form of mobile nanomachines capable of performing simple tasks, such as actuation and sensing. The computational part of nanomachines was modeled by variants of finite automata. Recently, a similar result was shown for nanomachines whose communication was inspired by so-called quorum sensing used by real bacteria [18]. The latter two results have been shown by simulating a counter automaton, probably computationally the least efficient universal computing device since it performs its computations using a unary representation of numbers.

The present paper makes a further step towards more general and computationally more efficient amorphous systems: we will consider amorphous systems with mobile processors communicating via a radio. The processors take a form of miniature finite-memory RAMs. The transition from systems with a static communication pattern to systems with an unpredictably changing communication pattern brings a number of new problems that have been encountered formerly neither in the classical theory of distributed systems nor in the previous models of amorphous systems. Namely, the situation is further complicated by a continuously changing communication topology in which new communication paths keep emerging while the old ones keep vanishing. Moreover, some processors may become temporarily unaccessible leading to problems with data consistency maintenance over the entire system at times when the nodes become again accessible. These changes call for novel approaches in the design of communicating protocol resulting in unusual time complexity estimation of the simulation algorithm: without additional assumptions on the nature of processor movements one cannot come with a better than a finite upper bound on the time complexity of this algorithm.

In its class of minimalist amorphous systems communicating over radio, our model of a flying amorphous computer presents the first model for which the computational universality in the efficient sense has been proven. Namely, when compared with the simulations from [16] and [18], where counter automata have been simulated, the current simulation deals with a simulation of a RAM which is a far more efficient universal computer than a counter automaton. From a (future) practical viewpoint, the model opens a door for exploitation of relatively efficient amorphous computing systems communicating via radio in an airborne medium or in a vacuum.

The structure of the paper is as follows. In Section 2 the definition of the flying amorphous computer is given along with the scenario of its use. In Section 3 we sketch the main ideas of simulation of a RAM computer without yet giving the details of a so-called setup phase in which the amorphous computer is preprocessed in order to be able to perform a simulation, and under the assumption that there is a rudimentary broadcast protocol available enabling a probabilistic communication among the nearest processors. The setup phase itself is described in Section 4, while the broadcast protocol in Section 5. Conclusions are in Section 6.

2   Flying Amorphous Computer

Definition 1. *A flying amorphous computer is a septuple $C=[N, A, P, r, s, T, v]$. The model has the following properties:*

(i) The computer consists of N nodes; there is one distinguished node called the base *node.*

(ii) Each node is modeled as a point in a square target area A. The initial positions of nodes in A are determined by a process P assigning a random position independently to each node.

(iii) The nodes are not synchronized but they operate at nearly the same clock speed. An operation taking time T on the fastest node takes at most $T(1 + \varepsilon_T)$ on the slowest node, for a fixed $\varepsilon_T > 0$.

Node properties:

(iv) A node has a fixed number of memory registers *of size s bits; initially, all registers contain zeros. There is a fixed subset of so-called* input and output registers *enabling message transmission among a node's neighbors. There also is a single register providing a real random number (truncated to s bits) between 0 and 1 upon request.*

(v) In each node there is a control unit *operating over all registers. The control unit operates according to a fixed program.*

Movement:

(vi) Initially, a non-zero random direction *vector u is assigned to each node. Thereafter, each node moves with a constant speed v in the direction of its vector u. A node does not know and cannot influence its direction vector.*

(vii) If a node is about to leave the target area A, then its direction vector is mirrored as a billiard ball bounced off a wall.

Communication properties:

(viii) Each node is equipped with a radio transceiver *by which it communicates with other nodes up to the distance r (called* communication radius*) using a shared radio channel; such nodes are called the* neighbors *of the node at hand.*

(ix) One message transmission takes time T.

(x) A node receives a message if exactly one of its neighbors sends that message during transmission time T. While sending, a node cannot receive a message. If two or more neighbors of a node are simultaneously sending a message, a collision occurs and the node receives no message.

(xi) A node cannot detect a collision, the radio receiver cannot distinguish the case when multiple neighbors send messages from the case when no neighbor sends a message.

The communication properties of a node are among the weakest possible and are enforced by requirements of the maximal technological simplicity of the underlying receiver.

In order to solve a computational task a flying amorphous computer operates according to the following scenario in which we assume the existence of an external operator. The scenario consists of three subsequent phases.

First, the operator deploys the nodes in the target area and in some way powers on the computer so that the nodes start moving.

In the second phase, the operator performs a setup of the computer. During the setup a unique address is assigned to each node and the input data of the computational task are loaded into the computer.

After the setup is finished, the computer's own computation can start. At that time, the operator may disconnect from the computer and the computer carries out the computation autonomously. After a sufficient time, the operator may reconnect to check the outcome of the computation.

3   Simulation

In order to prove the versatility of our model we prove its universality by showing that it can simulate a computation of a bounded RAM machine. We shall be using the standard model of a unit cost RAM, cf. [4] with registers of the same size s as the registers of our flying computer. The program, the input and the output data are also stored in these registers.

For the sake of clarity of our exposition, we first sketch the details of the simulation algorithm postponing the details how the previous phases of the computational scenario are implemented. We will merely assume that the computer has been properly formed in the setup phase and that there is a broadcast algorithm guaranteeing a message delivery between two concrete nodes in a finite time.

In order to start a simulation of a RAM with M registers we will require that the simulating flying amorphous computer also have at least M registers loaded with the same initial data as the RAM. This will happen in the setup phase to be described in Section 4.

Definition 2. *We say that a flying amorphous computer A is set up with respect to a RAM R with initial configuration $i_1, i_2, \ldots, i_M$ of R if A contains at least M nodes with unique addresses in the range 1 to M and if the j-th node of A contains the same data i_j as the register j in R does, for $j = 1, 2, \ldots, M$.*

The nodes of a flying amorphous computer communicate using broadcast from one node to its neighbors within their communication radius thus forming a multi-hop communication network. A broadcast operation is started at one node and delivers the message, possibly using several hops, to all nodes in a connected communication component. However, as the nodes are moving, not all nodes may be connected (perhaps through several intermediary nodes) to the originating node at times when needed. The message cannot be delivered to such nodes. We assume that such a situation can never occur for an infinitely long time. We will suppose that the nodes are moving in a such a way that it will never happen that two nodes remain forever in different connected components of the underlying communication graph.

Definition 3. *Let A be a flying amorphous computer, B a broadcasting protocol, and n_1, n_2 two nodes of A. Assume that in A message m is being repeatedly broadcast from node n_1 until it is successfully delivered to node n_2. We say that flying amorphous computer A with protocol B is* lively flying *if m is eventually delivered to n_2 after a finite amount of broadcast attempts.*

Note that in a lively flying computer the similar process of message delivery from n_1 to n_2 works also in the reverse direction, from n_2 to n_1. This can be used for a message delivery acknowledging.

We are ready to show how the simulation proceeds.

Theorem 1. *Let R be a RAM with M registers of size s, $M + 1 \leq N$. Let $I = i_1, i_2, \ldots i_M$ be the input data in R's registers. Let C be a computation on R taking time T_1 on input data I. Let A be a lively flying amorphous computer with broadcasting protocol B. Let the nodes of A except of the base node have memory capacity of at least $O(\log_2 M + \log_2 N + s)$ bits. Let A be set up with initial configuration I. Then A can simulate computation C in a finite time.*

Proof. Let us start with the memory requirements. Any node must hold an address in the range $[1..M]$ requiring $\lceil \log_2 M \rceil$ bits. Such a node must also hold a parameter k of a broadcasting protocol (cf. Section 5), which is of size $O(N)$ and, therefore, requires $O(\log_2 N)$ bits. The register also holds data of the same size as in R, requiring again s bits. In total, the space complexity of a node is $O(\log_2 M + \log_2 N + s)$ bits. Obviously, for a successful simulation it suffices if $N = M + 1$.

Now the amorphous computer can simulate the computation of R by simulating the individual instructions one after the other. During the simulation, the base node controls the computation and plays the role of the control unit of R. The values of registers of R are stored in other nodes of A.

For simulating an instruction the base node must be able to perform two kinds of operations. First, it must be able to simulate reading from and writing to memory registers what it does by communicating with the other nodes using broadcasts. Second, it has to simulate arithmetic computations, branching, and choosing of the next instruction to be processed. These operations are encoded in the internal fixed program of the base node.

For example, if A simulates instruction $A[1] + A[2] \rightarrow A[10]$, the base node firsts fetches the value of register 1 by communicating with a node with address 1, then the value of register 2. Then, it computes the addition and finally simulates storing of the result to register 10 by sending its contents to the node with address 10.

Obviously, a broadcast to some node may fail due to a temporary inaccessibility of that node. If, within a certain a priori given time interval, the base node does not receive an acknowledgement from the target node, the broadcast is repeated until an acknowledgement is received. Thanks to the amorphous computer being lively flying, at most a finite number of retries is required. It follows that one instruction can be simulated in finite time and, therefore, any finite computation can be simulated in finite time, too.

The computation ends with the *HALT* instruction in which case the base node enters a special halt state and stops processing of further instructions. The results are stored in the prescribed output nodes.   □

If the simulated program is too long to fit into the base node's memory, one can simulate a RASP machine (cf. [4]) in which the program is a part of the input data. The simulation is then similar to the previous RAM simulation, but now there is an extra read from a register in order to fetch the next instruction, which is then simulated.

Finally, note that a given instance of an amorphous computer has but a fixed number of nodes and so all computations performed on it are necessarily space-limited. Some constants used in the broadcast protocol (cf. Section 5) also depend on the size of the target area. Therefore, an amorphous computer presents a non-uniform computing model. However, the previous result shows that an infinite family of flying amorphous computers of increasing sizes has a universal computing power.

4   Setup

The setup has two phases: address assignment and input data loading.

Address assignment. The purpose of the address assignment process is to allocate unique addresses to the nodes of amorphous computer which are initially indistinguishable, all having zeroed memory. This process is controlled by the base node using the program from the left side of the following picture:

```
for a = 1 to M              |   procedure PickSingleNode()
   PickSingleNode()         |      broadcast(Initialize, 0)
   AssignAddress(a, h)      |      for i = 1 to h
                            |         broadcast(RandomGroup, i)
                            |         broadcast(WhichGroup?)
                            |         g = receive()
                            |         broadcast(Choose, g)
```

To pick a single node, the base node makes use of the algorithm from the right side of the previous picture. In this algorithm, procedure broadcast() represents a call of the broadcast protocol of the flying amorphous computer, which is described in the next section. In general, a broadcast can fail, but for the moment let us assume that the broadcast messages are always successfully delivered to all nodes. Namely, under this assumption it is easier to describe procedure PickSingleNode which makes use of the well-known method of probabilistic halving the set of candidates until a single one remains. The general case counting on the possibility of broadcast procedure failure is substantially more complicated since it must also handle the cases of temporary inaccessibility of some nodes and the problems of data consistency maintenance after the return of such nodes. This more involved case will be described later.

Procedure `PickSingleNode` initializes the nodes via the `Initialize` message. After receiving this message, all nodes still without an allocated address participate in the node selection procedure. In this procedure, the base node performs h so-called splitting rounds. During each round, all nodes randomly choose number 0 or 1 assigning themselves in this way to either of the two groups, 0 or 1. Then, all such nodes broadcast the number of the group they selected. From the received answers the base node randomly picks one group number and reports it back to all nodes. Only nodes that are in the selected group participate in the forthcoming round. It is expected that by each such split the number of candidate nodes is roughly halved. If value h is sufficiently high this algorithm selects exactly one node with high probability. The procedure `AssignAddress(a, h)` then assigns address a to a node that went through h rounds.

Theorem 2. *Let there be N nodes participating in the* `PickSingleNode()` *algorithm. The probability of selecting more than one node after h splitting rounds at is at most $N/2^h$.*

Proof. Consider the cardinality of the set of nodes after the last splitting round. After each round for each node the probability that another node remained in the same group is $1/2$. The probability that after h rounds a node still participates in the algorithm is $1/2^h$; therefore for all N nodes the probability that the cardinality of the remaining set is greater than 1 is, at most, $N/2^h$.   □

Now we return to the rather peculiar general case in which broadcasts in the address assignment phase can fail.

Recall that our goal selection of exactly one node with high probability is now complicated by the fact that there is no guarantee that a message sent to a certain subset of nodes will always reach all its members. However, for the correctness of the `PickSingleNode` algorithm it has been necessary that a single node is picked from among the nodes that have passed all splitting rounds.

For instance, if a node participating in a splitting round would not receive the `RandomGroup` message, then answering the subsequent `WhichGroup?` message that the node might later obtain would spoil the correctness of the entire algorithm. Namely, such a node could only provide an old value of g, not the current one. Thus, should a node miss a message, which is important for the correctness of selection process and which is addressed specifically to that node, the node must no longer participate in the algorithm.

In order to enable for a node to detect whether it has missed some round we make use of parameter `i` in `RandomGroup` message that was not important in the previous non-failing communication case. This parameter gives the number of the current splitting round. A node is allowed to participate in round $i+1$ if and only if the node has already passed successfully round i, i.e., if and only if it had first received message (`RandomGroup`,`i`) followed by (`Choose`,`g`), with g being the same as was the number of the node's chosen splitting group. In order to implement this idea it is enough for each node to remember the number of the last round in which it has participated. Then, upon receiving a (`RandomGroup`,`i`) message it can discover whether it has missed the previous message (i.e., the previous round) or not.

Upon detecting a missed message, the node at hand acts as if it was in the group that was not chosen, i.e., it stops participating in the next rounds of the splitting algorithm. This puts us on the safe side as far as the probability of selecting more than one node is concerned.

However, a premature termination of a node's further participation in the selection process could lead to an extreme case when all nodes stop participating eventually. Then, we end up with an empty set from which no node can be selected. In order to recognize this situation we introduce a mechanism of message acknowledgements. After performing a step in a computation, a node expected to perform such a step sends an acknowledging OK message back to the base node. The base node proceeds to the next step if and only if it hears at least one acknowledgement. Otherwise it repeats the current step. Technically, instead of performing, for example, a simple `broadcast(RandomGroup,i)`, the base node makes use of the following code (on the left side):

```
repeat                            |      repeat
   broadcast(RandomGroup,i)       |         broadcast(Write, a, x)
   m = receive()                  |         m = receive()
until m not empty                 |      until m not empty
```

With these changes, the probability of selecting exactly one node holds as in Theorem 2.

Input data loading. After the addresses were assigned to nodes but before starting the RAM simulation the input data must be stored into the respective nodes. The base node stores data x to a node with address a using the algorithm from the right side of the previous picture. Upon receiving the respective message, a node with address a stores x in its memory and responds with the OK message.

5 Communication

The communication capabilities of the individual nodes are as simple as possible. For instance, when a node transmits a message, it cannot determine if there are none, one, or several other nodes around it which might hear its message. Neither can a node determine if the sent message was successfully received by an other node. However, if the nodes adhere to a communication protocol providing a kind of coordination of the nodes' actions, a message can be transmitted from one node to other nodes with high probability even under these restrictive conditions.

The nodes communicate using broadcasting. A message originates in one node and is later broadcast to (potentially) all other nodes of the computer. The algorithm is relatively simple. The first node sends a message which is received by the node's immediate neighbors. These neighbors again send the message so that their neighbors receive it, etc. Note that whenever several nodes share a common neighbor, they should not send the message simultaneously since in such a case the message could not be delivered due to the collisions. In order to minimize the probability of collisions each node establishes a periodic sequence of time slots in which it sends the message randomly with a fixed probability p.

The protocol used by the nodes (the base node included) is described by the following code.

```
procedure broadcast(m,k)
while k <= B
    if random() < p then
        send(m,k+1)
        wait(2T)
    k := k+2
```

In the above procedure, calling send(m,k+1) causes one-shot sending of both message m and parameter $k+1$ from the node at hand. Parameter k of broadcast procedure keeps track of how many time slots have elapsed, in multiples of T, until the current moment. For the node where the message originates the starting value of k is 0. For other nodes, the current value of k is derived from the corresponding value in the lastly received message. The value of k is incremented by 2 on each time slot. Hencefore, all nodes stop broadcasting when k reaches the value of B, which is a global constant known in advance to all nodes of the whole flying amorphous computer. This will happen at about the same time; some small variation due to different clock speeds are possible.

The value of B influences how far from the originating node the message will spread. With too small a value, the message will not reach the nodes that are far away from the originating node; with too large a value, the excessive message sending after all nodes have already received the message just wastes time and energy.

We assume that the external operator is able to choose a reasonable value of constant B. Such a value should be selected, taking into account the number of nodes and their density in the target area. Asymptotically, the value rises as $O(\sqrt{N})$. In other words, it rises as the distance between the opposite corners of the target area if measured in units of r (the communication radius). The operator can find a reasonable value either by using a computer simulation of the system or by trying out several values on the actual flying amorphous computer.

Theorem 3. *Let all nodes of an amorphous computer be in a quiet state (i.e. not sending anything). Let X be a node that starts broadcasting a message using procedure* broadcast *at time t. Then, at time $t + BT(1 + \varepsilon_T)$ all nodes are again in a quiet state.*

Proof. The first node runs procedure broadcast with parameter k initially equal to 0. Each time through the loop the node waits for time $2T$ and the value of k is increased by 2. So, at most after time $BT(1 + \varepsilon_T)$ the node stops re-sending the message at hand. Transmitting a message takes time T. Therefore, whenever a neighbor node receives a message, value $k + 1$ corresponds to the running time of the procedure up to now, and hence also this node ends transmitting after at most $(B - k - 1)(1 + \varepsilon_T)$ time steps. By induction, all nodes end in time $BT(1 + \varepsilon_T)$. □

Note that in order to start a new broadcast it is necessary for all nodes to wait until the current broadcast ends. Namely, should a broadcast of two different messages have started simultaneously, then we cannot be sure which of the two messages would be delivered first. Therefore, the end of a current call to broadcast must be known to all nodes.

In order to guarantee the condition of non-concurrent broadcasts after the base node has started broadcasting a message, all other nodes must wait for time $B\varepsilon_T$ to be sure that in total time $B(1 + \varepsilon_T)$ has passed. Only thereafter a node can broadcast another message.

It is interesting to observe that there might be specific cases where a simultaneous broadcast cannot do any harm. For instance, in the address assignment protocol, when we wanted to know to which group some nodes belong, two different messages were sent at the same time from both groups. We did not care which one of the two messages was delivered since the groups had originally been selected randomly, anyway.

6   Conclusion

We have shown a computational universality of our model of flying amorphous computer. The nature of our results is quite unusual. In principle, a computation of a flying amorphous computer consists of two phases — of the preparation phase in which the computer is setup for the simulation, and the simulation proper.

The setup phase is of a probabilistic nature. As seen from Theorem 2, there is (a vanishing) probability that address allocation procedure will not end correctly, and there might appear the nodes with the same addresses. This will spoil the subsequent simulation. However, with overwhelming probability such a case will not happen and then the subsequent simulation algorithm will, in fact, be a deterministic algorithm: the result delivered by this algorithm will be correct with probability one. Once a flying amorphous computer is properly formed, it can be used for performing any simulation with the given number of nodes and its results will always be correct.

The next oddity of the simulation algorithm is its time complexity estimation in terms of finite time. This is due to the unpredictability of the communication paths formation. Interestingly, real computer simulations have revealed that within a flying amorphous computer a message is delivered to all nodes quite efficiently, in a wide range of velocities of node movements and parameters of the broadcast algorithm [9]. The simulations have also confirmed intuition that there is a tradeoff between the amount of nodes' mixing in a flying amorphous computer and the frequency of message repetition sendings in the broadcast protocol.

More research is needed in order to better understand and exploit the respective phenomena. Our results present but the first steps in this direction.

References

1. Abelson, H., et al.: Amorphous Computing. MIT Artificial Intelligence Laboratory Memo No. 1665 (August 1999)
2. Abelson, H., Allen, D., Coore, D., Hanson, C., Homsy, G., Knight Jr., T.F., Nagpal, R., Rauch, E., Sussman, G.J., Weiss, R.: Amorphous Computing. Communications of the ACM 43(5), 74–82 (2000)
3. Abelson, H., Beal, J., Sussman, G.J.: Amorphous Computing. Computer Science and Artificial Intelligence Laboratory, Technical Report, MIT-CSAIL-TR-2007-030 (June 2007)
4. Aho, A.V., Hopcroft, J.E., Ullman, J.D.: The Design and Analysis of Computer Algorithms. Addison-Wesley, Reading (1974)
5. Hoyle, F.: The Black Cloud, 219 p. Penguin Books (1957)
6. Kahn, J.M., Katz, R.H., Pister, K.S.: Next century challenges: mobile networking for "Smart Dust". In: Proceedings of the 5th Annual ACM/IEEE International Conference on Mobile Computing and Networking, MobiCom 1999, pp. 271–278. ACM, New York (August 1999)
7. Kahn, J.M., Katz, R.H., Pister, K.S.J.: Emerging Challenges: Mobile Networking for Smart Dust. Journal of Communications and Networks 2, 188–196 (2000)
8. Kurzweil, R.: The Singularity is Near, p. 652 pages. Viking Books, New York (2005)
9. Petrů, L.: Universality in Amorphous Computing. PhD Disseration Thesis, Dept. of Math. and Physics, Charles University, Prague (2009)
10. Petrů, L., Wiedermann, J.: A Model of an Amorphous Computer and Its Communication Protocol. In: van Leeuwen, J., Italiano, G.F., van der Hoek, W., Meinel, C., Sack, H., Plášil, F. (eds.) SOFSEM 2007. LNCS, vol. 4362, pp. 446–455. Springer, Heidelberg (2007)
11. Sailor, M.J., Link, J.R.: Smart dust: nanostructured devices in a grain of sand. Chemical Communications 11, 1375 (2005)
12. Vinge, V.: A Deepness in the Sky, 800 p. Tor Books (January 2000)
13. Warneke, B., Last, M., Liebowitz, B., Pister, K.S.J.: Smart Dust: communicating with a cubic-millimeter computer. Computer 34(1), 44–51 (2001)
14. Warneke, B., Atwood, B., Pister, K.S.J.: Smart dust mote forerunners. In: Proceedings of the 14th IEEE International Conference on Micro Electro Mechanical Systems, MEMS 2001, pp. 357–360 (2001)
15. Wiedermann, J., Petrů, L.: Computability in Amorphous Structures. In: Cooper, S.B., Löwe, B., Sorbi, A. (eds.) CiE 2007. LNCS, vol. 4497, pp. 781–790. Springer, Heidelberg (2007)
16. Wiedermann, J., Petrů, L.: Communicating mobile nano-machines and their computational power. In: Cheng, M. (ed.) NanoNet 2008. Lecture Notes of the Institute for Computer Sciences, Social Informatics and Telecommunications Engineering, vol. 3, pp. 123–130. Springer, Heidelberg (2009)
17. Wiedermann, J., Petrů, L.: On the Universal Computing Power of Amorphous Computing Systems. Theory of Computing Systems 46(4), 995–1010 (2009), http://www.springerlink.com/content/k2x6266k78274m05/fulltext.pdf
18. Wiedermann, J.: Nanomachine Computing by Quorum Sensing. In: Kelemen, J., Kelemenova, A. (eds.) Computation, Cooperation, and Life, Festschrifte. LNCS. Springer, Heidelberg (to appear, 2011)

Computation with Narrow CTCs[*]

A.C. Cem Say and Abuzer Yakaryılmaz

Boğaziçi University, Department of Computer Engineering,
Bebek 34342 İstanbul, Turkey
{say,abuzer}@boun.edu.tr

Abstract. We examine some variants of computation with closed time-like curves (CTCs), where various restrictions are imposed on the memory of the computer, and the information carrying capacity and range of the CTC. We give full characterizations of the classes of languages recognized by polynomial time probabilistic and quantum computers that can send a single classical bit to their own past. Such narrow CTCs are demonstrated to add the power of limited nondeterminism to deterministic computers, and lead to exponential speedup in constant-space probabilistic and quantum computation.

1 Introduction

It is known [AW09] that adding the capability of sending a polynomial number of bits through a closed timelike curve (CTC) to the past, so that it can be used as part of the input, to models as weak as constant-depth, polynomial-size Boolean circuits increases their computational power significantly, to match that of polynomial-space Turing machines. Interestingly, adding the same capability to a polynomial-time quantum computer results once again in the ability to solve precisely the problems in PSPACE, leading to Aaronson and Watrous' conclusion [AW09] that "CTCs make quantum and classical computing equivalent."

Since the information carrying capacity (*width*) of any CTC is finite, and the cost of building such a channel to the past may depend on its width critically, it is important to examine the power of computational models with "narrow" CTCs, i.e. those which are restricted to use a single-bit channel, regardless of the size of the input. The study of narrow CTCs [Bac04, AW09] has focused on polynomial-time computers as the core model until now, with results showing that classical computers augmented with narrow CTCs can recognize any language in BPP$_\text{path}$, whereas quantum computers with this capability can solve all problems in PP. Since it is not known whether any of the containments in P $\subseteq$ NP $\subseteq$ BPP$_\text{path}$ $\subseteq$ PP $\subseteq$ PSPACE is proper, we cannot presently say whether narrow CTCs are "useful" in these setups at all, and if so, whether they confer the same amount of power to classical and quantum models.

In this paper, we will complement the results mentioned above to provide full characterizations of the classes of languages recognized by probabilistic and

---

[*] This work was partially supported by the Scientific and Technological Research Council of Turkey (TÜBİTAK) with grant 108E142.

C.S. Calude et al. (Eds.): UC 2011, LNCS 6714, pp. 201–211, 2011.

quantum computers with narrow CTCs. It turns out that a CTC with a single classical bit provides a computer with precisely the power of postselection [Aar05].

The effects of restrictions on the core model of computation to be used inside the CTC on the power of the resulting setup were posed as open questions in [AW09]. In this regard, we consider real-time probabilistic and quantum finite automata (PFAs and QFAs), as well as several deterministic automaton models, as possible core models to be augmented with a narrow CTC. We show that narrow CTCs add the powers of limited nondeterminism or a certain kind of two-wayness to real-time automata. We are therefore able to prove that real-time PFAs, QFAs and deterministic pushdown automata (DPDAs) with narrow CTCs are strictly more powerful than their standard counterparts, whereas this addition does not increase the power of one-way deterministic finite automata (DFAs), or any deterministic two-way model. We also show that QFAs with narrow CTCs outperform their probabilistic counterparts, in contrast to the Aaronson-Watrous result on polynomial-time computation.

The rest of this paper is structured as follows. Section 2 summarizes previous work on the use of CTCs in computation. We define an alternative model of augmenting computers with CTCs in Section 3. Section 4 establishes the equivalence of the powers of postselection and narrow CTCs. In 5, we examine the effects of endowing weak deterministic machines with narrow CTCs. Section 6 is a conclusion.

2 Preliminaries

The existence of CTCs does not seem to be incompatible with the best available theories of spacetime. To cite just one example, an influential paper [MTY88] by Morris, Thorne, and Yurtsever describes how a technologically advanced civilization could first create a "wormhole", and then transform it to a "time machine" that can be used to send messages, or even people, backwards in time. The time machines of [MTY88] cannot be used to send anything to a time before their date of construction, say, d, and have a constant "range" of, say, T seconds, (determined by their builder at the beginning,) such that a message sent at any time point t $(t > d + T)$ is received at time point $t - T$. Once built, the machine can be used as many times as one wishes for such transmissions.

As noted by many authors of science fiction, a major problem with time travel is the "Grandfather Paradox," where a time traveler from the future prevents himself from traveling in the first place, leading to confusion about the state of the universe at the presumed time, say, A, of his arrival: He arrives if and only if he does not arrive. It was thought that Nature would prevent this logical inconsistency by simply not allowing time travel scenarios of that kind to be realized. Note that this argument assumes that the universe is supposed to be in exactly one, deterministic state, at all times. Probabilistic and quantum theories do not include this restriction, and David Deutsch [Deu91] showed that time travels to the past, including the above-mentioned scenario, would not lead to such problems if one just assumes that Nature imposes a *causal consistency condition* that the state x of the universe in the critical moment should be a fixed point of the operator f describing the evolution in the CTC, i.e. that $x = f(x)$. In the Grandfather Paradox scenario, Nature would "set" the state of the universe at time A to a distribution where the traveler arrives with probability $\frac{1}{2}$ to keep things consistent, as a "response" to the self-preventation action of the traveler.

As Deutsch noted, a computer which sends part of its output back in time to be used as part of its input can solve many computational problems much faster than what is believed to be possible without such loops. Bacon [Bac04] showed that NP-complete problems can be solved by polynomial-time computers with narrow CTCs. Aaronson and Watrous [AW09] proved, as mentioned in the introduction, that $AC^0_{CTC} = P_{CTC} = BQP_{CTC} = PSPACE_{CTC} = PSPACE$, where the subscript CTC under a class name indicates that the related machines have been reinforced by polynomial-width CTCs.

Let us review Aaronson and Watrous' model of quantum computation[1] with CTCs from [AW09]: A deterministic polynomial-time algorithm $\mathcal{A}$ takes an input w, and prints the description of a quantum circuit Q_w with rational amplitudes. Q_w acts on two registers of polynomial size. One register holds the information that is sent from the end of the computation in the future through the CTC, whereas the other one is a standard causality-respecting register, including a bit that will be used to report the output of the computation. The circuit is executed, with the CTC register set by Nature to some state satisfying the causal consistency condition described above, and the causality-respecting register initialized to all zeros, and the result is read off the output bit. A language L is said to be decided by such a CTC algorithm $\mathcal{A}$ if all members of L are accepted with high probability, and all nonmembers are rejected with high probability.

Note that this setup necessitates $\mathcal{A}$ to build a new CTC of the appropriate width for each different input w. This forces one [Deu91] to take the cost of this construction into account when analyzing the complexity, and the resources required may well scale exponentially in the width of the CTC.[2] The study of narrow CTCs, where this cost does not depend on the input length, is thus motivated.

3   Our Model

We will be considering several computation models that are augmented with the capability of sending a single classical bit of information from the time of the end of their execution back to the beginning.

We define a machine of type M_{CTC_1} as simply a machine of type M,[3] which has access to an additional bit in a so-called *CTC cell*. The CTC cell obtains

---

[1] The variant where the core model is classical is defined similarly.

[2] David Deutsch, personal communication.

[3] We will examine DFAs, PFAs, QFAs, both real-time and two-way DPDAs, time-bounded probabilistic and quantum Turing machines (PTMs and QTMs), and space-bounded deterministic Turing machines as core models. See [GHI67, Sip06, YS11b] for the standard definitions of these models.

its "initial" distribution from the future, according to the causal consistency condition. The program format of an M_{CTC_1} differs from that of an M so that it specifies the transitions to be performed for all possible combinations of not only the input symbol, internal state, etc., but also the CTC cell value. The set S of internal states of an M_{CTC_1} is defined as the union of three disjoint sets S_n, S_{p_0}, and S_{p_1}. The states in S_n are of the standard variety. When they are entered, the states in S_{p_1} (S_{p_0}) cause a 1 (0) to be sent back in time to be assigned to the CTC cell at the start of the execution. We assume that states in $S_{p_1} \cup S_{p_0}$ are entered only at the end of execution, (for real-time models, this is precisely when the machine is reading the end-marker symbol), and all states entered at that time are in $S_{p_1} \cup S_{p_0}$. Any number of members of S can be designated as accept states. The input string w is accepted if, for all stationary distributions of the evolution of the CTC bit induced by w, the machine accepts with probability at least $\frac{2}{3}$ with the CTC cell starting at that distribution. A string w is rejected if, for all stationary distributions of the evolution of the CTC bit induced by w, the machine rejects with probability at least $\frac{2}{3}$ with the CTC cell starting at that distribution. A language is recognized if all its members are accepted and all its nonmembers are rejected.

It is evident that any language recognized by any M_{CTC_1} according to our definition is also decided by some CTC algorithm á la the Aaronson-Watrous definition, described in Section 2. The motivation for the difference between the definitions is that the weakness of some of the core models we will use precludes us from performing any processing before using the CTC, and calculating or bounding the runtime, which determines the required "range" of the CTC beforehand. For more on this issue, see [SY11].

4   Postselection and Narrow CTCs

An important tool in the analysis of the capabilities of computers with narrow CTCs is the observation that one CTC bit endows any probabilistic or quantum core model with the power of postselection [Aar05]. This fact is already known, but since we have not seen it stated explicitly anywhere, we present a demonstration of it below.

Postselection is the capability of discarding all branches of a computation in which a specific event does not occur, and focusing on the surviving branches for the final decision about the membership of the input string in the recognized language. A formal definition of polynomial time computation with postselection can be found in [Aar05], where it was proven that PostBQP, the class of languages recognized by polynomial-time QTMs with postselection, equals the class PP. In the following, we present the analogous definition in the context of real-time computation with constant space.

A PFA (resp. QFA) $\mathcal{A}$ with postselection [YS11a] is simply an ordinary PFA (QFA) whose state set is partitioned into the sets of postselection accept, postselection reject, and nonpostselection states, and satisfies the condition that the probability that $\mathcal{A}$ will be in at least one postselection (accept or reject) state at the end of the processing is nonzero for all possible input strings. The overall acceptance and rejection probabilities of any input string w ($P^a_{\mathcal{A}}(w)$ and $P^r_{\mathcal{A}}(w)$, respectively) are calculated by simply discarding the computation paths ending at nonpostselection states, and performing a normalization so that the decision about the input is given by the postselection states:

$$P^a_{\mathcal{A}}(w) = \frac{p^a_{\mathcal{A}}(w)}{p^a_{\mathcal{A}}(w) + p^r_{\mathcal{A}}(w)} \quad \text{and} \quad P^r_{\mathcal{A}}(w) = \frac{p^r_{\mathcal{A}}(w)}{p^a_{\mathcal{A}}(w) + p^r_{\mathcal{A}}(w)}, \tag{1}$$

where $p^a_{\mathcal{A}}(w)$ and $p^r_{\mathcal{A}}(w)$, respectively, are the acceptance and rejection probabilities of w before the normalization.

Lemma 1. *Any language that can be recognized by a real-time automaton of type* $\mathrm{M} \in \{PFA, QFA\}$ *with postselection can be recognized by an* $\mathrm{M}_{\mathrm{CTC}_1}$.

Proof. Let $\mathcal{A}$ be the given automaton with postselection. We construct an $\mathrm{M}_{\mathrm{CTC}_1}$ $\mathcal{A}'$. The values 1 and 0 of the CTC bit are associated with acceptance and rejection, as explained below. $\mathcal{A}'$ imitates the behavior of $\mathcal{A}$ until the end of the input. If $\mathcal{A}$ ends up at a postselection state, $\mathcal{A}'$ halts with the same decision as $\mathcal{A}$ at that state. If $\mathcal{A}$ ends up at a nonpostselection state, $\mathcal{A}'$ simply reports the value it sees in the CTC bit as its decision. $\mathcal{A}'$ sends the value associated with its decision to the past as it halts.

The evolution of the CTC bit of $\mathcal{A}'$ for input w is described by the column stochastic matrix

$$\begin{pmatrix} 1 - p^a_{\mathcal{A}}(w) & p^r_{\mathcal{A}}(w) \\ p^a_{\mathcal{A}}(w) & 1 - p^r_{\mathcal{A}}(w) \end{pmatrix}, \tag{2}$$

whose only stationary distribution is

$$\begin{pmatrix} \frac{p^r_{\mathcal{A}}(w)}{p^a_{\mathcal{A}}(w) + p^r_{\mathcal{A}}(w)} \\ \frac{p^a_{\mathcal{A}}(w)}{p^a_{\mathcal{A}}(w) + p^r_{\mathcal{A}}(w)} \end{pmatrix}, \tag{3}$$

where the first and second entries stand for the probabilities of the values 0 and 1, respectively, meaning that $\mathcal{A}'$ recognizes $\mathcal{A}$'s language with the same error probability. $\qquad\square$

Let $\mathsf{BPP}_{\mathrm{CTC}_1}$ and $\mathsf{BQP}_{\mathrm{CTC}_1}$ denote the classes of languages recognized by polynomial time PTMs and QTMs with narrow CTCs using classical bits, respectively.[4] $\mathsf{BPP}_{\mathrm{path}}$ is the class of languages recognized by polynomial-time PTMs with postselection.

The results $\mathsf{BPP}_{\mathrm{path}} \subseteq \mathsf{BPP}_{\mathrm{CTC}_1}$ and $\mathsf{PP} \subseteq \mathsf{BQP}_{\mathrm{CTC}_1}$, that we alluded to in the introduction, are obtained using the link described above between CTCs and postselection.

We now present our main result that the power of postselection is all that a narrow CTC can confer on a computer. Let pPTM and pQTM denote polynomial-time PTMs and QTMs, respectively.

---

[4] Our definition of $\mathsf{BQP}_{\mathrm{CTC}_1}$ is different from the $\mathsf{BQP}_{\mathrm{CTC1}}$ given in [AW09], since Aaronson and Watrous consider a quantum bit sent through the CTC.

Lemma 2. *Any language that can be recognized by an* M_{CTC_1}*, where* M $\in$ *{PFA,QFA,pPTM,pQTM}, can be recognized by a machine of type M with post-selection.*

Proof. Let L be a language recognized by an M_{CTC_1} named $\mathcal{A}$. We start by constructing two machines of the standard type M, namely, $\mathcal{A}_0$ and $\mathcal{A}_1$, that simulate the computation of $\mathcal{A}$ by fixing 0 and 1 for the value of the CTC bit, respectively.

Let $p_{\mathcal{A}_i}^j(w)$ denote the probability that $\mathcal{A}_i$ will reach a configuration corresponding to sending the bit j to the past at the end of its computation when started on input w, where $i, j \in \{0, 1\}$. The CTC bit's evolution is described by

$$\begin{pmatrix} 1 - p_{\mathcal{A}_0}^1(w) & p_{\mathcal{A}_1}^0(w) \\ p_{\mathcal{A}_0}^1(w) & 1 - p_{\mathcal{A}_1}^0(w) \end{pmatrix}, \tag{4}$$

with stationary distribution

$$\begin{pmatrix} \frac{p_{\mathcal{A}_1}^0(w)}{p_{\mathcal{A}_0}^1(w) + p_{\mathcal{A}_1}^0(w)} \\ \frac{p_{\mathcal{A}_0}^1(w)}{p_{\mathcal{A}_0}^1(w) + p_{\mathcal{A}_1}^0(w)} \end{pmatrix}. \tag{5}$$

For $w \in L$, we therefore have the inequality

$$\frac{p_{\mathcal{A}_1}^0(w)}{p_{\mathcal{A}_0}^1(w) + p_{\mathcal{A}_1}^0(w)} p_{\mathcal{A}_0}^a(w) + \frac{p_{\mathcal{A}_0}^1(w)}{p_{\mathcal{A}_0}^1(w) + p_{\mathcal{A}_1}^0(w)} p_{\mathcal{A}_1}^a(w) \geq \frac{2}{3}. \tag{6}$$

We claim that, for all machine types M mentioned in the theorem statement, one can construct an instance of M, say, $\mathcal{A}'$, which will have two mutually exclusive collections of states, say, S_a and S_r, such that the probability that $\mathcal{A}'$ halts in S_a when started on input w is

$$\frac{1}{2} \left(p_{\mathcal{A}_1}^0(w) p_{\mathcal{A}_0}^a(w) + p_{\mathcal{A}_0}^1(w) p_{\mathcal{A}_1}^a(w) \right), \tag{7}$$

and the probability that $\mathcal{A}'$ halts in S_r is

$$\frac{1}{2} \left(p_{\mathcal{A}_1}^0(w) \left(1 - p_{\mathcal{A}_0}^a(w) \right) + p_{\mathcal{A}_0}^1(w) \left(1 - p_{\mathcal{A}_1}^a(w) \right) \right). \tag{8}$$

For instance, if M=pPTM, we first build two pPTMs, say, $\mathcal{A}_{10}$ and $\mathcal{A}_{01}$, for handling the two operands of the addition in Equation 7 by sequencing $\mathcal{A}_0$ and $\mathcal{A}_1$ to run on the input in the two different possible orders. $\mathcal{A}'$ simply runs $\mathcal{A}_{10}$ and $\mathcal{A}_{01}$, with probability $\frac{1}{2}$ each. Equation 8 is handled similarly.

If $\mathcal{A}$ is a real-time machine, the sequential processing described above is not permitted, and we instead perform tensor products of $\mathcal{A}_0$ and $\mathcal{A}_1$ to obtain the submachines $\mathcal{A}_{10}$ and $\mathcal{A}_{01}$.

Once $\mathcal{A}'$ is completed, we view it as a machine with postselection, by postselecting on the states in $S_a \cup S_r$ being reached at the end. S_a will be designated to be the set of accept states. S_r will be the reject states.

Performing the normalization described in Equation 1 on Equations 7 and 8 to calculate $\mathcal{A}'$'s probability of acceptance, one obtains precisely the expression depicted in Equation 6. The case of $w \notin L$ is symmetric. We conclude that $\mathcal{A}'$ recognizes L with exactly the same error probability as $\mathcal{A}$. $\qquad\square$

We have therefore proven

Theorem 1. *For any* $\mathrm{M} \in \{PFA, QFA, pPTM, pQTM\}$, $\mathrm{M}_{\mathrm{CTC}_1}$ *is equivalent in language recognition power to a machine of type* M *with postselection.*

Corollary 1. $\mathsf{BPP_{path}} = \mathsf{BPP_{CTC_1}}$, *and* $\mathsf{PP} = \mathsf{PostBQP} = \mathsf{BQP_{CTC_1}}$.

We can use Lemma 1 to demonstrate the superiority of real-time PFAs and QFAs with narrow CTCs over their standard versions, which can only recognize regular languages with bounded error: For a given string w, let $|w|_\sigma$ denote the number of occurrences of symbol σ in w. The nonregular language $L_{eq} = \{w \in \{a, b\}^* \mid |w|_a = |w|_b\}$ can be recognized by a PFA with postselection [YS11a]. As for quantum machines, the language $L_{pal} = \{w \in \{a, b\}^* \mid w = w^r\}$ is recognized by a QFA with postselection [YS11a]. L_{eq} is known [Fre81] to be recognizable by two-way PFAs at best in superpolynomial time [GW86], and the best known two-way QFA algorithm [AW02] for L_{pal} has exponential expected runtime. Furthermore, L_{pal} is known [DS92] to be unrecognizable by even two-way PFAs with bounded error, and no PFA with postselection can outperform a standard two-way PFA [YS11a], so we have established that finite-state quantum models with narrow CTCs outperform their probabilistic counterparts:

Corollary 2. *The class of languages recognized by* $QFA_{CTC_1}s$ *properly contains the class of languages recognized by* $PFA_{CTC_1}s$.

5 Weak Deterministic Models with Narrow CTCs

We adapt the argument used in [AW09] to prove that $\mathsf{P_{CTC}} \subseteq \mathsf{PSPACE}$ to state the following upper bounds for the powers of deterministic machines with narrow CTCs:

Theorem 2. *Let* $\mathcal{A}$ *be any machine of type* $\mathrm{M}_{\mathrm{CTC}_1}$ *that recognizes a language according to the definition in Section 3, where* M *is a deterministic model.* $\mathcal{A}$ *can be simulated by running two machines of type* M *in succession.*

Proof. We will assume that $\mathcal{A}$ never enters an infinite loop. (For a lengthier argument where we get rid of this assumption, see [SY11].) For a given input string w, let f_w be the mapping among probability distributions over the CTC bit realized by $\mathcal{A}$ when running on w. We have to find a distribution d, such that $f_w(d) = d$, and see how $\mathcal{A}$ responds to w when it starts with the CTC bit having that distribution.

We first run a machine $\mathcal{A}_1$ of type M obtained by fixing the CTC bit of $\mathcal{A}$ to 0 on input w. If this run ends at a state belonging to S_{p_0}, the collection of $\mathcal{A}$'s states that set the CTC bit to 0, we have found that $\begin{pmatrix} 1 \\ 0 \end{pmatrix}$ is a stationary distribution, and the response given to w by $\mathcal{A}_1$ is what $\mathcal{A}$ itself would announce if it were executed. If this first stage ends within S_{p_1}, then we run another machine $\mathcal{A}_2$ of type M, obtained by fixing the CTC bit of $\mathcal{A}$ to 1, on input w. Note that the only remaining possibilities for stationary distributions at this stage are $\begin{pmatrix} 0 \\ 1 \end{pmatrix}$ and $\begin{pmatrix} \frac{1}{2} \\ \frac{1}{2} \end{pmatrix}$, and in either case, $\mathcal{A}_2$'s response to w is certain to be identical to $\mathcal{A}$'s response, since $\mathcal{A}$ cannot have an error probability as big as $\frac{1}{2}$.   □

This construction can be realized by a two-way version of model M. It is known [She59] that two way DFA's are equivalent to their one-way versions.

Corollary 3. *One-way $DFA_{CTC_1}s$ recognize precisely the regular languages.*

Two-way DPDAs are more powerful than one-way DPDAs [GHI67]. Given a one-way $DPDA_{CTC_1}$ $\mathcal{A}$, we can apply the idea of Theorem 2 to obtain three DPDAs as follows: $\mathcal{A}_1$ is obtained by fixing the CTC bit to 0, and accepting if computation ends in a member of S_{p_0} that is also an accept state. $\mathcal{A}_2$ is obtained by fixing the CTC bit to 0, and accepting if computation ends in any member of S_{p_1}. $\mathcal{A}_3$ is obtained by fixing the CTC bit to 1, with no change to the accept states of $\mathcal{A}$.

Calling the languages recognized by these three machines L_1, L_2, and L_3, respectively, it is easy to see that the language recognized by $\mathcal{A}$ is $L_1 \cup (L_2 \cap L_3)$. We conclude that any language recognized by a one-way $DPDA_{CTC_1}$ can be expressed as the union of a deterministic context-free language (DCFL) with a language that is the intersection of two DCFLs.

To demonstrate that $DPDA_{CTC_1}s$ are actually more powerful than ordinary DPDAs, we will show that the capability of sending a finite number of bits to the past endows a machine with the power of limited nondeterminism.

The amount of nondeterminism used by a PDA can be quantified in the following manner [Her97]: The branching of a single move of a (nondeterministic) PDA is defined as the number of next configurations that are possible from the given configuration. The branching of a computation path of a PDA $\mathcal{N}$ is the product of the branchings of all the moves in this path. The branching of a string w accepted by $\mathcal{N}$ is the minimum of the branchings of the paths of $\mathcal{N}$ that accept w. Finally, the branching of $\mathcal{N}$ is the maximum of the branchings of the strings accepted by $\mathcal{N}$.

Fact 1. *The class of languages recognized by PDAs with branching k is the class of unions of k DCFLs.*

Theorem 3. *Any language that can be expressed as the union of two DCFLs can be recognized by a one-way $DPDA_{CTC_1}$.*

Proof. By Fact 1, we only need to show how to build a $DPDA_{CTC_1}$ that can simulate a given PDA $\mathcal{N}$ with branching 2.

We convert $\mathcal{N}$ to an equivalent PDA $\mathcal{N}'$, all of whose computational paths have branching exactly 2, by modifying the program so that for any computational path of $\mathcal{N}$ with branching greater than 2, $\mathcal{N}'$ simply scans the input until the end, and rejects without performing that excess branching. For every nondeterministic state of $\mathcal{N}'$, name the two outgoing branches 0 and 1. Convert $\mathcal{N}'$ to a $DPDA_{CTC_1}$ $\mathcal{A}$ which simulates $\mathcal{N}'$, selecting the ith branch if and only if it sees the value i in the CTC bit. At the end of the input, $\mathcal{A}$ sends the name of the current branch to the past if it is accepting the input. It sends the name of the other branch otherwise.

Table 1. Evolutions and stationary distributions of the CTC bit of $\mathcal{A}$

branch$_0$	branch$_1$	CTC transformation	Stationary distribution
Acc	Acc	$\begin{pmatrix} 1 & 0 \\ 0 & 1 \end{pmatrix}$	Any distribution
Acc	Rej	$\begin{pmatrix} 1 & 1 \\ 0 & 0 \end{pmatrix}$	$\begin{pmatrix} 1 \\ 0 \end{pmatrix}$
Rej	Acc	$\begin{pmatrix} 0 & 0 \\ 1 & 1 \end{pmatrix}$	$\begin{pmatrix} 0 \\ 1 \end{pmatrix}$
Rej	Rej	$\begin{pmatrix} 0 & 1 \\ 1 & 0 \end{pmatrix}$	$\begin{pmatrix} \frac{1}{2} \\ \frac{1}{2} \end{pmatrix}$

We consider the four possible cases corresponding to accept/reject responses of the two paths, and the associated evolutions of the CTC bit in Table 1. It is evident that $\mathcal{A}$ recognizes the language of $\mathcal{N}$ with zero error. $\qquad \square$

Since there exist languages (e.g. $\{a^i b^j c^k | i = j \text{ or } i = k\}$) that are not deterministic context-free, but which can be expressed as the union of two DCFLs, we conclude that the computation power of DPDAs is actually increased by the addition of a narrow CTC.

6 Concluding Remarks and Open Questions

We have examined the power of several computational models augmented by the capability of sending a single classical bit to the past. We have characterized the classes BPP_{CTC_1} and BQP_{CTC_1} in terms of classical conventional classes, and shown that real-time probabilistic and quantum finite automata, as well as deterministic pushdown automata, benefit from narrow CTCs. In [SY11], we establish that CTCs remain useful even if the information channel to the past has a small fixed range, and that narrow CTCs do not change the power of deterministic models with two-way access to the input string.

In an earlier paper [YS11a], we had shown that machines with postselection have precisely the same power as conventional machines that are able to reset their input head to the start of the input string and switch to the initial state, to restart the computation all over again [YS10]. The new link to narrow CTCs shows that postselection is indeed a profoundly interesting concept that requires further investigation.

Some open questions remain. To which extent can our results be generalized to CTCs with capacities of $k > 1$ bits? With more CTC bits, one can clearly implement more nondeterministic choices. With more nondeterminism, one can obtain more succinct finite automata for certain languages, and build PDAs with superior language recognition capability [Her97]. Assuming $\mathsf{BPP_{path}} \neq \mathsf{PSPACE} \neq \mathsf{PP}$, the equivalence of the power conferred by k-bit CTCs to that of postselection must break down at some point before $k = poly(n)$, where n is the length of the input. It would be interesting to clarify and quantify these relationships.

Acknowledgements

We thank David Deutsch, Scott Aaronson, Amos Ori, and Taylan Cemgil for their helpful answers to our questions. We are also grateful to the anonymous referees for their very useful suggestions.

References

[Aar05]   Aaronson, S.: Quantum computing, postselection, and probabilistic polynomial-time. Proceedings of the Royal Society A 461(2063), 3473–3482 (2005)

[AW02]   Ambainis, A., Watrous, J.: Two–way finite automata with quantum and classical states. Theoretical Computer Science 287(1), 299–311 (2002)

[AW09]   Aaronson, S., Watrous, J.: Closed timelike curves make quantum and classical computing equivalent. Proceedings of the Royal Society A 465(2102), 631–647 (2009)

[Bac04]   Bacon, D.: Quantum computational complexity in the presence of closed timelike curves. Physical Review A 70(032309) (2004)

[Deu91]   Deutsch, D.: Quantum mechanics near closed timelike lines. Physical Review D 44(10), 3197–3217 (1991)

[DS92]   Dwork, C., Stockmeyer, L.: Finite state verifiers I: The power of interaction. Journal of the ACM 39(4), 800–828 (1992)

[Fre81]   Freivalds, R.: Probabilistic two-way machines. In: Gruska, J., Chytil, M.P. (eds.) MFCS 1981. LNCS, vol. 118, pp. 33–45. Springer, Heidelberg (1981)

[GHI67]   Gray, J.N., Harrison, M.A., Ibarra, O.H.: Two-way pushdown automata. Information and Control 11, 30–70 (1967)

[GW86]   Greenberg, A.G., Weiss, A.: A lower bound for probabilistic algorithms for finite state machines. Journal of Computer and System Sciences 33(1), 88–105 (1986)

[Her97]   Herzog, C.: Pushdown automata with bounded nondeterminism and bounded ambiguity. Theoretical Computer Science 181, 141–157 (1997)

[MTY88]   Morris, M.S., Thorne, K.S., Yurtsever, U.: Wormholes, time machines, and the weak energy condition. Physical Review Letters 61(13), 1446–1449 (1988)

[She59]   Shepherdson, J.C.: The reduction of two–way automata to one-way automata. IBM Journal of Research and Development 3, 198–200 (1959)

[Sip06]   Sipser, M.: Introduction to the Theory of Computation, 2nd edn. Thomson Course Technology, United States of America (2006)

[SY11]   Say, A.C.C., Yakaryılmaz, A.: Computation with narrow CTCs. Technical Report arXiv:1007.3624 (2011)

[YS10]   Yakaryılmaz, A., Say, A.C.C.: Succinctness of two-way probabilistic and quantum finite automata. Discrete Mathematics and Theoretical Computer Science 12(4), 19–40 (2010)

[YS11a]   Yakaryılmaz, A., Say, A.C.C.: Probabilistic and quantum finite automata with postselection. Technical Report arXiv:1102.0666 (2011)

[YS11b]   Yakaryılmaz, A., Say, A.C.C.: Unbounded-error quantum computation with small space bounds. Information and Computation 279(6), 873–892 (2011)

Extended Watson-Crick L Systems with Regular Trigger Languages

David Sears and Kai Salomaa

School of Computing, Queen's University, Kingston, Ontario Canada K7L 3N6
{sears,ksalomaa}@cs.queensu.ca

Abstract. Watson-Crick Lindenmayer systems (L systems) add a control mechanism to ordinary L system derivations. The mechanism is inspired by the complementarity relation in DNA strings, and it is formally defined in terms of a trigger language (trigger, for short). In this paper we prove that Uni-Transitional Watson-Crick $E0L$ systems with regular triggers can recognize the recursively enumerable (RE) languages. We also find that even if the trigger is nondeterministically applied and the number of its applications can be unbounded then the computational power does not change. In the case where the number of applications of the trigger is bounded we find that the computational power lies within the $ET0L$ languages. We also find that Watson-Crick $ET0L$ systems where the number of complementary transitions is bounded by any natural number are equivalent in expressive power.

1 Introduction

Watson-Crick complementarity is a fundamental concept in DNA Computing. Several models abstracting and utilizing this concept have been formulated since Adleman's celebrated experiment [1]. One class of models is the family of Watson-Crick L systems introduced by V. Mihalache and A. Salomaa [6].

Informally, Watson-Crick L systems are L systems which consist of DNA-like alphabets, a complementary relation known as the Watson-Crick morphism and a set known as the trigger language (trigger, for short). DNA-like alphabets are alphabets which can be partitioned into two subalphabets of the same size: the purine and pyrimidine alphabets. The Watson-Crick morphism is a morphism which maps symbols in the purine alphabet to symbols in the pyrimidine alphabet and vice versa. Operationally, if the underlying L system is about to derive a string in the trigger language then in the next derivation step the complementary version of the string is derived instead, using the Watson-Crick morphism. The most commonly studied trigger language, referred to as the 'standard' trigger language, is the set of all strings where the number of pyrimidine symbols is greater than the number of purine symbols. Note that this language is context-free.

Studies had initially considered $D0L$ systems as the backbone of these models and the standard trigger [6,11,12,7]. Following these fundamental investigations other studies were conducted on variants of these systems.

C.S. Calude et al. (Eds.): UC 2011, LNCS 6714, pp. 212–223, 2011.

The computational power of these variants is investigated in [4,15,16,13,3]. In many instances it is found that these systems can recognize any arbitrary RE language. The *non-regular* standard trigger is used to establish these results.

Here we focus on Watson-Crick L systems with *regular* trigger languages. It is known that Watson-Crick $D0L$ systems with regular triggers define a restricted family of languages, and for this family even the language equivalence problem is decidable [5]. The situation changes dramatically when we allow variables and nondeterminism in the derivations. We show that Watson-Crick $E0L$ systems with regular triggers define all RE languages, and even some seemingly restricted variants of this model (like weak and uni-transitional systems) retain universal computational power.

In Sect. 2 the definitions and concepts needed to understand the results are recalled. In Sect. 3 it is shown that Uni-Transitional Watson-Crick $E0L$ systems are capable of recognizing the RE languages. In Sect. 4 a similar construction is provided which can be used to show how Watson-Crick $E0L$ systems where the Watson-Crick morphism is nondeterministically applied or not applied in derivations can recognize the RE languages. In Sect. 5 weaker variants of these systems are investigated whose computational power lies between the $E0L$ and $ET0L$ languages.

2 Preliminaries

Throughout this paper it is assumed that the reader is familiar with the basics of Formal Language Theory, in particular, L *Systems*. Below definitions are recalled in an attempt to fulfill the requirements needed for interpreting the results contained herein. For further information, refer to [9] or [10].

The empty string is denoted by ε. $alph(A)$ denotes the minimal/smallest alphabet such that arbitrary language A is a language over $alph(A)$. $\#_A b$ denotes the number of occurrences of elements in the set A which exist in string b.

An $E0L$ *system* is a 4-tuple $G = (\Sigma, h, \omega, \Delta)$ where Σ is an alphabet, $\Delta \subseteq \Sigma$ (Δ is called the *terminal* alphabet of G), h is a finite substitution on Σ (into the set of subsets of Σ^*) and ω, referred to as the *axiom*, is an element of Σ^*. Where convenient h, the finite substitution, will be represented by a set of rewriting productions. The *language of* G is defined by $L(G) = \bigcup_{i \geq 0} h^i(\omega) \cap \Delta^*$. Informally, an $ET0L$ system is an $E0L$ system employing a finite set of finite substitutions (called *tables*) rather than an individual finite substitution. The derivation step differs in that one of the tables is nondeterministically chosen and only rewriting productions within the chosen table are used.

The notion of Watson-Crick complementarity is only applicable to systems whose alphabet is *DNA-like*. A DNA-like alphabet is an alphabet with even cardinality $2n$, $n \geq 1$, where the letters are enumerated as follows: $\Sigma = \{a_1, \ldots, a_n, \overline{a_1}, \ldots, \overline{a_n}\}$. Thus, each of the non-barred letters a_i, $1 \leq i \leq n$, has its barred version $\overline{a_i}$. It is said that a_i and $\overline{a_i}$ are *complements*. The unbarred and barred subalphabets of a DNA-like alphabet are referred to as the purine and pyrimidine alphabets. Complementarity is formally expressed in terms of the *Watson-Crick morphism*. The Watson-Crick morphism (denoted by h_W) is a letter-to-letter endomorphism of Σ^* mapping each letter to its complementary letter. For example, for a in a purine alphabet, $h_W(a) = \bar{a}$ where $\bar{a}$ is in the corresponding pyrimidine alphabet. Similarly, for $\bar{a}$ in the pyrimidine alphabet, $h_W(\bar{a}) = a$ where a is in the purine alphabet. In this paper h_W is generalized over strings of arbitrary length, not just individual letters. That is, for a string $s = a_1 a_2 \ldots a_n$, $h_W(s) = h_W(a_1) h_W(a_2) \ldots h_W(a_n)$.

Consider a DNA-like alphabet denoted by Σ. The Watson-Crick morphism is invoked on a string in Σ^+ when the string is an element of a set known as the *trigger language* (trigger for short; also commonly denoted by TR). The trigger language is defined as a recursive subset of Σ^+. A trigger language is said to be sound if whenever $w \in TR$ then $h_W(w) \notin TR$. Note that the reverse case may not hold. It is important to note that in all subsequent investigations it is implicitly assumed that the systems being considered have sound trigger languages.

$E0L$ systems augmented with the Watson-Crick morphism and a trigger language are known as *Watson-Crick $E0L$ systems* [3,7,14]. A Watson-Crick $E0L$ system is a tuple (G, TR) where $G = (\Sigma, h, \omega, \Delta)$ is the underlying $E0L$ system utilized by (G, TR), Σ is a DNA-like alphabet, TR is the trigger language of the system and $\omega \in \Sigma^* \setminus TR$. The language recognized by (G, TR) consists of strings over Δ which can be obtained from the axiom (ω) in some number of derivation steps: $L((G, TR)) = \{s \in \Delta^* \mid \omega \Rightarrow^* s\}$. For the direct derivation step, unless otherwise specified, we consider the standard derivation mode: Let $w_1 = a_1 \cdots a_m$. Then $w_1 \Rightarrow w_2$ in standard derivation mode where

- $w_2 \in h(a_1) \cdots h(a_m)$ if $w_1 \notin TR$
- $w_2 \in h_W(a_1) \cdots h_W(a_m)$ if $w_1 \in TR$

Throughout the rest of the paper, if $w_1 \in TR$ then the next derivation step is represented as $w_1 \Rightarrow^{hw} w_2$. Worthy of note is that in previous investigations the definition of standard derivation mode differs slightly. In previous considerations, if a string is about to be derived which is in the trigger language then instead its Watson-Crick complement is derived. With the definition of standard derivation mode above, strings in the trigger language are allowed to be derived and in the following derivation step their Watson-Crick complement is derived. Considering either definition of standard derivation mode in this paper does not change the languages recognized by any of the considered systems since all Watson-Crick L systems in this paper have trigger languages which do not contain strings in the language that they recognize.

As an alternative direct derivation step, we introduce a new derivation mode called *weak derivation mode*. The weak derivation mode differs from the standard derivation mode defined above in the second condition: if $w_1 \in TR$ we set either $w_2 \in h(a_1) \cdots h(a_m)$ or $w_2 \in h_W(a_1) \cdots h_W(a_m)$.

Here an example is provided that illustrates the difference between standard and weak derivation modes: Consider Watson-Crick $E0L$ System $H = (G, TR)$ where $TR = \{a, b\}^*$ is the trigger language, G is the underlying $E0L$ system defined by $G = (\{S, a, b, \overline{S}, \overline{a}, \overline{b}\}, h, S, \{a, b\})$ where h is the set of productions defined as $\{S \rightarrow ab, a \rightarrow a, a \rightarrow aa, b \rightarrow b, b \rightarrow bb, \overline{S} \rightarrow \overline{S}, \overline{a} \rightarrow a, \overline{b} \rightarrow b\}$. Consider H to be operating in standard derivation mode. The only possible derivation in standard derivation mode is $S \Rightarrow ab \Rightarrow^{hw} \overline{ab} \Rightarrow ab \Rightarrow^{hw} \overline{ab} \Rightarrow \cdots$. The derivation sequence $ab \Rightarrow^{hw} \overline{ab} \Rightarrow ab$ will continue forever. The only terminal string which appears in this derivation sequence is ab and consequently the language recognized by H when operating in standard derivation mode is $\{ab\}$. Notice that the trigger language is sound. Now consider H to be operating in weak derivation mode. Many derivations become possible: $S \Rightarrow ab \Rightarrow^{hw} \overline{ab} \Rightarrow \cdots$, $S \Rightarrow ab \Rightarrow a^2 b^2 \Rightarrow a^4 b^4 \Rightarrow a^8 b^8 \Rightarrow \cdots$, $S \Rightarrow ab \Rightarrow aab \Rightarrow aaab \Rightarrow \cdots$, $S \Rightarrow ab \Rightarrow^{hw}$ $\overline{ab} \Rightarrow ab \Rightarrow abb \Rightarrow aabbb \Rightarrow \cdots$, etc. Whenever a string is derived in TR the system can always nondeterministically choose to continue deriving using the underlying $E0L$ system rather than applying h_W in the next derivation step, yielding different derived strings. The language recognized by H when operating in weak derivation mode is $\{a^n b^m \mid n \geq 1, m \geq 1\}$.

We provide some example abbreviations for systems considered in this paper. $E0L(REG)$ refers to Watson-Crick $E0L$ systems with regular triggers. $E0L(REG,$ uni) refers to such systems which are uni-transitional, that is, each derivation of a terminal word can apply the Watson-Crick morphism at most once [13,14]. $E0L(REG, weak)$ refers to systems with regular triggers operating instead in weak derivation mode. $E0L(REG, weak, uni)$ refers to these weak derivation mode systems which are at most uni-transitional. $ET0L(REG, weak, bounded)$ refers to such systems where instead $ET0L$ systems are considered as the underlying generative mechanism and the number of complementary transitions is bounded by a constant [14].

3 Uni-transitional Watson-Crick $E0L$ systems

In this section it is proven that Uni-Transitional Watson-Crick $E0L$ systems [13] with regular trigger languages are sufficiently powerful to generate all RE languages. We know from [8] that one-sided *Penttonen-normal-form (PNF)* grammars recognize all RE languages. The productions of a PNF grammar are restricted to be of the forms

$$AB \rightarrow AC, \ A \rightarrow BC, \ A \rightarrow a, \ A \rightarrow \varepsilon \tag{1}$$

where A, B, C are nonterminals and a is a terminal.

In what follows a construction of a Watson-Crick $E0L$ system with a regular trigger is provided which recognizes the language of an arbitrary PNF grammar. The system constructed simulates directly all context-free productions of the given PNF grammar. Additionally, two new sets of special symbols are introduced and these symbols are used to simulate the context-sensitive productions. The regular trigger is specified such that it recognizes 'bad' simulations (produced by nondeterminism) involving the special symbols. 'Good' simulations are not captured by the trigger.

3.1   Construction

Let $G = (\Sigma, V, S, P)$ be a Penttonen-normal-form grammar with productions of P as in (1). Here Σ is the terminal alphabet, V the set of nonterminals and S the start nonterminal. Let $\hat{V} = \{\hat{v} \mid v \in V\}$, $(V \times V) = \{(a,b) \mid a,b \in V\}$. Let $H = (K, TR)$ be an $E0L(REG)$ system where K is the underlying $E0L$ system and TR is a regular trigger language. Here $K = (\Sigma \cup V \cup \hat{V} \cup (V \times V) \cup \overline{\Sigma} \cup \overline{V} \cup \overline{\hat{V}} \cup \overline{(V \times V)}, P_H, S, \Sigma \cup \overline{\Sigma})$ where all subalphabets of K are mutually disjoint and the set of productions P_H consist of the following rules:

- For any $a \in \Sigma$, $a \to a$ is in P_H
- Always when $A \to BC$, $A \to a$, or $A \to \varepsilon$ is in P where $A, B, C \in V, a \in \Sigma$, the same production is in P_H
- For any $A \in V$, P_H contains the productions $A \to A$, $A \to \hat{A}$, $\hat{A} \to A$ where $\hat{A} \in \hat{V}$
- For any $B, C \in V$, P_H contains the productions $B \to (B,C)$, $(B,C) \to C$
- For any $a \in \overline{\Sigma} \cup \overline{V} \cup \overline{\hat{V}} \cup \overline{(V \times V)}$, $a \to a$ is in P_H

Let $X = \Sigma \cup V \cup \hat{V} \cup (V \times V)$. The trigger language TR for the $E0L(REG)$ system H, is defined as:

$$TR = (\ X^* \hat{V} X^* \ \cup \ X^*(V \times V)X^*\ ) -$$
$$\{\ ac(d,e)b \mid a,b \in (\Sigma + V)^*, \ c \in \hat{V}, \ d,e \in V, \ (d,e) \in (V \times V), (cd \to ce) \in P\}$$

It is easy to verify that the trigger TR is sound.

3.2   Proof of Language Equivalence

Let G, H, TR, P, P_H be as defined in Sect. 3.1. Here we show that the language recognized by context-sensitive grammar G is equal to the language recognized by system H.

First we establish that derivations of G can be simulated by H. It is immediate that a derivation step of G involving a context-free production can be directly simulated in H, due to the definition of P_H. Next we show that also any context-sensitive production in G can be simulated by H.

Lemma 3.2.0.1. *Let* $a = pABq$, $b = pACq$ *where* $A, B, C \in V$, $p, q \in (\Sigma + V)^*$ *and let* $AB \to AC \in P$. *If* $a \Rightarrow_G b$ *then* $a \Rightarrow_H^* b$.

Proof. For any $i \in \Sigma \cup V$, $i \to_H i$ by definition of P_H. So $p \Rightarrow_H p$ and $q \Rightarrow_H q$. $A \to_H \hat{A}$, $B \to_H (B,C)$ by definition of P_H. So $pABq \Rightarrow_H p\hat{A}(B,C)q$. $\hat{A} \to_H A$, $(B,C) \to_H C$ by definition of P_H. So $p\hat{A}(B,C)q \Rightarrow_H pACq$. So $pABq \Rightarrow_H p\hat{A}(B,C)q \Rightarrow_H pACq$ and thus $a \Rightarrow_H^* b$.   □

Using the above observations it easy to show by induction that any derivation in G can be directly simulated in H, and hence:

Lemma 3.2.0.2. $L(G) \subseteq L(H)$

We need the property that the language $L(H)$ cannot contain symbols of $\overline{\Sigma}$.

Lemma 3.2.0.3. $L(H) \subseteq \Sigma^*$

Proof. Consider the derivation of a terminal string u in the system H. If h_W is applied in the derivation of u then the string produced after applying h_W contains a symbol in $\overline{V}$ (due to the definition of TR), a nonterminal, for which there exists only a self-rewriting production in P_H. $alph(TR) \cap \{\overline{V}\} = \varnothing$ so the trigger will not be applied in any subsequent derivations. This contradicts the fact that a terminal string is derived. On the other hand, if h_W is not applied in the derivation then u clearly cannot contain elements of $\overline{\Sigma}$.   □

The following lemma establishes that, for any $x, y \in (\Sigma + V)^*$, if x directly derives y in H then x derives y in G via a sequence of derivation steps where a context-free production is applied at each step. Here $\Rightarrow^?$ means derives in 0 or 1 steps.

Lemma 3.2.0.4. *For any* $x, y \in (\Sigma + V)^*$, *if* $x \Rightarrow_H y$ *then* $x \Rightarrow_G^* y$.

Proof. Recall that the only productions of H with both sides in $(\Sigma + V)^*$ are productions that simulate context-free productions of G. Hence we can write $x = a_1 \cdots a_n$ where $a_i \in \Sigma \cup V$, $y = b_1 \cdots b_n$ where $b_i \in (\Sigma + V)^*$ such that $a_i \Rightarrow_H b_i$ where $1 \leq i \leq n$. Note that $a_i \Rightarrow_H b_i$ iff $a_i \Rightarrow_G^? b_i$ where $1 \leq i \leq n$ by definition of P, P_H. So $a_i \Rightarrow_G^? b_i$ where $1 \leq i \leq n$. If this is the case then $(b_1 \cdots b_{i-1}) \cdot (a_i \cdots a_n) \Rightarrow_G^? (b_1 \cdots b_i) \cdot (a_{i+1} \cdots a_n)$ where $1 \leq i \leq n$.   □

The following proves that if a string $x \in (\Sigma + V)^*$ derives a string of the form $r\hat{A}(B,C)s$ in system H via a sequence of simulations of context-sensitive productions then x derives a string of the form $rACs$ in grammar G. This will be useful when proving that strings derived over the alphabet $(\Sigma + V)^*$ by sequences of simulations of context-sensitive productions in H can also be derived in G.

Lemma 3.2.0.5. *For any* $x \in (\Sigma + V)^*$, *if* $x = w_0 \Rightarrow_H \cdots \Rightarrow_H w_n = r\hat{A}(B,C)s$ *where* $r, s \in (\Sigma + V)^*$, $\hat{A} \in \hat{V}$, $(B,C) \in (V \times V)$, $w_{i-1} \Rightarrow_H w_i$ *are valid simulations of context-sensitive productions* ($w_i \in (\Sigma + V)^* \hat{V}(V \times V)(\Sigma + V)^*$ *and* $w_i \notin TR$) *and* $1 \leq i \leq n$ *then* $x \Rightarrow_G^+ rACs$.

Proof. **Base Case:** Consider $x \Rightarrow_H r\hat{A}(B,C)s$. Now x must contain substring AB as the only production for $\hat{A} \in \hat{V}$ in P_H is $A \to \hat{A}$ and the only production for $(B,C) \in (V \times V)$ in P_H is $B \to (B,C)$. So x must be of the form $pABq$ where $A, B \in V$, $p, q \in (\Sigma + V)^*$ since $x \in (\Sigma + V)^*$. It must be the case that $p \Rightarrow_H r$ and $q \Rightarrow_H s$. If this is the case then $pABq \Rightarrow_G^* rABq \Rightarrow_G^* rABs$ by Lemma 3.2.0.4. $AB \to AC$ must be a production in G since $r\hat{A}(B,C)s \notin TR$ and by definition of TR. Then $rABs \Rightarrow_G rACs$. So $x = pABq \Rightarrow_G^* rABq \Rightarrow_G^* rABs \Rightarrow_G rACs$. So $x \Rightarrow_G^+ rACs$.

   Inductive Step: Let $x = w_0 \Rightarrow_H^{n+1} w_{n+1} = t\hat{D}(E,F)u$ where $t, u \in (\Sigma + V)^*$, $\hat{D} \in \hat{V}$, $(E,F) \in (V \times V)$ and $w_{i-1} \Rightarrow_H w_i$ are valid simulations of context-sensitive productions ($w_i \in (\Sigma + V)^* \hat{V}(V \times V)(\Sigma + V)^*$ *and* $w_i \notin TR$) where $1 \leq i \leq n + 1$. Then $x \Rightarrow_H^n r\hat{A}(B,C)s$ and so $x \Rightarrow_G^* rACs$ by the Inductive Hypothesis. r or s must contain substring DE as the only production for $\hat{D} \in \hat{V}$ in P_H is $D \to \hat{D}$ and the only production for $(E,F) \in (V \times V)$ in P_H is $E \to (E,F)$.

1. Consider the case when s contains substring DE: s must be of the form $vDEw$ where $v, w \in (\Sigma + V)^*$ since $s \in (\Sigma + V)^*$. If this is the case then $r\hat{A}(B,C)s = r\hat{A}(B,C)vDEw \Rightarrow_H r'ACv'\hat{D}(E,F)w' = t\hat{D}(E,F)u$ where $r \Rightarrow_H r'$, $\hat{A} \Rightarrow_H A$, $(B,C) \Rightarrow_H C$, $v \Rightarrow_H v'$, $w \Rightarrow_H w'$ and $r', v', w' \in (\Sigma + V)^*$ since $t\hat{D}(E,F)u \in (\Sigma + V)^*\hat{V}(V \times V)(\Sigma + V)^*$. $rACs = rACvDEw \Rightarrow_G^*$ $r'ACvDEw$ by Lemma 3.2.0.4. $r'ACvDEw \Rightarrow_G^* r'ACv'DEw$ by Lemma 3.2.0.4. $r'ACv'DEw \Rightarrow_G^* r'ACv'DEw'$ by Lemma 3.2.0.4. $r'ACv'DEw' \Rightarrow_G$ $r'ACv'DFw' = tDFu$ since $DE \to DF$ is a production in G by $t\hat{D}(E,F)u \notin TR$ and definition of TR. So $x \Rightarrow_G^+ tDFu$ by transitivity of $\Rightarrow_G^*$.

2. Consider the case when r contains substring DE: The proof follows symmetrically for this case. □

The following lemma shows that if a string y in $(\Sigma + V)^*$ is derivable from x in $(\Sigma + V)^*$ via a sequence of simulations of context-sensitive productions in system H then y is also derivable from x in grammar G.

Lemma 3.2.0.6. *For any $x, y \in (\Sigma + V)^*$, if $x = w_0 \Rightarrow_H \cdots \Rightarrow_H w_n = u \Rightarrow_H y$ and $w_{i-1} \Rightarrow_H w_i$ are valid simulations of context-sensitive productions ($w_i \in (\Sigma + V)^*\hat{V}(V \times V)(\Sigma + V)^*$ and $w_i \notin TR$) where $1 \leq i \leq n$ then $x \Rightarrow_G^+ y$.*

Proof. Assume $x = w_0 \Rightarrow_H \cdots \Rightarrow_H w_n = u \Rightarrow_H y$ for any $x, y \in (\Sigma + V)^*$ and $w_{i-1} \Rightarrow_H w_i$ are valid simulations of context-sensitive productions. u is of the form $r\hat{A}(B,C)s$ where $r, s \in (\Sigma + V)^*$, $\hat{A} \in \hat{V}$, $(B,C) \in (V \times V)$. $u = r\hat{A}(B,C)s \Rightarrow_H r'ACs' = y$ where $\hat{A} \Rightarrow_H A$, $(B,C) \Rightarrow_H C$, $r \Rightarrow_H r'$, $s \Rightarrow_H s'$ and $r', s' \in (\Sigma + V)^*$ since $y \in (\Sigma + V)^*$. $x \Rightarrow_G^+ rACs$ by Lemma 3.2.0.5, $x = w_0 \Rightarrow_H \cdots \Rightarrow_H w_n = r\hat{A}(B,C)s$ and $w_{i-1} \Rightarrow_H w_i$ are valid simulations of context-sensitive productions. $rACs \Rightarrow_G^* r'ACs$ by Lemma 3.2.0.4. $r'ACs \Rightarrow_G^* r'ACs' = y$ by Lemma 3.2.0.4. So $x \Rightarrow_G^+ y$ by transitivity of $\Rightarrow_G^*$. □

Lemma 3.2.0.7. *For any $x, y \in (\Sigma + V)^*$, if $x \Rightarrow_H^n y$ where $n \geq 0$ then $x \Rightarrow_G^* y$.*

Proof. **Base Case:** The case $n = 0$ follows by the reflexivity of $\Rightarrow^*$.

Inductive Step: Assume $x \Rightarrow_H^n z \Rightarrow_H^m y$ where $x, y, z \in (\Sigma + V)^*$, $m \geq 1$ and y is the first occurrence of a string in $(\Sigma + V)^*$ in a derivation from z. $x \Rightarrow_G^* z$ by the Inductive Hypothesis. Consider the possibilities for m:

- If $m = 1$ then $z \Rightarrow_G^* y$ by Lemma 3.2.0.4 and so $x \Rightarrow_G^* y$.
- If $m > 1$ then $z \Rightarrow_H^m y$ can be rewritten as $z \Rightarrow_H w_1 \Rightarrow_H \cdots \Rightarrow_H w_m = y$ and $w_i \notin (\Sigma + V)^*$ where $1 \leq i < m$. Consider the possibilities for w_j where $1 \leq j < m$:

  1. If there exists a string $w_j \in TR$ where $1 \leq j < m$ then $\#_{\overline{\hat{V} \cup \overline{(V \times V)}}} w_{j+1} \geq 1$.

     Only self-rewriting productions exist for symbols $\hat{V}$ and $\overline{(V \times V)}$. This implies that no subsequent string will be derived in $(\Sigma + V)^*$, which contradicts the assumption that $z \Rightarrow_H^m y$. So it must be that $w_i \notin TR$ where $1 \leq i < m$.

  2. If there exists a string w_j where $\#_{\overline{V} \cup \overline{\hat{V} \cup \overline{(V \times V)}}} w_j \geq 1$ and $1 \leq j < m$ then no subsequent string in $(\Sigma + V)^*$ will be derived since only self-rewriting

productions exist for symbols $\overline{V}$, $\hat{V}$ and $\overline{(V \times V)}$. This contradicts the assumption that $z \Rightarrow_H^m y$. So it must be that w_i does not contain any symbols in $\overline{V} \cup \hat{V} \cup \overline{(V \times V)}$ where $1 \leq i < m$.

3. Assume that there exists a string w_j where $\#_{\overline{\Sigma}} w_j \geq 1$ and $1 \leq j < m$. Symbols in $\overline{\Sigma}$ can only be produced by applications of h_W since there are no productions in P_H which derive them. This implies that the string appearing immediately before w_j in the derivation is in TR. However, as proved in 1., if there exists a $w_k \in TR$ where $1 \leq k < m$ then this contradicts the assumption that $z \Rightarrow_H^m y$. So it must be that w_i does not contain any symbols in $\overline{\Sigma}$ where $1 \leq i < m$.

Thus $\#_{\overline{\Sigma} \cup \overline{V} \cup \hat{V} \cup \overline{(V \times V)}} w_i = 0$ where $1 \leq i < m$ by 2 and 3 above. $w_i \in (\Sigma + V)^* \hat{V} (V \times V)(\Sigma + V)^*$ since $w_i \notin (\Sigma + V)^*$, $\#_{\overline{\Sigma} \cup \overline{V} \cup \hat{V} \cup \overline{(V \times V)}} w_i = 0$ and $w_i \notin TR$ by 1 above where $1 \leq i < m$. Then $z \Rightarrow_G^+ y$ by this fact, $z \in (\Sigma + V)^*$, $w_i \notin TR$ by 1 above where $1 \leq i < m$ and Lemma 3.2.0.6. Thus $x \Rightarrow_G^+ y$.

So, in general, for both cases where $m = 1$ and $m > 1$, $x \Rightarrow_G^* y$.   □

Lemma 3.2.0.8. $L(H) \subseteq L(G)$

Proof. By Lemma 3.2.0.7, for any $y \in (\Sigma + V)^*$, if $S \Rightarrow_H^* y$ then $S \Rightarrow_G^* y$. Thus the claim follows by Lemma 3.2.0.3.

Now we can state the main result of this section:

Theorem 3.2.0.9. *Any RE language can be generated by a E0L(REG, uni) system.*

Proof. Any RE language can be generated by a PNF context-sensitive grammar (1), see Theorem 4 of [8]. The $E0L(REG)$ system defined in Sect. 3.1 was shown to be equivalent to an arbitrary grammar of this form by Lemmas 3.2.0.2 and 3.2.0.8. Clearly the constructed $E0L(REG)$ system is uni-transitional.   □

Note that the Watson-Crick E0L system used in the proof of Theorem 3.2.0.9, in fact, has the property that no successful derivation of a terminal word can use the Watson-Crick morphism. This type of systems are a special case of uni-transitional systems and they are called *blocking* systems in [14].

4   Weak Derivation Watson-Crick *E0L* Systems

In this section a construction is provided which can be used to show that Watson-Crick *E0L* systems with regular trigger languages operating in weak derivation mode are sufficiently powerful to generate all RE languages. The language recognized by the system in the construction is also the same if the system operates in standard derivation mode. The fundamental difference between this construction and the construction in Sect. 3 is that in order to simulate a context-sensitive derivation h_W must be applied. That is, if a string is derived with symbols in $\hat{V} \cup (V \times V)$ which is not in the trigger language (not a valid simulation of a context-sensitive derivation) then no terminal strings will be derived as these symbols will only rewrite themselves. As a consequent of this, another important difference arises: there is no upper bound placed on the number of applications of h_W in any given derivation if the system operates in either weak or standard derivation mode since there is no assumed bound on the number of context-sensitive derivations for any arbitrary derivation in any one-sided context-sensitive grammar. The proof of language equivalence found in [14] Sect. 4.2 is very similar to the proof used in Sect. 3.

Let $G = (\Sigma, V, S, P)$ be a one-sided context-sensitive grammar whose productions in P are of the forms in (1). Let $\hat{V} = \{\hat{v} \mid v \in V\}$, $(V \times V) = \{(a,b) \mid a, b \in V\}$. Let $I = (J, TR2)$ be an $E0L$(REG) system where J is the underlying $E0L$ system and $TR2$ is the regular trigger language. $J = (\Sigma \cup V \cup \hat{V} \cup (V \times V) \cup \overline{\Sigma} \cup \overline{V} \cup \widehat{\overline{V}} \cup \overline{(V \times V)}, P_I, S, \Sigma \cup \overline{\Sigma})$ where all subalphabets of J are mutually disjoint and P_I is the following production set:

- For any $a \in \Sigma$, $a \to a$ is in P_I
- Always when $A \to BC$, $A \to a$, or $A \to \varepsilon$ is in P where $A, B, C \in V, a \in \Sigma$, the same production is in P_I
- For any $A \in V$, P_I contains the productions $A \to A$, $A \to \hat{A}$, $\hat{A} \to \hat{A}$, $\widehat{\overline{A}} \to A$ where $\hat{A} \in \hat{V}$, $\widehat{\overline{A}} \in \widehat{\overline{V}}$
- For any $B, C \in V$, P_I contains the productions $B \to (B, C)$, $(B, C) \to (B, C)$, $\overline{(B, C)} \to C$
- For any $\overline{A} \in \overline{V}$, $\overline{A} \to A$ is in P_I
- For any $\overline{a} \in \overline{\Sigma}$, $\overline{a} \to a$

The trigger $TR2$ is defined as the regular set:

$$TR2 = \{ \, ac(d, e)b \mid a, b \in (\Sigma + V)^*, \ c \in \hat{V}, \ d, e \in V, \ (d, e) \in (V \times V), cd \to ce$$
$$\text{is a production rule in grammar } G \, \}$$

Theorem 4.0.1. *Any RE language can be generated by a $E0L$(REG, weak) system.*

Proof. Proven in [14] Theorem 4.2.1 for the construction outlined above.   □

5   Watson-Crick *ET0L* Systems with Bounded Complementary Transitions

In this section it is demonstrated that the family of languages recognized by $E0L$(REG, weak, uni) systems are a subset or equal to the $ET0L$ languages. This result is also used to show that $ET0L$ languages are equal to the $ET0L$(REG, weak, bounded) languages. It is strongly believed that the languages of these systems are in fact properly contained in the $ET0L$ languages but this has not been proven conclusively.

The following theorem provides a construction which can convert any $E0L$(REG, weak, uni) system H into an equivalent $ET0L$ system $\tilde{H}$. A table, T_2 is provided in $\tilde{H}$ which contains the productions of the underlying $E0L$ system used in H. T_2 is used to derive terminal strings. Two more tables, T_1 and T_3, are provided which are used to simulate derivations where in the last step the Watson-Crick morphism is applied. T_1 contains the same productions as the underlying $E0L$ system of H but where each symbol from the original alphabet of H is represented by nonterminals which have two state symbols attached to them corresponding to states in a Deterministic Finite Automata which recognizes the trigger language used in H. T_3 writes the sentential forms produced by T_1 to their complementary versions from which rewriting can only be performed in T_3 iff the string represented by the state-encoded string in the previous derivation step is in the trigger language. The idea of representing the Trigger Language as a Deterministic Finite Automata (DFA) and then using the states of this DFA to encode nonterminals used in the simulation of derivations of strings in the Trigger Language is inspired by [2].

Theorem 5.0.1. *There exists an algorithm that, given an $E0L$(REG, weak, uni) system H, produces an equivalent $ET0L$ system $\tilde{H}$.*

Proof. Let K be a language recognized by an $E0L$(REG, weak, uni) system denoted by H. Let $H = (G, TR)$ where $G = (\Sigma, h, \omega, \Delta)$ and TR is recognized by DFA $D = (Q, \Sigma, \delta, s_0, F)$. K is recognized by $ET0L$ system $\tilde{H} = (\Sigma \cup \hat{\Sigma} \cup \overline{\hat{\Sigma}} \cup \{\beta\}, T, \omega, \Delta)$ where $\hat{\Sigma} = \{< s_1, a, s_2 > \mid a \in \Sigma, s_1, s_2 \in Q\}$, $\overline{\hat{\Sigma}} = \{\overline{a} \mid a \in \hat{\Sigma}\}$, all subalphabets are mutually disjoint and T consists of the following tables:

- T_1 which consists of productions
  - $\omega \to < s_0, a_1, s_1 > \cdots < s_{k-1}, a_k, s_k >$ where $a_1, \ldots, a_k \in \Sigma$, $s_0, \ldots, s_k \in Q$, $\omega \to a_1 \ldots a_k$ in G, and $s_k \in F$
  - $< s_1, a, s_2 > \to < s_1, a_1, q_1 > \cdots < q_{k-1}, a_k, s_2 >$ where $a, a_1, \ldots, a_k \in \Sigma$, $s_1, s_2, q_1, \ldots, q_{k-1} \in Q$ and $a \to a_1 \ldots a_k$ in G
  - $< s_1, a, s_2 > \to \varepsilon$ where $a \in \Sigma$, $s_1, s_2 \in Q$, $a \to \varepsilon$ in G and $s_1 = s_2$
  - $a \to \beta$ where a is any symbol in $alph(\tilde{H})$ with no production above
- $T_2 = h \cup \{a \to \beta \mid a \in \hat{\Sigma} \cup \overline{\hat{\Sigma}} \cup \{\beta\}\}$
- T_3 which consists of productions
  - $< s_1, a, s_2 > \to \overline{< s_1, a, s_2 >}$ where $s_1, s_2 \in Q$, $a \in \Sigma$
  - $\overline{< s_1, a, s_2 >} \to \overline{a}$ where $s_1, s_2 \in Q$, $a, \overline{a} \in \Sigma$, $h_W(a) = \overline{a}$, $h_W(\overline{a}) = a$ and $\delta(s_1, a) = s_2$
  - $a \to \beta$ where a is any symbol in $alph(\tilde{H})$ with no production above

$L(H) \subseteq L(\tilde{H})$:
All strings derived in H without an application of h_W can be derived by applying the productions of T_2. All strings derived in H that have h_W applied in one of their derivations can be derived first in T_1 as a sequence of symbols $< s_{1_{a_1}}, a_1, s_{2_{a_1}} > \cdots < s_{1_{a_n}}, a_n, s_{2_{a_n}} >$ where $a_1, \ldots, a_n \in \Sigma$, $s_{1_{a_1}}, \ldots, s_{1_{a_n}}, s_{2_{a_1}}, \ldots$, $s_{2_{a_n}} \in Q$, $a_1 \ldots a_n \in TR$, $\omega \Rightarrow^* a_1 \ldots a_n$ in H and $\delta(s_{1_{a_i}}, a_i) = s_{2_{a_i}}$ where $1 \leq i \leq n$.

Each symbol in this string can immediately derive its Watson-Crick complement via an application of T_3. Useful productions (productions which do not directly derive β) for these complementary symbols are provided in T_3 and in no other table. T_3 is then applied which writes these $\overline{< s_1, a, s_2 >}$ symbols to $\overline{a} \in \Sigma$. From here useful derivations can only occur in T_2 which is the same $E0L$ system as the underlying $E0L$ system of H.

$L(\tilde{H}) \subseteq L(H)$:

Consider any string $s \in L(\tilde{H})$. That is, $\omega \Rightarrow^*_{\tilde{H}} s$ where $s \in \Delta^*$. Symbols in Δ are derivable only from productions in T_2 or the following production set in T_3:
$\{ \ \overline{< s_1, a, s_2 >} \ \rightarrow \ \overline{a} \ \text{where} \ s_1, s_2 \in Q, \ a, \overline{a} \in \Sigma, \ h_W(a) = \overline{a}, \ h_W(\overline{a}) = a \ \text{and} \ \delta(s_1, a) = s_2 \ \}$. This implies that one of two derivations are possible for s from ω in $\tilde{H}$:

$$\omega \Rightarrow^*_{T_2} s \quad \text{or} \quad \omega \Rightarrow^*_{T_1} \hat{a} \Rightarrow_{T_3} \overline{\hat{a}} \Rightarrow_{T_3} \overline{a} \Rightarrow^*_{T_2} s$$

where $a, \overline{a} \in \Sigma^*$, $\hat{a} \in \hat{\Sigma}^*$, $\overline{\hat{a}} \in \overline{\hat{\Sigma}}^*$. Consider the first case. s is derived by using productions in T_2 iff it is derived by using productions in h. So if s is derived solely by productions in T_2 then $\omega \Rightarrow^*_H s$. Consider the second case. $\omega \Rightarrow^*_{T_1} \hat{a}$ iff $\omega \Rightarrow^*_h a$. Assume $a \notin TR$. If this is the case then $\overline{\hat{a}}$ contains symbols of the form $\overline{< s_1, b, s_2 >}$ where $s_1, s_2 \in Q$, $b \in \Sigma$ and $\delta(s_1, b) \neq s_2$. All such symbols have no useful productions (are only rewritten to the trap symbol β) in T_1, T_2, T_3. This implies that $\overline{\hat{a}}$ does not derive $\overline{a}$ in T_3, a contradiction. So a must be in TR. So if $\omega \Rightarrow^*_{T_1} \hat{a} \Rightarrow_{T_3} \overline{\hat{a}} \Rightarrow_{T_3} \overline{a}$ then $\omega \Rightarrow^*_h a \Rightarrow^{hw} \overline{a}$. $\overline{a} \Rightarrow^*_{T_2} s$ iff $\overline{a} \Rightarrow^*_h s$. Thus, if $\omega \Rightarrow^*_{T_1} \hat{a} \Rightarrow_{T_3} \overline{\hat{a}} \Rightarrow_{T_3} \overline{a} \Rightarrow^*_{T_2} s$ then $\omega \Rightarrow^*_h a \Rightarrow^{hw} \overline{a} \Rightarrow^*_h s$.

Thus $L(H) = L(\tilde{H})$ by the two arguments above. The truth of the statement of the Theorem follows. $\square$

Note that Theorem 5.0.1 can be extended to show that $E0L$(REG, weak) systems where h_W is applied at most n $(n \geq 1)$ times in any derivation can be simulated by $ET0L$ systems. Nonterminals of the form $< s_i, a, s_{i+1} >$ used in Theorem 5.0.1 can be generalized to the form $< (s_1, s_2, \ldots, s_m), a, (s_1', s_2', \ldots, s_m') >$ where $1 \leq m \leq n$. Here $(s_1, s_2, \ldots, s_m)$ and $(s_1', s_2', \ldots, s_m')$ form tuples of states. The states s_1 and s_1' for the complements of such nonterminals would be used by table T_3 to judge whether the string represented at the previous derivation step is in fact in the trigger language of the original $E0L$(REG, weak) system. If such a string containing nonterminals of the form $\overline{< (s_1, s_2, \ldots, s_m), a, (s_1', s_2', \ldots, s_m') >}$ can be rewritten usefully by T_3 (without producing instances of the trap symbol β) then all nonterminals are rewritten to the form $< (s_2, \ldots, s_m), \overline{a}, (s_2', \ldots, s_m') >$. In a simulation of a valid derivation requiring up to n applications of the Watson-Crick morphism this process would repeat until T_3 rewrites a string in the form Σ^* as its original definition in Theorem 5.0.1.

This idea can be extended further to show that $ET0L$(REG, weak, bounded) languages are contained within the $ET0L$ languages. This same idea can be used so that, instead of just T_1 simulating derivations of strings in the trigger language, several tables could be used. This set of tables simulate the productions of each table in an arbitrary $ET0L$ system according to the encoding and derivation of complementary symbols used in T_1. Also, instead of just T_2, all tables from the underlying $ET0L$ system are included. $ET0L$ systems are by definition $ET0L$(REG, weak, bounded) systems where the trigger language is disjoint with any sentential form derived by the system. Thus, the following Corollary is immediate:

Corollary 5.0.2. $L(ET0L) = L(ET0L(REG, weak, bounded))$

References

1. Adleman, L.M.: Molecular Computation Of Solutions To Combinatorial Problems. Science 266, 1021–1024 (1994)
2. Bar-Hillel, Y., Perles, M., Shamir, E.: On Formal Properties of Simple Phrase-Structure Grammars. Zeitschrift für Phonetik, Sprachwissenschaft und Kommunikationsforschung 14, 143–177 (1961)
3. Csima, J., Csuhaj-Varjú, E., Salomaa, A.: Power and size of extended Watson-Crick L systems. Theor. Comput. Sci. 290(3), 1665–1678 (2003)
4. Csuhaj-Varjú, E.: Computing by networks of Watson-Crick D0L systems. In: Ito, M. (ed.) Algebraic Systems, Formal Languages and Computation. RIMS Kokyroku 1166, Research Institute for Mathematical Sciences, pp. 43–51. Kyoto University, Kyoto (August 2000)
5. Honkala, J., Salomaa, A.: Watson-Crick D0L systems with regular triggers. Theor. Comput. Sci. 259(1-2), 689–698 (2001)
6. Mihalache, V., Salomaa, A.: Lindenmayer and DNA: Watson-Crick D0L Systems. Bulletin of the EATCS 62 (1997)
7. Mihalache, V., Salomaa, A.: Language-Theoretic Aspects of DNA Complementarity. Theor. Comput. Sci. 250(1-2), 163–178 (2001)
8. Penttonen, M.: One-Sided and Two-Sided Context in Formal Grammars. Information and Control 25(4), 371–392 (1974)
9. Rozenberg, G., Salomaa, A.: The Mathematical Theory of L systems. Academic Press, New York (1980)
10. Rozenberg, G., Salomaa, A. (eds.): Handbook of Formal Languages. Word, Language, Grammar, vol. 1. Springer-Verlag New York, Inc., New York (1997)
11. Salomaa, A.: Turing, Watson-Crick And Lindenmayer. Aspects Of DNA Complementarity. In: Calude, C., Casti, J., Dinneen, M.J. (eds.) Unconventional Models of Computation, pp. 94–107. Springer, Heidelberg (1997)
12. Salomaa, A.: Watson-Crick Walks and Roads on D0L Graphs. Acta Cybern 14(1), 179–192 (1999)
13. Salomaa, A.: Uni-transitional Watson-Crick D0L systems. Theor. Comput. Sci. 281(1-2), 537–553 (2002)
14. Sears, D.: The Computational Power of Extended Watson-Crick L Systems. Master's thesis, School of Computing, Queen's University, Kingston, Ontario, Canada (2010), http://hdl.handle.net/1974/6224
15. Sosík, P.: D0L System + Watson-Crick Complementarity = Universal Computation. In: Margenstern, M., Rogozhin, Y. (eds.) MCU 2001. LNCS, vol. 2055, pp. 308–320. Springer, Heidelberg (2001)
16. Sosík, P.: Universal computation with Watson-Crick D0L systems. Theor. Comput. Sci. 289(1), 485–501 (2002)

Computing with Planar Toppling Domino Arrangements

William M. Stevens

Oxford University Department of Psychiatry,
Warneford Hospital, Oxford, OX3 7JX
william@stevens93.fsnet.co.uk

Abstract. A method for implementing Boolean logic functions using arrangements of toppling dominoes is described. Any desired combinational function can be implemented. A circuit constructed using this method has no timing or order constraints on its inputs and requires no out-of-plane bridges for passing one line of dominoes over another. Since it is built using toppling dominoes, a circuit can be used only once.

Keywords: Logic circuit, dual-rail logic, one-shot logic, non-electronic logic, domino.

1   Introduction

There are several different reasons for studying domino computation. There are didactic reasons that have nothing directly to do with science: for people interested in the physical basis of computing it is interesting and enjoyable to build computing primitives from simple materials that are easy to manipulate and which do not require elaborate tools to construct. Domino computers have been technologically possible for many thousands of years, but only since the advent modern mathematics and digital computing theory has anybody had the insight and motivation to both build one and tell other interested people about it.

Just as pure mathematicians strive to explore and understand all areas of mathematics, regardless of their practical utility, so some scientists apply the same philosophy to physical phenomena. Even those who have a different philosophy and who believe that all scientific endeavour should be directed towards practical outcomes must admit that there are well known examples in mathematics of ideas that seemed entirely useless at the time that they were first conceived of and developed, but which later turned out to be fundamental to theoretical science, which is in turn fundamental to several modern technologies. Domino systems are one-shot, collision-based systems, so studying their information processing potential could lead to insights into other systems that fall into one or both of these categories.

A one-shot computing system is a system that can be configured to carry out a computation, but it can only perform the computation once. A collision-based system is one in which mobile objects or patterns interact with one another to

C.S. Calude et al. (Eds.): UC 2011, LNCS 6714, pp. 224–233, 2011.
© Springer-Verlag Berlin Heidelberg 2011

carry out information processing operations. Examples of collision-based systems include the systems based on the Belousov-Zhabotinsky (BZ) reaction that have been devised by Gorecki et al. [1], and Fredkin and Toffoli's billiard ball model [2], which is also a one-shot system. A comprehensive collection of work related to collision-based computing systems can be found in [3].

1.1 Previous Work

The idea of using interacting dominoes to carry out Boolean operations has occurred to many people, as a web search around the subject area will reveal.

O'Keefe [4] seems to have been the first to explore the area in some detail and publish his findings. O'Keefe described a system of logic gates made from dominoes in which a travelling disturbance on a line of dominoes represents a Boolean value of 1, and the absence of a disturbance represents a value of 0. Figure 1 shows a NOR gate in O'Keefe's system. The relative timing of the inputs to this gate is critical to its correct functioning. In the NOR gate, the A and B inputs to the gate must be applied before the auxiliary 1 input has been applied so that they have time to prevent the auxiliary 1 input from causing a value of 1 to be output at C.

Fig. 1. A NOR gate in O'Keefe's system

O'Keefe did not restrict his dominoes to a two-dimensional surface, he allowed bridges so that one line of dominoes could cross another without interference. O'Keefe points out that constraining a system to planar geometry requires that the flow of information in a domino system must be carefully managed so as to avoid signals being blocked by barriers of de-energised dominoes.

1.2 Challenges

This paper addresses two of the challenges that O'Keefe's work raises.

- Can a toppling domino logic system be devised that has no timing constraints on its inputs?
- Can arbitrary combinational functions be implemented in a planar system of dominoes, without requiring bridges to allow one domino line to cross another?

2   Domino Interactions and Boolean Algebra

There are a large number of ways that dominoes can be arranged. This paper makes use of conventional arrangements in which all dominoes initially stand upright, spaced apart from one another, and topple over in the course of propagating a disturbance. There are two different transitions that an individual upright domino can undergo: it can topple in one direction, or it can topple in the opposite direction. This paper does not make user of the finer details of the physics of toppling waves of dominoes, but interested readers can find out about some of these in the paper by Wagon et al. [5].

In this paper two domino configurations are defined that will be used exclusively as the basis for constructing other arrangements. These configurations are shown in Figure 2, along with symbols that will be used in schematic diagrams of more complex arrangements later on.

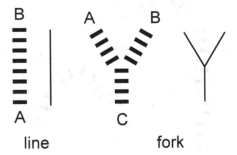

Fig. 2. Elementary domino configurations

The first is the *line* configuration. When A topples towards B then A will cause B will topple in the same direction as A. When B topples towards A then A will topple in the same direction as B. If A and B both topple towards one another at the same time then the two toppling wavefronts will collide and nothing further will happen. Once a domino is toppled it can no longer take part in any further interaction.

The second is the *fork* configuration. This configuration has three lines of dominoes meeting at a junction, dominoes at the external ends of these three lines are labelled A,B and C. When C topples toward the junction, then it will cause both A and B to topple away from the junction. When A topples toward the junction it will cause C to topple away from the junction, but B will be unaffected. When B topples toward the junction it will cause C to topple away from the junction, but A will be unaffected. It is possible that two or three lines leading into the fork configuration topple simultaneously. Clearly if all three of A, B and C topple towards the junction simultaneously there can be no resulting output. If both A and B topple toward the junction at the same time this will cause C to topple away from the junction. If A and C (or equivalently, B and C) topple toward each other at approximately the same time then whether or not B will topple is dependent on the precise timing and location of the resulting collision. Clearly if the collision happens nearer to C than to A then B will not topple. If the collision happens nearer to A than to C then B will topple. If the collision happens in the vicinity of the fork junction then whether or not B will topple is unpredictable.

The word 'signal' refers to a toppling domino wavefront. So the phrase 'a signal is applied to an input of a mechanism', means that the domino located at that input is toppled in the direction that will cause a toppling wavefront to travel into the mechanism. The phrase 'a signal emerges from an output of a mechanism' means that the domino located at the output is toppled by a wavefront emerging from the mechanism.

2.1   One Way Line

A one way line can be constructed using two forks as shown in Figure 3. A signal entering the configuration from A will split at fork F into two signals that will collide with each other, preventing any signal from emerging at B. In the other direction, a signal entering the configuration from B will pass through fork G, through F and then emerge from A. Correct operation of the one way line requires that the rate of signal propagation along any two domino lines is similar, otherwise the collision that is required after a signal from A splits at fork F may happen in the wrong location. Figure 3 also shows the schematic symbol that will be used for a one way signal line.

2.2   Single Line Crossover

A single line crossover is a mechanism with two inputs A and B and two outputs A' and B' which are crossed over topologically. A signal may arrive either at input A or at input B. A signal arriving at A will be passed to output A', a signal arriving at B will be passed to B'. This definition of a single line crossover does not specify what will happen if a signal arrives at both inputs: in this paper the single line crossover will only be used in situations where it is guaranteed that only one of the inputs will ever receive a signal.

A pair of forks can be used as the basis of a single line crossover, with one way lines on each input to prevent signals from one input propagating back to the other input. Figure 4 shows the single line crossover and the schematic symbol that will be used to represent it. A total of six forks are used in a single line crossover.

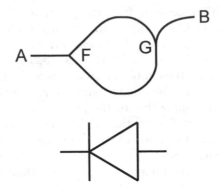

Fig. 3. A one way line

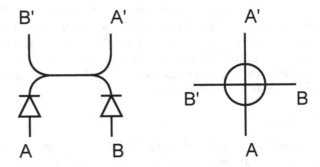

Fig. 4. A single line crossover

2.3   Both Mechanism

A *both* mechanism has two inputs and one output. Only when signals have arrived at both inputs will it generate an output. It is not immediately obvious that this mechanism can be implemented using only the fork and line configurations. Figure 5 shows the both mechanism and the schematic symbol that will be used to represent it.

If a signal is applied to A well before a signal is applied to B, then the one of the two paths that emanate from the fork near input B will be interrupted, permitting a signal derived from B to emerge at C. If a signal is applied to B well before a signal is applied to A then the same process happens, with A and B exchanged. A situation that might occur if both A and B are applied close together in time is that in which A prevents B from propagating to C, but B also prevents A from propagating to C. To prevent this situation from occurring, the length of the path T must be long enough so that if a signal from B reaches fork F before a signal from A, then a signal from B will reach fork G before a signal from A. A total of 17 forks are used in the both mechanism.

In part because the both mechanism uses one way lines, and in part because of the timing constraint in the previous paragraph, correct operation of the both

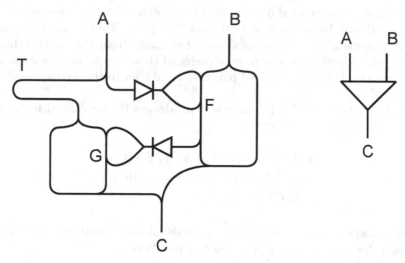

Fig. 5. A both mechanism

mechanism requires that the rate of signal propagation along any two domino lines is similar.

2.4 Dual-Rail Boolean Algebra

The mechanisms described in the previous sections can be mapped onto some of the constituents of a Boolean algebra. The way in which we choose to map signals onto Boolean values means that it is impossible to construct a mechanism to correspond to the negation operation within this Boolean algebra. Using von Neumann's double-line trick, introduced in [6], this incomplete Boolean algebra can be used to construct a complete *dual-rail* Boolean algebra that does have a negation operation. Because we will be dealing simultaneously with two different Boolean algebras, one constructed from the other, we will need to be clear at every point which one we are referring to. Let us call the incomplete Boolean algebra (without negation) B, and the complete Boolean algebra that we construct B'. All of the constituents of B' will be marked with an apostrophe.

The familiar definition of a two-element Boolean algebra is used, consisting of a set containing two elements $\{0, 1\}$ and three operations AND, OR and NOT defined in the conventional way.

We use the passage of a single signal along a propagation path at any time to represent the element 1 in algebra B, and the non-occurrence of a signal on a path at any time to represent the element 0. The both mechanism corresponds to the AND operation in algebra B. The fork mechanism corresponds to the OR operation in algebra B. Given the way that we have chosen to represent the elements 0 and 1, we cannot construct a mechanism that corresponds to the NOT operation in algebra B because detecting 'the non-occurrence of a signal on a path at any time' cannot be done in a finite time. We would need to wait forever to be sure that an event was never going to happen.

Algebra B' is constructed from algebra B as follows. The two elements $\{0', 1'\}$ of algebra B' are defined as $0' = (1, 0)$ and $1' = (0, 1)$. That is, each element in algebra B' is made using an ordered pair of elements from algebra B. When we use a letter, such as x, to refer to an element of B' we use subscripts x_0 to refer to the first member of the ordered pair and x_1 to refer to the second member of the ordered pair.

The AND', OR' and NOT' operations of algebra B' are defined in terms of the constituents of B as follows:

$$x \text{ AND' } y = (x_0 \text{ OR } y_0, x_1 \text{ AND } y_1)$$
$$x \text{ OR' } y = (x_0 \text{ AND } y_0, x_1 \text{ OR } y_1)$$
$$\text{NOT' } x = (x_1, x_0)$$

It is straightforward to show that these definitions satisfy the conventional definitions for conjunction, disjunction and negation.

With this definition of the NOT' operation, negation of a value x in algebra B' can be realised simply by crossing over the signal paths along which the signals representing x_0 and x_1 travel. This can be accomplished using the single line crossover of Figure 4.

In addition to allowing a negation operation to be realised topologically by crossing over propagation paths, the other advantage that the dual-rail system has is that it removes any need to be concerned with the relative timing of signals, and so allows asynchronous circuits to be constructed. When the occurrence or non-occurrence of a signal is used to represent a Boolean value, without any other signal for timing reference, then it is not possible to know how long to wait before deciding that the signal had not occurred. In the dual-rail system used here we can expect either a signal corresponding to 0', or a signal corresponding to 1' to arrive at each dual-rail output once the circuit has finished operating. We will not be left in any state of ambiguity over whether the circuit has finished or not.

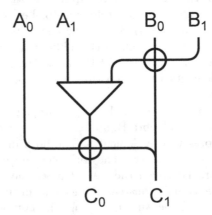

Fig. 6. A dual-rail NAND gate

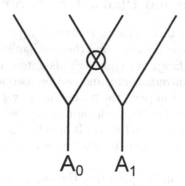

Fig. 7. A dual-rail fanout mechanism

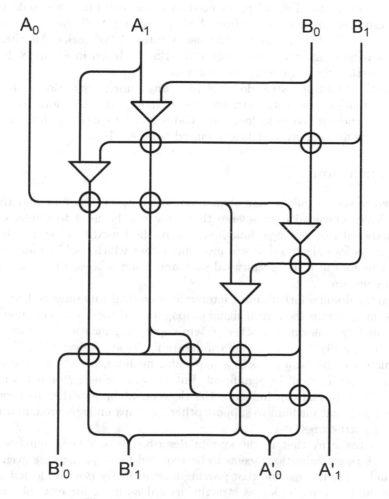

Fig. 8. A dual-rail crossover mechanism

3   Logic Functions and Planar Crossover

We are now in a position to be able to construct a planar, dual-rail, one-shot, asynchronous Boolean logic system using the mechanisms introduced above. A dual-rail NAND gate is shown in Figure 6. This gate uses a total of 30 forks.

The NAND gate is a universal gate, and can be used as the basis for realising any Boolean function. To complete the proof that any Boolean function can be realised it must be shown that dual-rail signals can be split into two, so that a single output can feed into several gates. It must also be shown that dual-rail signals can be routed from one gate to another without any restriction, perhaps crossing each other.

Dual-rail signal paths can be split using the configuration shown in Figure 7, which makes use of 8 forks.

When two dual-rail signal paths need to cross each other we could use the planar crossover network made from AND gates and NOT gates described by Dewdney in [7]. This method would use a total of 266 forks. An alternative planar arrangement, which uses only 140 forks, is shown in Figure 8. It is not known whether this arrangement is minimal.

Since the dual-rail system does not have any timing restrictions, the precise length or route of any path between gates is unimportant, so long as the path starts and ends in the right locations. Gates can be spaced as far apart as is necessary to fit as many signals as required between them.

4   Conclusion

This paper shows that a pair of domino interactions, the fork and the line, are logically universal in the sense that they can be used to implement any combinational Boolean logic function. The method used to show this involved taking a Boolean algebra that was incomplete, but which could be implemented using dominoes in a straightforward way, and using it to construct a complete Boolean algebra.

Since the domino fork and line interactions are logically universal, subject to the assumption that the rate of signal propagation along any two domino lines is approximately uniform, any other system supporting identical or similar interactions is also likely to be a good candidate for logical universality. The direct construction of Boolean gates in a particular medium, based on these interactions, may not in itself be significant, but knowing whether or not a medium is logically universal can influence the decision about whether to investigate the ability of that medium to support other, perhaps undiscovered, information processing structures.

It is noteworthy that in the system described here a large number of elementary fork configurations seems to be required for implementing even simple Boolean logic functions, and that proving how arbitrary Boolean logic functions can be implemented is not as straightforward as it is, for example, in relay circuits. It seems that within planar domino systems, asynchronous Boolean functions are relatively complex phenomena.

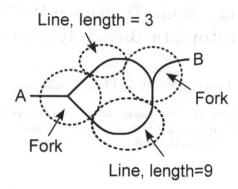

Fig. 9. A one way line split into four regions

A possible route for further research in this area is to formalise and to gener-alise the concepts used in this paper. One way that this could begin is illustrated in Figure 9. Here a one way line is split into four regions, each of which contains either a fork or a line, and has timing parameters related to line lengths. The response of a single region to all possible combinations of input events could be described by formalising the descriptions given in section 2. Composition rules could then be formulated that would permit the behaviour of a mechanism to be deduced from the behaviour of its constituent mechanisms. Working out the details of a formal system of this type is likely to be simpler in a domino system, where each event can only occur once and where each region can undergo only a finite number of state transitions, than in the more general case where events can occur any number of times and where regions may return to a previous state. Similar schemes have been used in the domain of asynchronous circuit design and also in process calculi, but (to my knowledge) none of these have been ex-tensively used for describing, modelling or understanding information processing in event-propagating media such as domino systems, collision-based systems and spiking neural networks.

References

1. Gorecki, J., Gorecka, J.N.: Multi-argument logical operations performed with ex-citable chemical medium. J. Chem. Phys. 124(084101) (2006)
2. Fredkin, E., Toffoli, T.: Conservative Logic. J. Theo. Phys. 21(3,4), 219–253 (1982)
3. Adamatzky, A. (ed.): Collision Based Computing. Springer, London (2002)
4. O'Keefe, S.: Implementation of Logical Operations on a Domino Substrate. Int. J. Unconventional Computing 5(2), 115–128 (2009)
5. Wagon, S., Briggs, W., Becker, S.: The Dynamics of Falling Dominoes. The UMAP Journal 26(1), 35–47 (2005)
6. Von Neumann, J.: Probabilistic Logics and the Synthesis of Reliable Organisms from Unreliable Components. In: Shannon, C., McCarthy, J. (eds.) Automata Studies, pp. 43–98. Princeton University Press, Princeton (1956)
7. Dewdney, A.K.: Logic circuits in the plane. ACM SIGACT News 10(3), 38–48 (1979)

Increasing Fault-Tolerance in Cellular Automata-Based Systems

Luděk Žaloudek and Lukáš Sekanina

Faculty of Information Technology, Brno University of Technology,
Božetěchova 2, 612 66 Brno, Czech Republic
{izaloude,sekanina}@fit.vutbr.cz

Abstract. In the light of emergence of cellular computing, new cellular computing systems based on yet-unknown methods of fabrication need to address the problem of fault tolerance in a way which is not tightly connected to used technology. This may not be possible with existing elaborate fault-tolerant cellular systems so we strive to reach simple solutions. This paper presents a possible solution for increasing fault-tolerance in cellular automata in a form of static module redundancy. Further, a set of experiments evaluating this solution is described, using triple and quintuple module redundancy in the automata with the presence of defects. The results show that the concept works for low intensity of defects for most of our selected benchmarks, however the ability to cope with errors can not be intuitively deduced as indicated on the example of the majority problem.

Keywords: cellular automata, fault tolerance, static module redundance, TMR, cellular computing, rule 30, Game of Life, Byl's Loop.

1 Introduction

With today's unending need for faster and ubiquitous computation, the von Neumann architecture is being pushed to its limits. As new problems emerge [2], more and more computation is performed in parallel and even new computing paradigms emerge. One of these new paradigms is what Sipper calls cellular computing [14].

Cellular computing is based on the model of cellular automaton (CA) and its three key features: simplicity, locality of interactions and massive (or vast) parallelism. The CA model is flexible and capable of performing extensive range of computational tasks (overview of most notable may be found in [14]). Also, the CA model keeps rising in importance in several other areas including simulation of biological, chemical or social phenomena.

Since cellular computing is based on a model rather than a physical piece of hardware, it is not limited in the technology, by which it could be fabricated or synthesized. As a relatively new paradigm and possible future solution to (some of) our computational needs, it could be implemented not only in silicon (where there are existing examples of cellular-based hardware [5,11]).

C.S. Calude et al. (Eds.): UC 2011, LNCS 6714, pp. 234–245, 2011.

However novel approaches to digital hardware fabrication like chemical synthesis or inkjet printing are not yet refined and possibly prone to defects and transient errors more than the most common CMOS technology. In order to address this problem, methods of fault-tolerance need to be introduced. Existing cellular systems [5,11] possess fault-tolerance mechanisms which arguably significantly complicate the underlying hardware. Of the simpler methods of fault-tolerance the most prominent are module redundancy and error correcting codes. The goal of this paper is to demonstrate the feasibility of module redundancy on several distinct examples of CA tasks.

Section 2 introduces the cellular automaton model together with its notable modifications, properties and problems. One of the problems, specifically fault-tolerance in CA, is discussed in the remainder of Section 2 and current methods of fault-tolerance in CA are outlined. In Section 3 our solution is proposed. Section 4 describes sample CA problems which are then used in an experimental setup. Results of aforementioned experiments are provided in Section 5. Section 6 concludes the paper.

2   Cellular Automata

2.1   General Description of Cellular Automata

A cellular automaton is a d–dimensional grid of cells, where each cell is a finite automaton. The cells operate according to their *local transition functions*. In the most common scenario, the cells work synchronously – a new state of every cell is calculated from its previous state and the previous states of the cell's 'neighbors' at each time step. If all automata of the grid are identical (i.e., the automata operate according to the same local transition function – *rule*) the automaton is called *uniform*, otherwise it is *non-uniform*. By *configuration* of the cellular automaton we mean the states of all the cells at a given moment. The global behavior is captured in the *global transition function*, which defines a transformation from one configuration to the next configuration of the cellular automaton, in case of synchronous automaton the transition between steps. The sequence of configurations, determined by the global transition function, represents the *computation* of the cellular automaton.

Theoretically, the CA model assumes an infinite number of cells. However, in the case of practical applications the number of cells is finite. In that case, it is necessary to define the *boundary conditions*, i.e. the setting of the boundary cells' neighborhood. One of the states is also usually used as *quiescent* or inactive state. A convention is that when a quiescent cell has an entirely quiescent neighborhood, it will remain quiescent at the next time step.

The simplest one-dimensional uniform cellular automaton, with only two states and nearest neighbors neighborhood $N = \{-1, 0, 1\}$ (only left and right neighbor cells together with the cell itself are relevant for the local transition function[1]), can exhibit very complex behavior [18]. Each such CA is uniquely de-

---

[1] We also denote this parameter as neighborhood radius r and in the nearest neighborhood case we would write that $r = 1$.

fined by a mapping $\{0, 1\}^N \rightarrow \{0, 1\}$. Hence there are 2^8 such cellular automata, each of which is uniquely specified by the (transition) rule i ($0 \leq i < 256$). The transition functions of such automata then can be called as *rule 0, rule 1, rule 2* etc.

With two-dimensional cellular automata, the neighborhood is usually comprised of five or nine cells (see Fig. 1). The central cell is normally included into its own neighborhood. In the 5-neighbor (von Neumann) model, the individual transition rules would adopt the form $CTRBL \rightarrow C'$ where $CTRBL$ specifies the states of the Center, Top, Right, Bottom, and Left positions of the neighborhood's present state and C' represents the next state of the center cell.

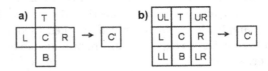

Fig. 1. a) 5-cell von Neumann's neighborhood, **b)** 9-cell Moore's neighborhood

The applications of cellular automata include many scientific areas, especially the modeling of complex (biological, chemical, computational, etc.) phenomena. Because the number of cells is typically very large, the cellular automaton can model a massively parallel computational system where some useful computation can *emerge* only on the base of local interactions. The properties of cellular automata have been investigated by means of analytic as well as experimental methods [8,15,16,17,18].

In order to obtain specific behaviors, the concept of cellular automata has been modified in several ways. Examples include *non–uniform* CA [15], *non–local* CA [10], *asynchronous* CA [3] or CA *with global control* [13].

2.2   Notable Inherent Properties and Problems

The cellular automaton model also exhibits several notable inherent properties that set it apart from conventional computational models: *massive parallelism*, *locality* of cellular interactions and *simplicity* of basic components (cells) [15,14]. This was mentioned in Section 1, however the terms need to be clarified. Note these properties are especially important for our case where every cell would be physically implemented (no simulation considered).

Massive parallelism is something rarely seen even in today's supercomputers. When talking about it in CA, we mean vast numbers of cells on much larger scale than the numbers of processing elements usually found in conventional parallel systems. The cell figures in CA are often expressed by exponential notation 10^X.

Locality of interactions means total absence of global control. Transitions between states are based on nearest neighborhood. Best known examples of such neighborhood are von Neumann's and Moore's neighborhood for 2D CA.

Simplicity refers to cells as basic computational elements. Instead of depending on complex components similar to modern processors (e.g. Pentium), a CA cell should be easily constructed as a simple electronic circuit or using chemical synthesis.

These properties put together are key to *emergent behavior* mentioned in Subsection 2.1. However, these features also require certain prerequisites, which we usually assume when modeling cellular automata in software: perfect clock synchronization (for synchronous CA) and error-free implementation platform. The two prerequisites turn into problems when we attempt to implement CA physically.

In particular, perfect clock synchronization may be unachievable when considering truly massive cellular arrays due to different lengths of clock wiring in different parts of the cellular array and the spatial requirement for such wiring. Moreover, since cellular automata are one of possible future solutions to problems of conventional computational architectures [14], we cannot assume that the clocking problem will not encounter further complications when migrating cellular architectures to different materials and/or fabrication methods (with respect to current CMOS technology). One of possible solutions and probably the most feasible is the employment of asynchronous CA mentioned in Subsection 2.1. With recent achievements in the field of delay-insensitive circuits, techniques employing asynchronous CA were demonstrated [12]. It is arguable if such functionality may be duplicated in different technology such as chemical synthesis. In addition, severe modification of local transition function (rules) with respect to synchronous version of implemented applications is required when applying the asynchronous CA paradigm.

The second problem is virtual impossibility of attaining error-free implementation platform for cellular-based computing machines. Not only do we encounter a variety of transient errors (caused by cross talk, noise etc.) and transient faults (e.g. temporary hardware element failures due to ambient radiation), with decreasing component sizes and the dawn of nanotechnology in mind, we must also deal with permanent defects originating in fabrication or arising during the lifetime of the hardware. Such defects may not be detected (or even detectable) after the fabrication and the CA model should be prepared to deal with them independently of hardware.

Unlike solving problems with synchronization, attaining fault-tolerance is the subject of this paper, which is discussed in more detail in Subsection 2.3.

2.3   Fault-Tolerance in Cellular Automata

In recent years several methods of fault-tolerance in CA have been implemented or proposed. Those which are employed in existing cellular-based hardware [5,11] added new layers of functionality and also complexity. Cells are formed into supercells, routing mechanisms circumvent defective elements and positional information and whole sets of functionality are passed to each cell/element. Although such systems possess fault-tolerant functionality, the overhead cost is extremely high in terms of hardware complexity.

Other, much simpler mechanisms were introduced to the CA model, which were inspired by more conventional digital circuit solutions [6,12]. Most prominent of such solutions are module redundancy and error detecting and correcting codes.

When utilizing ECC in CA, the fault-tolerance deals with errors that occur in the cell memory, i.e. its state information or its transition function look-up table, if that is implemented in such way. Example of cellular automaton with error detection and correction is [12].

3　Proposed Approach

Generally, static module redundancy which is our main concern in this paper is based on backing up the subsystem's function by multiplying the modules which compute it. The results of each module are then compared in a voting element which produces the final result by majority voting [9]. Best practice is to use odd number of modules to insure no ties occur. The most commonly used form of static modular redundancy in digital circuits is TMR (triple). Other possible approaches to module redundancy include dynamic redundancy and/or reconfiguration and combinations of previously mentioned. Of course, such methods increase the complexity of the implementation due to additional multiplexing, voting and/or reconfiguring elements. This paper is further limited to static NMR so we do not complicate the fabrication process of cellular systems which should be cheap and easily produced in vast numbers.

An example of a general analysis of reliability of TMR in quantum-dot CA may be found in [6]. However, we believe that distinct types of CA tasks may behave differently depending on their complexity and number of active cells in presence of errors. The approach to such discovery is presented in following paragraphs.

Since we work with CA as a possible model for future computing systems, we cannot predict which technologies or materials will be used in actual cellular system fabrication. Because of that, transient errors are not considered in our analysis. The reason is apparent possibility of difference in transient error sensitivity in various technologies which could be used in cellular system's fabrication (e.g. chemical synthesis, DNA-based scaffolding, ink-printed circuits etc.).

We also limit our analysis to small standalone defects. Such defects could occur during production or during the lifetime of the application and may not be easily detectable like common large-scale burst defects which destroy hundreds of elements in tightly localized patterns. Since we assume such environment and implementation platform, we can assume that usage of TMR or 5-module redundancy (5MR) should be sufficient.

The design of each cell is straightforward and was roughly outlined at the beginning of this section. Each cell contains three (TMR) or five (5MR) modules, which have common inputs from the cell's neighborhood. Each module computes the local transition function F of the CA and all module's results are provided to the majority voting element. The voting element then passes the dominating result as the next state of the cell. In theory, TMR can mask one error and 5MR can mask two errors without system failure. Graphical representation of one CA cell with TMR is depicted in Fig. 2.

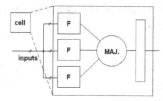

Fig. 2. Inside a CA cell with TMR. Triple modules take *inputs* from neighborhood and compute the local transition function *F*. Results are then voted on by *MAJ.* element which passes the prevailing result to register as the cell's new state.

4   The Experiments

This section describes how the experiments were conducted. First the benchmark tasks are outlined, later the experiment's conditions are recounted.

4.1   The Benchmark Problems

In order to evaluate the modified CA's behavior in the presence of defects, four common CA computational tasks were selected.

The first task is in a 1D CA and it is commonly known as *rule 30*. The designation originates from Wolfram's classification scheme [18] and denotes the number which comes from the new states' values when the transition rule set is sorted by neighborhood values in descending order as shown on Fig. 3. Note that in this case, the CA uses only the simplest 3-cell (r = 1) neighborhood. The rule produces chaotic aperiodic behavior and may be used for e.g. random number generation [18].

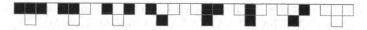

Fig. 3. The sequence of individual transition rules in descending order can form a number from the resulting states. In this case we can transform binary 00011110 into decimal 30 which explains why this transition function is called *rule 30*.

Fig. 4 a) shows several steps in a 1D CA with *rule 30*, where the initial configuration is *seeded* with one active state in the middle.

Another common 1D CA task, which is often used as a benchmark [1,15], is the problem of *majority* (or alternatively known as *density*). The goal of density computation is a classification of the initial CA configuration by determining whether there is a majority of 0s or 1s. Majority could be easily solved from a global view, however a CA works only with local interactions. In this case, CA uses 7-cell (r = 3) neighborhood and the result should be known in number of steps equal to double the width of the CA. The final result is identified by the final configuration of the CA, where the cellular space is completely filled with the dominant state (Fig. 4 b).

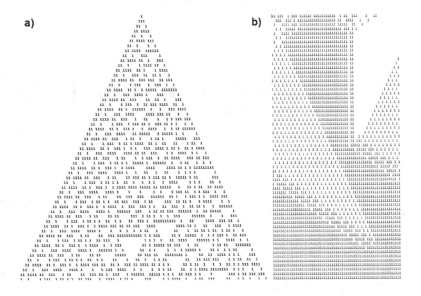

Fig. 4. a) Evolution of rule 30, **b)** successful computation of majority (rule taken from [1]). Note that white spaces represent the cells in state 0.

It is not possible to solve the majority problem for all possible configurations but researchers strive to find the best possible solution e.g. by evolutionary approaches. Their success is measured in the percentage of successful runs on a set of randomly generated initial configurations [15]. For our experiments we selected one of the better known solutions which is capable of correct classification of 82.326% configurations [1]. The rule can be encoded in hexadecimal number by the sequence FFAAFFAAFFAAFFAAA0AA00A0A0AA00A0.

Third task that was selected for our experiments is commonly known as Conway's *Game of Life* [7]. It is well known artificial life simulation with a 2-state 2D CA with von Neumann neighborhood (5 cells - center and 4 immediate neighbors). Rules for Life can be briefly described as following:

1. Any live cell with fewer than two live neighbors dies, as if caused by under-population.
2. Any live cell with two or three live neighbors lives on to the next generation.
3. Any live cell with more than three live neighbors dies, as if by overcrowding.

4. Any dead cell with exactly three live neighbors becomes a live cell, as if by reproduction.

With these simple rules the behavior can be surprisingly complex and several notable types of structures within the cellular space emerge, including *oscillators* and *spaceships*. One of such spaceships is called a glider: a structure capable of translating trough cellular space by nontrivial (as in multi-step) transformations. The glider is used in the experiments described in this paper and its translation is depicted in Fig. 5.

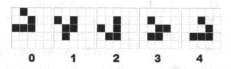

Fig. 5. The translation of the *glider* spaceship in the Game of Life in 4 steps.

Final benchmark task used in our experiments is the self-replication of the so-called *Byl's Loop* [4]. Byl's Loop uses von Neumann neighborhood and apart from the other mentioned benchmarks, it uses 6 states. The loop is capable of replicating itself in 26 steps from its initial configuration and continues to do so until it runs of cellular space. Fig. 6 a) shows the replication of Byl's Loop in 5-step jumps and Fig. 6 b) shows the cellular space after 11 generations of loop replication. The replication mechanism is very carefully connected to the shape of the loop and even minuscule alterations in its configuration cause failure of the replication process.

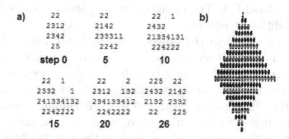

Fig. 6. a) Single Byl's Loop's replication, b) cellular space after 11 generation of replications

From the description of the benchmarks, it is clear that tasks with increasing complexity were selected in order to provide the experiments with a breadth of different problems representing wide area of cellular automata applications.

4.2   The Experimental Setup

Each experiment used CA with cyclic neighborhood. That means that each edge cell is connected to the cell on the opposite edge. Size of the CA was always the same: 225 cells, i.e. 15x15 for 2D and 1x225 for 1D. The reason was to provide similar conditions for all experiments.

All initial configurations were carefully set so the maximum possible area of the CA could be used during the 100 steps each CA performed. In the case of rule 30, the initial configuration was seeded by two cells in state 1 on positions 64 and 147 which created the image of two triangles connecting together at the bottom. For the Game of Life computation, three gliders were set in column above each other from the upper left corner of the CA (each glider in the position shown in the 0 frame of Fig. 5) separated by space of two quiescent cells, so the gliders would travel trough the whole CA during the 100-step run. The Byl's Loop initial configuration (see Fig. 6) was set with one loop in the upper left corner of the cellular space so there would be 8 replicated loops after the 100-step run, filling roughly 8/9 of the available cellular space.

The defects in each run were generated on the level of redundant modules. Note that three levels of fault-tolerance were simulated: (i) no redundance (single module per cell), (ii) TMR and (iii) 5MR. Each module could be damaged with the probability ranging from 0,5% to 15% with uniform distribution. Damaged module manifests itself as dead – it always generates state 0. Each case was run 500 times.

A correct run occurs when the CA's configuration after defined number of steps is identical to the configuration of the same CA run without errors for the defined number of steps. Other definitions of correctness are not considered.

5   Results

The tables below are grouped by task type and in order of ascending redundancy of the simulated CA. The tables show the percentage of correct runs with different probabilities of defects.

Special case not shown in the tables is the majority task, which was not so extensively tested as the other benchmarks. The reason was that the used rule shifts the information about density trough the CA (as seen in Fig. 4 b)). The parameters for the few runs were also different – the CA was only 64 cells long as in Fig. 4. If there is a failing 'dead' cell in a CA running the majority task, 0 states are shifted from its position and the result turns in favor of 0. There was no point in testing the majority since most of the results ended in failure. On the other hand, the evolution of rule 30 may be distorted by defective cells but it still retains at least part of its shape because the failing cells do not distribute their 0 states so aggressively.

Many of the tables show results of 99% or 98%. Because each experiment was run 500 times, some freak probability cases always caused few of the runs to have several defective adjacent modules. Such occurrences may not be likely, but still possible.

Table 1. Selected results for all benchmarks without fault tolerance. Each single number was obtained from 500 runs.

No FT	Rate of correct runs up to 30, 50, 80 and 100 steps [%]											
Prob. of defects [%]	30	50	80	100	30	50	80	100	30	50	80	100
	Rule 30				Game of Life				Byl's Loop			
0.5	55	42	33	33	56	42	38	38	83	73	56	48
0.7	42	28	22	22	49	43	31	31	75	62	45	34
1	31	16	11	11	32	15	12	12	67	49	30	23
1.5	16	7	4	4	19	7	6	6	53	34	20	12
2	6	1	0	0	11	3	3	3	50	26	12	8
2.5	0	0	0	0	3	0	0	0	38	16	6	2
3	0	0	0	0	0	0	0	0	30	10	3	0

Table 2. Selected results for all benchmarks with TMR. Each single number was obtained from 500 runs.

TMR	Rate of correct runs up to 30, 50, 80 and 100 steps [%]											
Prob. of defects [%]	30	50	80	100	30	50	80	100	30	50	80	100
	Rule 30				Game of Life				Byl's Loop			
0.5	99	98	98	98	98	97	97	97	99	99	99	97
0.7	98	97	96	96	98	97	97	97	99	99	98	97
1	95	94	93	93	97	95	94	94	98	98	96	95
1.5	91	88	86	86	92	88	88	88	97	96	93	92
2	86	79	76	76	89	83	82	82	94	92	87	85
2.5	78	69	64	64	82	73	72	72	93	88	82	77
3	68	60	54	54	74	63	62	62	87	83	74	69
4	56	41	33	33	61	49	46	46	75	59	42	34
5	55	40	33	33	42	27	23	23	75	59	42	34
8	10	4	2	2	15	5	4	4	48	27	11	6
10	3	0	0	0	3	0	0	0	33	14	3	1
15	0	0	0	0	0	0	0	0	8	1	0	0

Table 3. Selected results for all benchmarks with 5MR. Each single number was obtained from 500 runs.

TMR	Rate of correct runs up to 30, 50, 80 and 100 steps [%]											
Prob. of defects [%]	30	50	80	100	30	50	80	100	30	50	80	100
	Rule 30				Game of Life				Byl's Loop			
0.5	99	99	99	99	99	99	99	99	99	99	99	99
1	99	99	99	99	99	99	99	99	99	99	99	99
1.5	99	99	99	99	99	99	99	99	99	99	99	99
2	98	97	97	97	98	98	98	98	99	99	99	98
2.5	97	96	95	95	98	98	98	98	99	98	98	97
3	96	95	93	93	97	95	95	95	99	99	98	98
4	94	89	87	87	92	88	87	87	96	95	94	92
5	87	81	78	78	89	81	79	79	95	92	87	84
6	79	68	64	64	84	77	76	76	92	87	80	76
7	72	58	51	51	42	73	61	58	88	81	71	66
8	58	47	38	38	59	46	43	43	77	64	49	40
10	33	18	11	11	39	25	22	22	74	57	41	33
12.5	13	3	1	1	16	7	5	5	53	33	16	11
15	3	0	0	0	4	0	0	0	36	15	4	2

6   Conclusion

The results give the indication that static NMR in cellular automata can work with the presence of low defect rate. There was no surprise when 5MR showed substantially better results than TMR. It is arguable which method would suffice if any. Some sources [12] claim that future materials based on nanotechnology could have defect rate up to 10%. However, such claims are at best questionable since we can't analyze future materials and even if some of them already exist, it is quite possible that the fabrication process would improve significantly when or if cellular systems are actually implemented. It is also possible that the defect rate would be even higher.

Another observation which should be considered is the difference between the benchmark task's results. The tables show that some tasks are more sensitive to errors caused by defects than others. Especially Byl's Loop shows lower sensitivity. However, the majority problem indicates that it cannot work with defects at all which was a little surprising. From the intuitive viewpoint, one could assume that few defects could not make such a difference with a configuration that has overwhelming majority of one state. Evidently, some CA tasks work in ways which make them extremely sensitive to permanent errors.

The initial configurations were designed so the runs would use most of the cellular space but those configurations also need to be taken into account, especially with the rule 30 and Byl's Loop which don't use most of the cellular space until late in the computation. If those tasks were configured differently, they could possibly show similar behavior to the majority problem.

In conclusion, it is safe to say that although cellular automata look like uniform arrays of elements, the nature of computations running in them makes analysis of their error susceptibility an assignment dependent on said computations and that such analyses can not be generalized. Moreover the fault-tolerance in cellular systems can be increased by very simple and straightforward manner which NMR undoubtedly is. NMR could also be independent on hardware platform since grouping and connecting several modules in each cell should be much easier task than adding multiple layers of closely connected complex 'circuitry'.

For future work, we plan to model cellular systems with additional types of errors, possibly in GPU accelerated cellular automata simulator which we successfully implemented [19]. The defects could be modelled differently e.g. by permanent state of 1 or by random state. Other issue is the mathematical reliability model for TMR, which is quite straightforward, however we feel the need to include the CA's task so the model would make sense for the purpose of designing fault-tolerant CA applications. Experiments described in this paper also inspired us to investigate more thoroughly, how the presence of errors would change the behavior of known CA, especially their membership in Wolfram Classes [17].

Acknowledgements. This work was partially supported by the grant Natural Computing on Unconventional Platforms GP103/10/1517, the FIT grant FIT-11-S-1 and the research plan Security-Oriented Research in Information Technology, MSM0021630528.

References

1. Andre, D., Bennett III, F.H., Koza, J.R.: Discovery by genetic programming of a cellular automata rule that is better than any known rule for the majority classification problem. In: Proceedings of the First Annual Conference on Genetic Programming, GECCO 1996, pp. 3–11. MIT Press, Cambridge (1996)
2. Beckett, P., Jennings, A.: Towards nanocomputer architecture. In: Proceedings of the Seventh Asia-Pacific Conference on Computer Systems Architecture, CRPIT 2002, pp. 141–150. Australian Computer Society, Inc., Australia (2002)
3. Bersini, H., Detour, V.: Asynchrony induces stability in cellular automata based models. In: Brooks, R.A., Maes, P. (eds.) Artificial Life IV, pp. 382–387. MIT Press, Cambridge (1994)
4. Byl, J.: Self-reproduction in small cellular automata. Physica D 34(1-2), 295–299 (1989)
5. Durbeck, L., Macias, N.: The cell matrix: An architecture for nanocomputing. Nanotechnology 12(3), 217–230 (2001)
6. Dysart, T., Kogge, P.: System reliabilities when using triple modular redundancy in quantum-dot cellular automata. In: IEEE International Symposium on Defect and Fault Tolerance of VLSI Systems, DFTVS 2008, pp. 72 –80 (2008)
7. Gardner, M.: The fantastic combinations of john conway's new solitaire game 'life'. Scientific American 223, 120–123 (1970)
8. Garzon, M.: Models of massive parallelism: analysis of cellular automata and neural networks. Springer, London (1995)
9. Lala, P.K. (ed.): Self-checking and fault-tolerant digital design. Morgan Kaufmann Publishers Inc., San Francisco (2001)
10. Li, W.: Phenomenology of nonlocal cellular automata. Journal of Statistical Physics 68, 829–882 (1992), doi:10.1007/BF01048877
11. Mange, D., Sipper, M., Stauffer, A., Tempesti, G.: Towards Robust Integrated Circuits: The Embryonics Approach. Proceedings of IEEE 88(4), 516–541 (2000)
12. Peper, F., Lee, J., Abo, F., Isokawa, T., Adachi, S., Matsui, N., Mashiko, S.: Fault-tolerance in nanocomputers: A cellular array approach. IEEE Transactions on Nanotechnology 3(1), 187–201 (2004)
13. Sekanina, L., Komenda, T.: Global control in polymorphic celular automata. Journal of Cellular Automata 6(4), 1–21 (2011)
14. Sipper, M.: The emergence of cellular computing. Computer 32(7), 18–26 (1999)
15. Sipper, M.: Evolution of Parallel Cellular Machines: The Cellular Programming Approach. LNCS, vol. 1194. Springer, Heidelberg (1997)
16. Toffoli, T., Margolus, N.: Cellular Automata Machines: A New Environment for Modeling. MIT Press, Cambridge (1987)
17. Wolfram, S.: Cellular Automata and Complexity: Collected Papers. Addison-Wesley, Reading (1994)
18. Wolfram, S.: A New Kind of Science. Wolfram Media, Inc., Champaign (2002)
19. Zaloudek, L., Sekanina, L., Simek, V.: Accelerating Cellular Automata Evolution on Graphics Processing Units. International Journal on Advances in Software 3(1), 294–303 (2010)

Author Index